A Practical Approach to Advanced Mathematical Modelling in Civil Engineering

A Practical Approach to Advanced Mathematical Modelling in Civil Engineering

MOHAMMAD HEIDARZADEH

University of Bath, Bath, United Kingdom

THEODOSIOS K. PAPATHANASIOU

Aston University, Birmingham, United Kingdom

YURUI FAN

Brunel University, Uxbridge, United Kingdom

HAMID BAHAI

Brunel University, Uxbridge, United Kingdom

OXFORD
UNIVERSITY PRESS

OXFORD
UNIVERSITY PRESS

Great Clarendon Street, Oxford, OX2 6DP,

United Kingdom

Oxford University Press is a department of the University of Oxford.
It furthers the University's objective of excellence in research, scholarship,
and education by publishing worldwide. Oxford is a registered trade mark of
Oxford University Press in the UK and in certain other countries

Published in the United States of America by Oxford University Press
198 Madison Avenue, New York, NY 10016, United States of America

British Library Cataloguing in Publication Data

Data available

Library of Congress Control Number: 2024947742

ISBN 9780198854241
ISBN 9780198854258 (pbk.)

DOI: 10.1093/9780191888656.001.0001

Printed and bound by
CPI Group (UK) Ltd, Croydon, CR0 4YY

The manufacturer's authorised representative in the EU for product safety is Oxford University Press España S.A.
of El Parque Empresarial San Fernando de Henares, Avenida de Castilla, 2 – 28830 Madrid (www.oup.es/en or
product.safety@oup.com). OUP España S.A. also acts as importer into Spain of products made by the manufacturer.

Preface

Motivation for writing this book

This book draws upon extensive experience gained by the authors from teaching engineering mathematics across various universities in the UK. Through this experience, we identified a significant gap in the availability of textbooks tailored to the needs of engineering students and professionals, particularly those emphasizing practical applications and ready-to-use computer code. Specifically, this gap became apparent during the supervision of final-year undergraduate and MSc students for their dissertations. It became evident that many of these students lacked the necessary knowledge and mathematical tools to effectively complete their dissertations. This book aims to fill that gap by providing comprehensive coverage of engineering mathematics for civil engineers, with a focus on its direct relevance and applicability to real-world civil engineering applications. By providing practical examples drawn from diverse subdisciplines within civil engineering, supplemented with MATLAB code, the book serves as an all-in-one reference textbook for teaching engineering mathematics to civil engineers and supporting students to successfully complete their dissertations and other research works.

Content of the book

The book begins with an exploration of the mathematical modelling process in Chapter 1, which serves as the foundation of engineering mathematics. This critical aspect is covered through several real-world examples. Recognizing a gap in the existing literature where this topic is not systematically addressed, we have attempted to introduce a methodical, step-by-step approach accompanied by a variety of practical examples. Subsequent chapters are dedicated to specific disciplines within civil engineering: structural engineering (Chapter 2), geotechnical engineering (Chapter 3), coastal engineering (Chapter 4), Water and Environmental Engineering (Chapter 5), and transportation engineering (Chapter 6). Appendices A and B serve as valuable resources, offering supplementary materials designed to support readers in two key areas: basic engineering mathematics and MATLAB programming.

Readership and how to use the book

This book is intended for a readership comprising final-year undergraduate, MSc, and PhD students of civil engineering and mechanical engineering, as well as practising engineers. The incorporation of advanced mathematical models such as machine learning, deep learning, and finite difference methods, complemented by corresponding MATLAB code, enhances the book's value for practising engineers.

To use the book as a textbook for university teaching, we suggest employing the entire volume for either a semester-long or a year-long module (course) to ensure students acquire sufficient knowledge across various subdisciplines of civil engineering. Familiarity with diverse disciplines is crucial for students, given the potential for their work in any specialization upon graduation. However, due to time constraints, it may not be feasible to cover all the examples within each chapter over a single semester or year; hence, we advise focusing on selected topics from each chapter. The book contains 62 MATLAB scripts which are available at the following website: http://www.oup.com/AdvancedMathematicalModelling.

Mohammad Heidarzadeh
Theodosios K. Papathanasiou
Yurui Fan
Hamid Bahai
London, United Kingdom
May 2024

Contents

List of Figures

Chapter 1
Mathematical Modelling

1.1 Mathematical models and modelling

It is common to develop mathematical models for physical phenomena or systems in order to study them. Mathematical equations, concepts, and relationships are used for developing models using various variables that characterize different aspects of a phenomenon such as dimensions, material, speed, density, elasticity, and other parameters. Conservation and constitutive laws are used to develop mathematical models, and are discussed in detail in this chapter. There are many advantages in developing mathematical models for real-world phenomena, such as (Bender 2000):

- Models help to simplify complicated phenomena to their most basic elements and thus are essential for understanding them and discovering their various features.
- Models allow the development of predictive tools for real-world systems, such as models for tsunami waves that help to forecast their effects.
- Models offer cost and time saving for studying a real-world phenomenon because they allow the phenomenon to be tested in a relatively short time and at a low cost. Optimization of real-world phenomena is possible using models, and thus models contribute towards saving resources.
- Mathematical models are powerful tools for testing hypotheses explaining real-world phenomena in a systematic and scientific way.

An example of a mathematical model is shown in Figure 1.1 where a multistorey building is modelled using a mass–spring–damper system in which a series of masses m are connected by springs with stiffness k and dampers with damping coefficient c. Each storey of the building is modelled using one mass (e.g., m_1), one spring (e.g., k_1), and a damper (e.g., c_1).

The process of applying mathematical equations and relationships to represent a real-world phenomenon using a model is called 'mathematical modelling'. Led by the increased application of computers in science and technology over the past few decades, mathematical modelling has become popular and has been used in various fields, including engineering, biology, medical sciences, sociology, and economics. In engineering, and particularly in civil engineering, mathematical modelling has been widely used, for instance in areas such as the seismic design of structures, hydrological modelling, flood modelling, slope stability analysis, traffic flow modelling, sensor-based modelling for structural health monitoring, water quality modelling, pavement design, project scheduling, and many other areas.

A Practical Approach to Advanced Mathematical Modelling in Civil Engineering. Mohammad Heidarzadeh et al.,
Oxford University Press. © Mohammad Heidarzadeh, Theodosios K. Papathanasiou, Yurui Fan, Hamid Bahai (2025).
DOI: 10.1093/9780191888656.003.0001

(a)

(b)

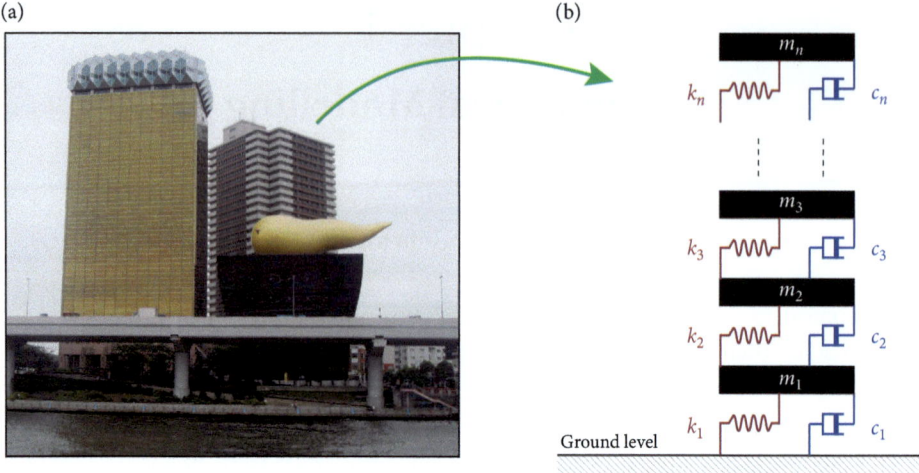

Figure 1.1 (a) Two tall multistorey buildings in central Tokyo (Japan). (b) The representative model for a multistorey building using a system of masses m, springs (with stiffness k), and dampers (with damping coefficient c).

1.2 Building a mathematical model

According to Bender (2000), building a mathematical model requires imagination in the first place, supplemented with mathematical skills. The process of building a mathematical model can be divided into the following steps:

- **Step 1: Problem formulation**

  The problem needs to be defined clearly and the objectives of the model creation should be articulated. In other words, it should be clear what the model is expected to do. For example, referring to Figure 1.1, the mathematical model for the multistorey building is expected to calculate oscillations of the building, its damping performance, and lateral drifts of the storeys, among other objectives.

- **Step 2: Outlining the problem and conceptualization**

  At this stage, the problem is broken down into its parts, and the interactions among various parts are identified. It should be established how various parts influence one another. Referring to the example in Figure 1.1, the key parts of the problem are the masses of the building storeys, and the stiffness and damping properties of the columns and inter-storey walls. The masses are connected to each other using columns with certain stiffness and damping effects.

- **Step 3: Assumptions**

  An essential part of every mathematical model are the assumptions made to simplify the problem. The real world is complicated, and it is infeasible to take all the details into a model. The forces and properties that are insignificant are identified and ignored. As an example, for the case of the multistorey building of Figure 1.1, we assume that the total weight of each storey is represented by a concentrated weight of m_i.

- **Step 4: Parametrization and variables**

  Now is the time to define the variables and parameters that shape the final model. The base units are usually length (L), time (T), and mass (M), and many variables are defined using them, grouped into three classes of geometric (such as length, width, and volume), kinematic (such as velocity and acceleration), and dynamic (such as pressure, stress, and momentum) parameters. For the multistorey building of Figure 1.1, some of the parameters are shown in the figure (m_1, k_1, c_1, and so on).

- **Step 5: Deriving the equations**

  By defining the variables in the previous step, the equation or relationship among the variables can be derived using fundamental principles and empirical data. The incorporation of conservation and constitutive laws is essential in the derivation of the equations. For example, for the multistorey building of Figure 1.1, the equation can be derived by applying the principle of static equilibrium of a mass–spring–damper system, which indicates that all forces applied to the mass are in equilibrium. Such forces include the inertia forces due to the weights, forces applied in the springs, forces in the dampers, and external forces such as wind or earthquake.

- **Step 6: Simulation and analysis**

  The derived mathematical model can be used to study the behaviour of the real-world phenomenon under various conditions by changing the range of input parameters and environmental forces. By interpreting the results, the phenomenon can be understood.

- **Step 7: Testing and validation**

  The model needs to be validated through testing its performance against actual real-world data and benchmark examples. This will help to understand the capabilities, accuracy, and limitations of the model. The example model shown in Figure 1.1 can be validated by actual data collected through planting instruments such as drift meters in the building and monitoring the building drifts for a period of a few weeks or months under the impact of winds or earthquakes.

1.3 Constructing a mathematical model: An example

Here, we practise the above steps by developing a mathematical model for a system of a single-storey building (Figure 1.2). The following steps are taken sequentially:

Step 1: Problem formulation

The example in Figure 1.2 requires the development of a mathematical model that can calculate the lateral drift of the single-storey building along the x axis. Free vibration of the system is requested in this example, and thus the external force is zero, that is, $f(t) = 0$.

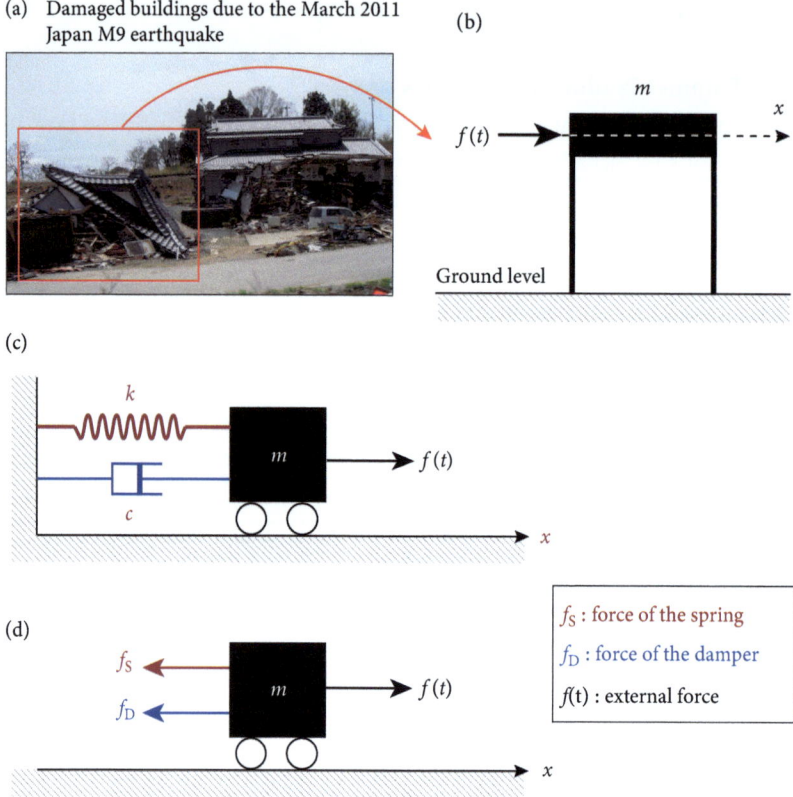

Figure 1.2 (a) A single-storey building damaged during the 11 March 2011 Japan M9 earthquake in the northeast of Japan. (b) Sketch representing a single-storey building. (c) The representative model for this single-storey building using a system of masses (m), springs (with stiffness k), and dampers (with damping coefficient c). (d) Free-body diagram showing all the forces acting on the mass of the single-storey building.

Step 2: Outlining the problem and conceptualization

The problem shown in Figure 1.2 involves a mass (m), a spring with stiffness k, and a damper with damping coefficient c. These elements are connected so that the mass movement is controlled by the spring and the damper. They are in continuous interaction among themselves. The stiffer the spring, the smaller the movement of the mass, and vice versa.

Step 3: Assumptions

Several assumptions are made here (Figure 1.2), among which are: the single-storey building is assumed to be made of rigid elements; the entire mass of the floor and the walls is considered as a concentrated single mass; changes in the density and material properties are ignored and thus a single type of material is assumed.

Step 4: Parametrization and variables

Referring to Figure 1.2, the variables of this problem are:

m: mass of the single-storey building
k: stiffness coefficient of the spring
c: damping coefficient of the damper
x: displacement of the mass
t: time
f_S: force exerted by the spring
f_D: force exerted by the damper
$f(t)$: the external force exerted on the mass, such as wind or earthquake.

Step 5: Deriving the equations

Newton's second law is applied:

$$\sum F = ma \tag{1.1}$$

where F is force, m is mass, and a is acceleration. Considering the various forces applied to the mass (Figure 1.2d), Eq (1.1) can be rewritten as

$$f(t) - f_S - f_D = ma \tag{1.2}$$

The acceleration of the mass (a) is defined as the second derivative of the displacement with respect to time. Therefore, it is expressed as $a = \frac{d^2x}{dt^2}$. In physics, we understand that the forces from the spring and damper are given by $f_S = kx$ and $f_D = c\frac{dx}{dt}$, respectively. Therefore, we substitute these terms into Eq (1.2):

$$m\frac{d^2x}{dt^2} + c\frac{dx}{dt} + kx = f(t) \tag{1.3}$$

Step 6: Simulation and analysis

Equation (1.3) is a second-order ordinary differential equation that can be solved readily. We assume that the external force is absent initially, $f(t) = 0$, and thus the free vibration of the system is solved here. It is also assumed that the mass of the storey is $m = 10$ kg, the damping coefficient is $c = 10$ kg/s, and the stiffness of the spring is $k = 90$ kg/s^2. Assume that the system starts its vibration with an initial drift of $x = 0.16$ m, and its initial velocity is zero. With these data, the equation of motion takes the following form:

$$10\frac{d^2x}{dt^2} + 10\frac{dx}{dt} + 90x = 0 \tag{1.4}$$

To solve second-order ordinary differential equations (ODEs) with the general form

$$\frac{d^2x}{dt^2} + a\frac{dx}{dt} + bx = 0 \tag{1.5}$$

we need to form the characteristic equation as follows:

$$\lambda^2 + a\lambda + b = 0 \tag{1.6}$$

and calculate $\Delta = \sqrt{a^2 - 4b}$. Depending on the value of Δ, solutions of Eq (1.5) will take one of the following forms:

- If $\Delta > 0$, then Eq (1.6) has two real answers, λ_1 and λ_2, and the solution will take the form $x(t) = z_1 e^{\lambda_1 t} + z_2 e^{\lambda_2 t}$.
- If $\Delta = 0$, then Eq (1.6) has one real answer of λ and the solution will be $x(t) = (z_1 + z_2 t) e^{\lambda t}$.
- If $\Delta < 0$, then Eq (1.6) has two imaginary answers of $\lambda_1 = p + qi$ and $\lambda_2 = p - qi$, and the solution will be of the form $x(t) = e^{pt}(z_1 \cos qt + z_2 \sin qt)$.

The constants z_1 and z_2 are obtained using the specific initial and boundary conditions of the problem.

To solve the mathematical model represented by Eq (1.4), it is rearranged as follows:

$$\frac{d^2x}{dt^2} + \frac{dx}{dt} + 9x = 0 \tag{1.7}$$

The characteristic equation becomes: $\lambda^2 + \lambda + 9 = 0$, which gives two answers: $\lambda_1 = -0.5 + 2.96\,i$ and $\lambda_2 = -0.5 - 2.96\,i$. Therefore, the solution of the equation of motion becomes

$$x(t) = e^{-0.5t}(z_1 \cos 2.96t + z_2 \sin 2.96t) \tag{1.8}$$

Now, the two constant values of z_1 and z_2 need to be calculated. The problem stated that the drift is $x = 0.16$ m at the beginning $(t = 0)$. By applying this initial condition to Eq (1.8), we have $z_1 = 0.16$. The other condition states that the initial velocity is zero, which means $\frac{dx}{dt}\big|_{t=0} = 0$. By taking the first derivative of Eq (1.8) relative to time, and applying the latter condition, we have $z_2 = 0.027$. Therefore, the solution is complete, and the final equation is derived as follows:

$$x(t) = e^{-0.5t}\,[\,0.16\,\cos(2.96t) + 0.027 \sin(2.96t)\,] \tag{1.9}$$

A plot of this solution is shown in Figure 1.3, demonstrating that the initial displacement is 0.16 m at $t = 0$, followed by cycles of positive and negative drifts (i.e., drifts of the building to the left and right) where the maximum drift decreases with time. Eventually, Figure 1.3 shows that the oscillations disappear after around 8 s. The MATLAB script in Code 1.1 can be used to plot the oscillations shown in Figure 1.3. This and the other MATLAB scripts presented in this chapter can be downloaded from the following website: http://www.oup.com/AdvancedMathematicalModelling.

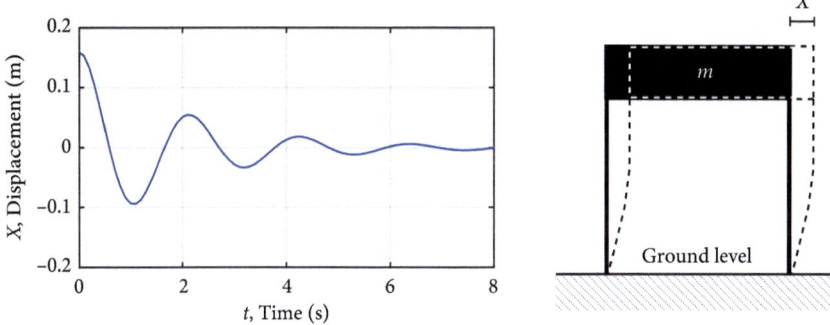

Figure 1.3 Time history plot of the oscillations of a single-storey building with input data including mass $m = 10$ kg, stiffness of the spring $k = 90$ kg/s^2, and damping coefficient $c = 10$ kg/s. This figure is generated using the MATLAB script in Code 1.1.

Step 7: Testing and validation

Normally, validation of a mathematical model is done through an experimental or empirical dataset. For the case of the example shown in Figure 1.2 with a solution in Eq (1.9), an accelerometer can be placed on the building for a certain period of time, such as a week or a month, to collect experimental data to validate the model.

Code 1.1 *The MATLAB code to plot the time history of the motion equation of a single-storey building with $m = 10$ kg, $k = 90$ kg/s^2, and $c = 10$ kg/s. This MATLAB code (and other MATLAB scripts from this chapter) can be downloaded from the following website: http://www. oup.com/AdvancedMathematicalModelling. This code produces Figure 1.3.*

```
% This code plots oscillations of a
% single degree of freedom system (m, k, c)
clc; clear; close all;
set(0,'DefaultAxesFontsize',16);
%
m=10; % mass of the system (kg)
k=90; % stiffness of the spring (kg/s^2)
c=10; % damping coefficient of damper (kg/s)
%
t=0:0.1:10;
x=exp(-0.5.*t).*(0.16.*cos(2.96.*t)+...
0.027.*sin(2.96.*t));
%
subplot('position',[0.2 0.2 0.3 0.35]);
plot(t,x,'b', 'LineWidth',1.5);
xlim([0 8]);ylim([-0.2 0.2]);
xlabel('t, Time (s)'); ylabel('x, Displacement (m)');
set(gca,'linewidth',1.5);
grid on;
%
% End of the code %
```

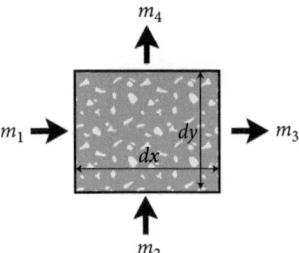

Figure 1.4 A soil element in an incompressible medium.

1.4 Conservation principles

Conservation principles play critical roles in the analysis and understanding of physical phenomena and in the development of mathematical models. Conservation principles are basic principles describing the balance and preservation of certain physical quantities such as mass, energy, momentum, area, volume, and other parameters (Kreyszig 2011; Bender 2000). Some of the main conservation principles are described in this section.

1.4.1 Conservation of mass (continuity law)

The law of conservation of mass dictates that within a closed system, the total mass remains unaltered throughout time. In other words, within a closed system, mass remains constant, with no creation or destruction. It is also known as the 'continuity law'. In the context of geotechnical engineering, referring to the soil element in Figure 1.4, the sum of the mass of a fluid with density ρ entering the control volume with dimensions $dx \times dy$ should be equivalent to the combined mass exiting the control volume and any mass that accumulates within it. Referring to Figure 1.4, and assuming a situation with an incompressible medium with steady flow, the amount of mass flow entering the element (m_{in}) is equal to the mass flow leaving the cell (m_{out}):

$$m_{in} = m_{out} \tag{1.10}$$

$$(m_1 + m_2) = (m_3 + m_4) \tag{1.11}$$

1.4.2 Conservation of energy

The principle of energy conservation asserts that energy can undergo conversion from one form to another but cannot be created or destroyed. For example, in the context of structural engineering, conservation of energy is applied in the design of structural motion control systems due to environmental loading such as wind and earthquake. Most structural motion control systems, such as base isolation and tuned mass dampers, are based on structural motion damping where the oscillations due to wind or earthquake

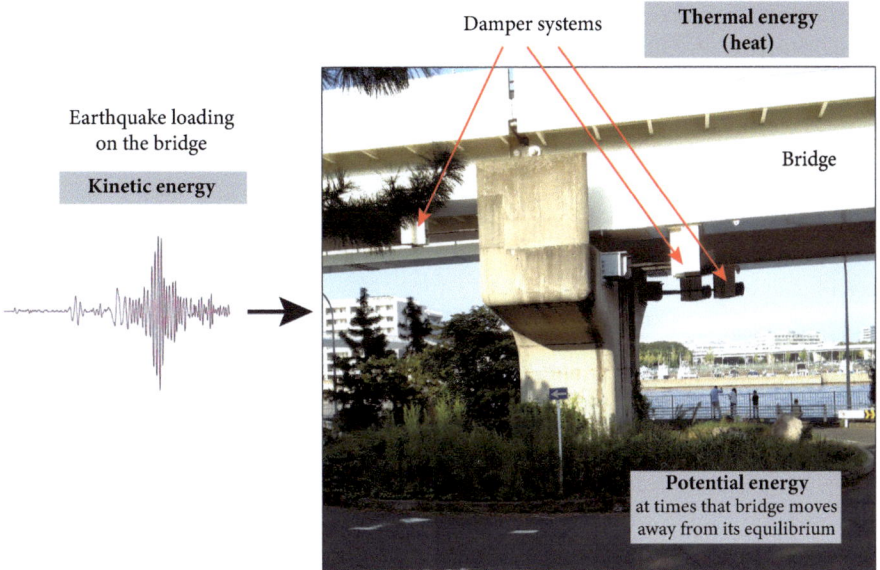

Damper systems

Thermal energy (heat)

Earthquake loading on the bridge

Kinetic energy

Bridge

Potential energy at times that bridge moves away from its equilibrium

Figure 1.5 Conservation of energy in a bridge in Japan where the kinetic energy from earthquake oscillations is transformed to thermal energy (heat) in dampers and potential energy as the structure moves away from its equilibrium.

are dissipated. In this process, the input kinetic energy from earthquake or wind, which causes excessive oscillations, is absorbed and converted to potential energy and heat in the dampers.

For the bridge shown in Figure 1.5, the earthquake loading imparts kinetic energy (KE) to the structure in the form of structural vibrations and oscillations. Part of this energy is absorbed by the dampers and is consequently transformed to thermal energy (heat), which is called dissipated energy (DE) here. In the meantime, part of the kinetic energy is converted to potential energy (PE) when the structure moves away from its equilibrium position. The total energy (TE) in the system is given by the following equation:

$$TE = KE + PE - DE \qquad (1.12)$$

According to the conservation of energy, the total energy in the system remains constant during the whole oscillation process as energy converts from one form to another.

1.4.3 Conservation of momentum

The principle of momentum conservation states that within a closed system, the total momentum remains unchanged as time progresses. In physics, momentum (M) is defined as the product of mass (m) and its velocity (v):

$$M = m \times v \qquad (1.13)$$

The principle of momentum conservation is the outcome of Newton's second law. Assume that two objects with masses m_1 and m_2 interact in a closed system. Based on Newton's third law, the forces on each other (F_1 and F_2) have the same magnitude but in opposite directions:

$$F_1 = -F_2 \tag{1.14}$$

According to Newton's second law, $F = ma$, where a is acceleration $\left(a = \frac{dv}{dt}\right)$. Therefore, Eq (1.14) can be rewritten as

$$m_1 \frac{d(v_1)}{dt} = -m_2 \frac{d(v_2)}{dt} \tag{1.15}$$

As mass is a constant value, the two masses m_1 and m_2 can be moved into the parentheses, and Eq (1.15) is rearranged to the following form:

$$\frac{d(m_1\,v_1)}{dt} + \frac{d(m_2\,v_2)}{dt} = 0 \tag{1.16}$$

This equation can be rearranged further:

$$\frac{d}{dt}(m_1\,v_1 + m_2\,v_2) = 0 \tag{1.17}$$

Equation (1.17) implies that the total momentum of a system of two masses (m_1 and m_2) remains the same over time (e.g., between t_1 and t_2). In another words, Eq (1.17) implies that

$$(m_1\,v_1 + m_2\,v_2)_{t_2} = (m_1\,v_1 + m_2 v_2)_{t_1} \tag{1.18}$$

1.5 Initial value problems (IVP)

In the realm of mathematics, an initial value problem (IVP) pertains to the task of finding a solution to a differential equation while taking into account an initial condition, which is usually at the initial time. Therefore, an IVP has two components: an ODE and an initial value such as $(t_0,\,y_0)$ (Kreyszig 2011). The initial condition helps to determine the value of the unknown function in the ODE. Within this framework, the solution to an IVP reveals the evolution of the system from its initial state as specified by the initial value. As an example, the motion of the system of a single-storey building discussed earlier in this chapter is an IVP. Some other examples of IVPs in civil and environmental engineering applications are:

- **Structural dynamics**: As seen in the previous section, IVPs can be used to predict structural motion under various loadings such as wind and earthquake.
- **Transportation engineering**: IVPs are applied in traffic flow modelling and predicting traffic congestion.

- **Groundwater flow**: Movement of water in aquifers is usually modelled using IVPs.
- **Hydrology**: Phenomena such as rainfall, runoff, and flood-related flows in rivers and dam reservoirs can be modelled using IVPs.
- **Pollution transport**: The propagation of pollution in rivers and reservoirs is an IVP problem.
- **Landslide prediction**: Potential landslides and the failure probabilities can be modelled as IVPs.
- **Sediment transport**: Movement of sediments in rivers, coasts, and estuaries can be described using IVPs.

As a numerical example of an IVP, we employ an initial value problem to simulate the temporal changes in pollution concentration within a reservoir. Water reservoirs, such as the one shown in Figure 1.6, serve as primary sources for domestic and industrial water supply. Hence, it is essential to maintain continuous vigilance over their water quality.

In the following, the seven steps discussed in Section 1.2 are employed to develop a mathematical model for the temporal changes in pollution concentration in a dam reservoir.

Step 1: Problem formulation

The objective of this problem (Figure 1.6) is to develop a mathematical model for the temporal changes of pollution concentration, $C(t)$, in the dam reservoir by taking into account the initial concentration of the pollution, $C(t = 0)$, the water discharge rate into the reservoir (Q), and the total volume of reservoir water (V). The function $C(t)$ represents pollution concentration changes over time.

Step 2: Outlining the problem and conceptualization

Pollution concentration in a body of water, such as a dam reservoir, depends on a few parameters such as the initial concentration of the pollution, $C(t = 0)$, the input water discharge rate into the reservoir (Q), and the total volume of reservoir water (V).

Figure 1.6 Photograph showing the reservoir of the Cheddar dam in the UK. The parameters used for modelling temporal changes of pollution concentration are given in the top-right corner.

For instance, as Q increases, pollution concentration will decline more rapidly. Moreover, a greater reservoir water volume will lead to a more rapid decrease in pollution concentration.

Step 3: Assumptions

In the example shown in Figure 1.6, the following assumptions are made: (i) The output water discharge rate from the reservoir is assumed to be zero. (ii) The total reservoir water volume (V) is assumed to be constant and therefore the increase of water volume due to the input discharge (Q) is neglected. (iii) It is assumed that the input discharge rate (Q) is constant over time.

Step 4: Parametrization and variables

Referring to Figure 1.6, the variables of this problem are:

V: total reservoir water volume
Q: input water discharge rate into the reservoir
C: concentration of pollution in the reservoir at any given time
$C(t = 0)$: initial concentration of pollution
P: volume of the pollution in the reservoir at any given time
t: time.

Step 5: Deriving the equations

Considering the above parameters, the pollution concentration in the reservoir at any time, C, can be given by the following equation:

$$C = \frac{P}{V} \tag{1.19}$$

where P is the volume of the pollution in the reservoir at time t, and V is total reservoir water volume. The derivative of Eq (1.19) with respect to time is taken, as follows:

$$\frac{dC}{dt} = \frac{1}{V} \frac{dP}{dt} \tag{1.20}$$

At the time interval dt, we understand that $dP = -Q \cdot C \, dt$. By substituting this expression into Eq (1.20), we obtain

$$\frac{dC}{dt} = -\frac{1}{V} \frac{dt \, (Q \cdot C)}{dt} \tag{1.21}$$

This equation is rearranged to

$$V \frac{dC}{dt} + Q \, C = 0 \tag{1.22}$$

Equation (1.22) yields the variation of pollution concentration (C) in the reservoir at any time.

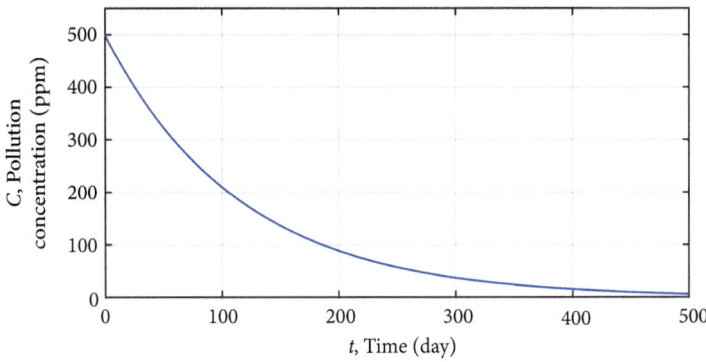

Figure 1.7 Time history plot of the temporal variation of pollution concentration in a dam reservoir. The total reservoir water volume is $V = 25 \times 10^6$ m³, the input discharge rate is $Q = 2.48$ m³/s, and the initial concentration of pollution in the reservoir water is $C(t = 0) = 500$ ppm. This figure is generated using Code 1.2.

Step 6: Simulation and analysis

As a numerical example for the variation of pollution concentration in a dam reservoir (Eq 1.22), we assume a total water volume of $V = 25 \times 10^6$ m³, an input discharge rate of $Q = 2.5$ m³/s into the reservoir, and an initial concentration of pollution in the reservoir water of $C(t = 0) = 500$ ppm (parts per million). Using these input parameters, Eq (1.22) becomes

$$25 \times 10^6 \frac{dC}{dt} + 2.5 \, C = 0 \tag{1.23}$$

This is a simple ODE whose solution is of exponential type as follows:

$$\frac{dC}{C} = -10^{-7} \, dt \tag{1.24}$$

$$C = K_1 \, e^{-10^{-7}t} \tag{1.25}$$

The constant K_1 is determined using the initial value of the problem, which is $C(t = 0) = 500$ ppm. This results in $K_1 = 500$. Therefore, the final solution of the ODE is

$$C = 500 \, e^{-10^{-7}t} \tag{1.26}$$

The solution for the variation of pollution concentration in the reservoir is plotted in Figure 1.7 using the MATLAB script in Code 1.2. According to Figure 1.7, it takes more than 100 days for the pollution concentration to reach 200 ppm, and over 500 days are required for the complete removal of the pollution from the reservoir.

Step 7: Testing and validation

Validation of this mathematical model, addressing the temporal variation of pollution concentration in a dam reservoir, can be done by taking samples of water quality at different intervals and measuring pollution concentration. Those measured values can be compared with the theoretical values predicted by Eq (1.26) and Figure 1.7 for validation.

Code 1.2 *The MATLAB code to plot the temporal variation of pollution concentration in a dam reservoir. The total reservoir water volume is $V = 25 \times 10^6\ m^3$; the input discharge rate is $Q = 2.48\ m^3/s$; and the initial concentration of pollution in the reservoir water is $C(t = 0) = 500\ ppm$. This code, which produces Figure 1.7, can be downloaded from the following website: http://www.oup.com/ AdvancedMathematicalModelling.*

```
% This code plots variations of pollution
% concentration in a reservoir
clc; clear; close all;
set(0,'DefaultAxesFontsize',16);
%
V=25*10^(6); % reservoir water volume (m^3)
Q=2.5; % input discharge of reservoir (m^3/s)
c0=500; % initial pollution concentration at t=0 (ppm)
%
t=0:3600:500*24*3600; % time for 500 days
C=500.*exp(-1.*10^(-7).*t);
%
subplot('position',[0.2 0.2 0.45 0.45]);
plot(t./(24.*3600),C,'b', 'LineWidth',1.5);
xlim([0 500]); ylim([0 550]);
xlabel('t, Time (day)'); ylabel('C, Pollution concentration (ppm)');
set(gca,'linewidth',1.5);
grid on;
%
% End of the code %
```

1.6 Boundary value problems (BVP)

A BVP consists of an ODE and specified boundary conditions. In simpler terms, a BVP involves the search for a function that not only satisfies a given differential equation within a defined domain but also complies with particular conditions set at the boundaries of that domain. The difference between an IVP and a BVP is that an IVP involves a single initial condition within the domain, while a BVP specifies conditions at multiple points or boundaries. Examples of applying BVPs in civil engineering applications are:

- **Geotechnical engineering**: BVPs are used in geotechnical engineering to analyse foundations and soil masses under various loading conditions. The boundary conditions involve the supports of the foundations and the applied loads.
- **Structural engineering**: A wide range of problems are of the BVP type in structural engineering such as analysis of structural elements like beams and columns. Analyses of structures such as bridges, buildings, dams, foundations, and roads are based on BVPs, where the boundary conditions include the applied loads and various types of structural supports.

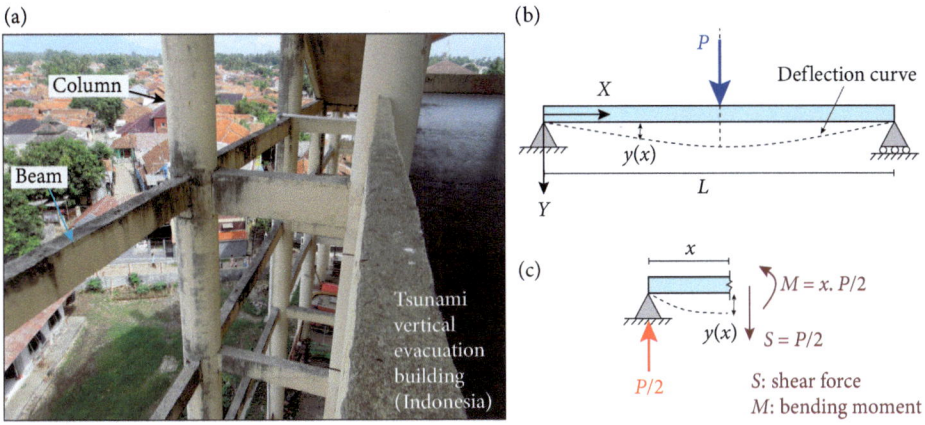

Figure 1.8 (a) Photo showing a tsunami vertical evacuation building in Indonesia where a beam and a column are marked. (b) Sketch of a simply supported beam of length L with a point load (P) at its centre. The function $y(x)$ is the deflection of the beam at distance x from its left end. (c) The free-body diagram of the left side of the beam showing the shear (S) and moment (M) at distance x from the left end of the beam.

- **Hydraulic engineering**: Problems such as the design of water supply networks, open channels, and sewage systems are basically BVPs where the boundary conditions are hydraulic heads and discharges at different points of the system, such as the entry and exit points.
- **Coastal engineering**: Sediment transport in coastal areas and estuaries are BVPs where sediment concentrations at boundaries are used as boundary conditions. Most of the design in coastal engineering involves BVPs where wave conditions at boundaries are specified at certain points or boundaries to calculate the heights and dimensions of structural elements such as seawalls and breakwaters.

An example of a BVP in civil engineering is the deflection of a simply supported beam which is under a point load (P) at its centre (Figure 1.8). The beam is composed of a roller support on one end and a hinge support on the other end. Beams serve as essential structural elements in every building, where the calculation of their deflections under point and distributed loads is an integral part of structural analysis.

For the beam shown in Figure 1.8b, the deflection curve under the concentrated load P is symmetric, with its maximum value occurring at the midpoint of the beam where the concentrated load is applied. Therefore, we only need to solve for the deflection in the first half of the beam. This problem has three boundary conditions, as listed in Table 1.1: deflection is zero at the two ends of the beam, and the slope of the deflection curve is zero in the middle of the beam.

From structural analysis, we know that the deflection of a beam, $y(x)$, can be calculated using the following ODE (Kreyszig 2011):

$$\frac{d^2y(x)}{dx^2} = -\frac{M(x)}{EI} \tag{1.27}$$

Table 1.1 Various boundary conditions for the boundary value problem of the deflection, $y(x)$, of a beam of length L under a concentrated load at its centre. The beam is a simply supported beam composed of a roller support on one end and a hinge support on the other end (see Figure 1.8).

Number	Location (x)	Boundary condition	Comment
1	$x = 0$	$y(x) = 0$	At the left end, the deflection is zero
2	$x = L/2$	$dy/dx = 0$	In the middle of the beam, the slope of the deflection curve is zero
3	$x = L$	$y(x) = 0$	At the right end, the deflection is zero

where x is the distance from the left end of the beam (Figure 1.8), E is Young's modulus of elasticity of the material of the beam, $M(x)$ is the bending moment at distance x, and I is the moment of inertia of the beam's cross section. Considering the free-body diagram in Figure 1.8c, the reaction of the left-hand side support is $P/2$. Therefore, the bending moment at distance x can be calculated as

$$M(x) = \frac{P}{2} x \tag{1.28}$$

By substituting the bending moment in Eq (1.27), we get

$$\frac{d^2 y(x)}{dx^2} = -\frac{P x}{2 EI} \tag{1.29}$$

Two integrations of this equation result in the following equation:

$$y(x) = -\frac{1}{12 EI} P x^3 + C_1 x + C_2 \tag{1.30}$$

where C_1 and C_2 are constants that can be determined by applying the boundary conditions listed in Table 1.1. From boundary condition 1, we achieve $C_2 = 0$. Boundary condition 2 results in $C_1 = \frac{1}{16 EI} P L^2$. Therefore, the equation for the deflection of the beam takes the following form:

$$y(x) = -\frac{1}{12 EI} P x^3 + \left(\frac{1}{16 EI} P L^2 \right) x \tag{1.31}$$

After rearranging, the equation becomes

$$y(x) = \frac{P x}{48 EI} \left[3L^2 - 4x^2 \right] \tag{1.32}$$

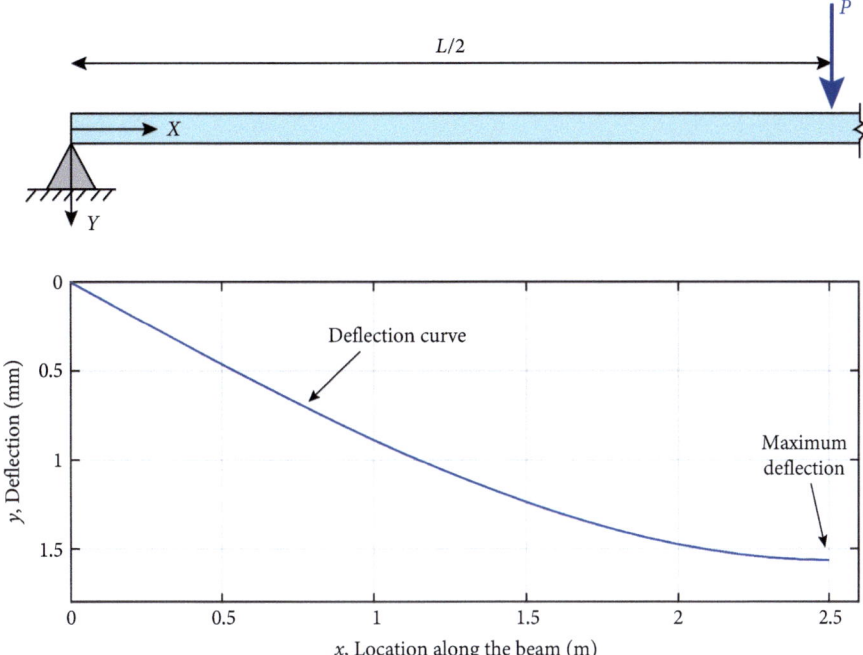

Figure 1.9 Plot of the deflection curve of a beam of length 5 m (L = 5 m), a point load of P = 75 kN, Young's modulus of E = 2.5 × 10^7 kN/m^2 for concrete, and I = 0.005 m^4. Note that the deflection of half of the beam is plotted here. This figure is generated using Code 1.3.

The maximum deflection of the beam at the midpoint (y_{max}) is obtained by substituting $x = L/2$ in Eq (1.32), which results in

$$y_{max} = \frac{PL^3}{48\,EI} \tag{1.33}$$

Assuming a length of 5 m for the beam (L = 5 m), a point load of P = 75 kN, a Young's modulus of E = 2.5 × 10^7 kN/m^2 for concrete, and I = 0.005 m^4, the deflection curve of the beam is plotted in Figure 1.9 using the MATLAB script in Code 1.3. This code can be downloaded at: http://www.oup.com/AdvancedMathematicalModelling.

Code 1.3 *MATLAB code for plotting the deflection curve of a beam with length of 5 m (L = 5 m), a point load of P = 75 kN, a Young's modulus of E = 2.5 × 10^7 kN/m^2 for concrete, and I = 0.005 m^4. Note that the deflection of half of the beam is produced here. This code generates Figure 1.9.*

```
% This code plots the deflection
% of a beam under a point load (P)
clc; clear; close all;
set(0,'DefaultAxesFontsize',16);
```

Code 1.3 *Continued*

```
%
L=5; % length of beam (m);
P= 75; % point load (KN);
E=2.5.*10.^(7); % Young's modulus (KN/m^2);
I=0.005; % moment inertia of cross section(m^4);
%
x=0:0.1:2.5; % plot deflection of half of beam
y=P.*x.*(3.*(L.^2)-(4.*x.^2))./(48.*E.*I).*1000;
% deflection is reported in millimeters
%
subplot('position',[0.2 0.2 0.55 0.45]);
plot(x,y,'b', 'LineWidth',1.5);
set(gca, 'YDir', 'reverse')
xlim([0 2.6]); ylim([0 1.8]);
xlabel('x, Location along the beam (m)');
ylabel('y, Deflection (mm)');
set(gca,'linewidth',1.5);
grid on;
%
% End of the code %
```

1.7 Constitutive laws

Mathematical modelling was discussed in the previous sections of this chapter with several real-world examples. As mentioned above, constitutive laws form the basis for developing mathematical models; we have used some of them earlier in this chapter. Constitutive laws are equations or relationships that describe the behaviour of a particular material under various conditions and loads. A few common constitutive laws for civil engineering applications are presented in this section.

1.7.1 Hooke's law

Hooke's law, also known as the law of elasticity, states that there is a linear relationship between the deformation and the force (or load) in elastic materials or within the elastic limit of any material. When these conditions are met, the object reverts to its original shape and size once the load is removed. This law is commonly applied to concrete and steel in civil engineering. If a specific material is subjected to a stress level σ under a particular load resulting in a corresponding strain of ε, Hooke's law applies within its elastic limit as follows (Salsa 2016):

$$\sigma = E\,\varepsilon \tag{1.34}$$

where σ is stress, ε is strain, and E is the Young's modulus (or modulus of elasticity).

As an application of Hooke's law, here we calculate the deformation of a concrete column in a concrete building with cross section A and Young's modulus E (Figure 1.10). The stress applied to this column and the corresponding strain are calculated as follows:

$$\sigma = \frac{P}{A} \tag{1.35}$$

and

$$\varepsilon = \frac{\Delta L}{L} \tag{1.36}$$

where ΔL is the deformation of the column under the load P (Figure 1.10). By combining Eqs (1.34), (1.35), and (1.36), the deformation can be expressed as

$$\Delta L = \frac{PL}{EA} \tag{1.37}$$

The deformation of the column in Eq (1.37) is plotted in Figure 1.11 assuming a square cross section for a concrete column with dimensions 0.3 m × 0.3 m (cross-sectional area of $A = 0.09$ m^2), a length of 3 m for the column ($L = 3$ m), a Young's modulus of $E = 2.5 \times 10^7$ kN/m^2 for concrete, and by changing the concentrated force in the domain $0 - 200$ kN. The MATLAB script in Code 1.4 can be used to generate Figure 1.11. Note that the results of Eq (1.37) and Figure 1.11 are valid only within the elastic limit of the column.

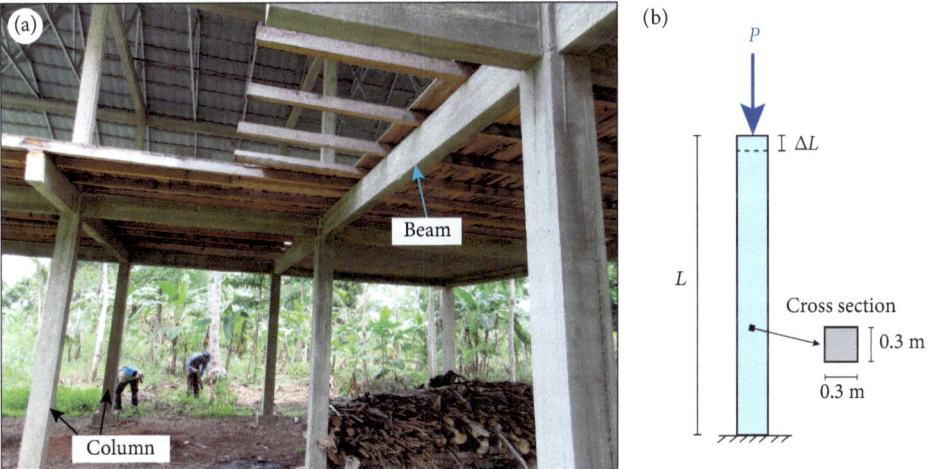

Figure 1.10 (a) Photo showing a column in a residential concrete building under construction in Indonesia. (b) Sketch of a column with length L under the concentrated load P.

Code 1.4 *MATLAB code for plotting the deformation of a column of length 3 m ($L = 3$ m) as a function of loading, under a point load in the range of $P = 0 - 200$ kN, a Young's modulus of $E = 2.5 \times 10^7$ kN/m^2, and a cross-sectional area of $A = 0.09$ m^2. This code generates Figure 1.11.*

```
% This code plots the deformation of a
% column under a point load (P)
clc; clear; close all;
set(0,'DefaultAxesFontsize',16);
%
L=3; % length of beam (m);
% P is point load (KN);
E=2.5.*10.^(7); % Young's modulus (kN/m^2);
A=0.09; % cross section area (m^2);
%
P=0:1:200; % point load P variations (kN)
deltaL=(P.*L./(E.*A)).*1000;
% deformation is reported in millimeters
%
subplot('position',[0.2 0.2 0.55 0.45]);
plot(P,deltaL,'b', 'LineWidth',1.5);
xlim([0 210]); ylim([0 0.3]);
xlabel('P, Point load (kN)');
ylabel('DeltaL, Deformation (mm)');
set(gca,'linewidth',1.5);
grid on;
%
% End of the code %
```

1.7.2 Darcy's law

Darcy's law is a fundamental law in hydrogeology and soil mechanics that governs the flow of water through porous media such as soil and rock. According to Das (2019), Darcy's law states that the rate of flow through porous media (Q) is directly correlated with the hydraulic gradient ($\Delta h/Z$, where Δh is the hydraulic head difference between two

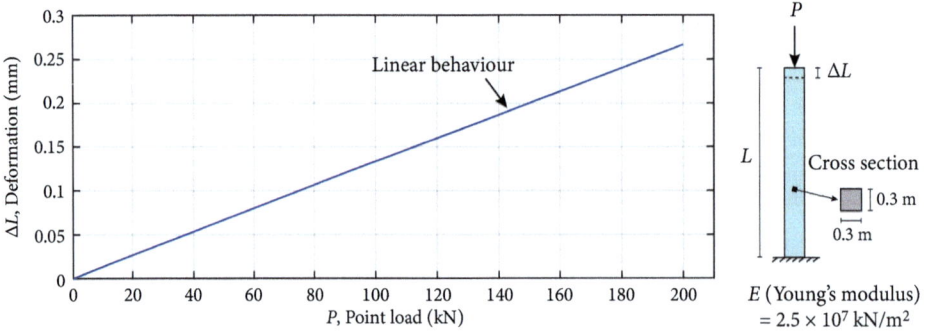

Figure 1.11 Plot of the deformation of a column of length 3 m ($L = 3$ m) as a function of loading, under a point load in the range of $P = 0 - 200$ kN, a Young's modulus of $E = 2.5 \times 10^7$ kN/m², and a cross-sectional area of $A = 0.09$ m². This figure is produced using Code 1.4.

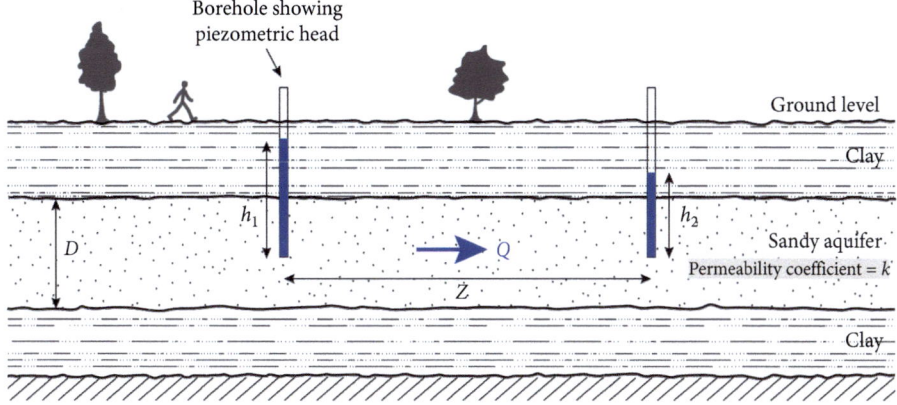

Figure 1.12 Sketch showing a sandy aquifer confined by two relatively impermeable clay layers. The two boreholes give the piezometric water heads at two locations.

points in the porous media as shown in Figure 1.12, and Z is the length of the flow path between the two points) and the hydraulic conductivity of the medium (k, also known as permeability coefficient):

$$Q = k A \frac{\Delta h}{Z} \tag{1.38}$$

where Q is the discharge, k is the hydraulic conductivity of the medium (also known as the permeability coefficient), A is the cross-sectional area of the porous medium, Δh is the hydraulic head difference, and Z is the length of the flow path (Figure 1.12). The value of k depends on the type of the materials. For example, it is approximately 5×10^{-3} m/s for medium sand, but it is around 1×10^{-9} m/s for silty clay (Das 2019). Darcy's law can also be expressed in terms of water flow velocity through soil (v) as follows:

$$v = k \frac{\Delta h}{Z} \tag{1.39}$$

1.7.3 Mohr–Coulomb failure criterion

The Mohr–Coulomb (MC) failure criterion describes the failure of soil materials under various conditions. This theory states that the failure is produced by a combination of normal (σ) and shear (τ) stresses (Das 2019), as described by the following equation:

$$\tau = c + \sigma \tan \phi \tag{1.40}$$

where τ is the shear strength on the failure plane, σ is the normal stress on the failure plane, c is the cohesion of the soil, and ϕ is the angle of friction of the soil (or internal

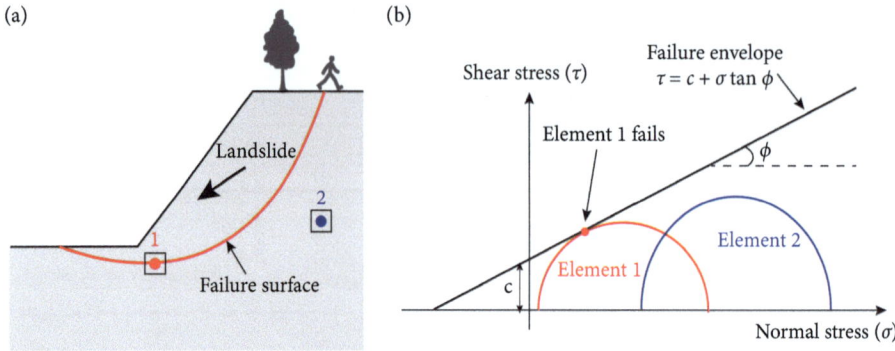

Figure 1.13 (a) Sketch of a slope where a failure surface is marked with two soil elements, 1 and 2. Soil element 1 is obtained from the failure surface, while element 2 is taken from an unaffected part of the slope. (b) Graphical representation of the Mohr–Coulomb failure criterion and the stress condition of the two soil elements.

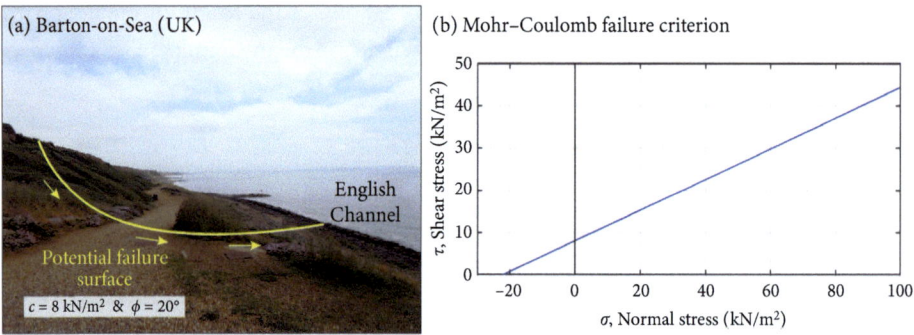

Figure 1.14 (a) Photo of a slope in Barton-on-Sea, UK. (b) The Mohr–Coulomb failure criterion for a potential slope failure assuming cohesion $c = 8$ kN/m^2 and internal friction angle $\phi = 20°$.

friction angle). Equation (1.40) is shown graphically in Figure 1.13b, where the MC failure envelope is a straight line. In Figure 1.13b, we assume that the slope fails along the curve marked in red. This curve is named the 'failure surface'. Two soil elements are marked in this slope (Figure 1.13): element 1 is located on the failure surface while element 2 belongs to an unaffected part of the slope. According to the Mohr–Coulomb failure criterion, the stress conditions in element 1 result in a state where the combination of its normal and shear stresses touches the failure envelope, whereas this is not the case for element 2 (see Figure 1.13).

As an application of the Mohr–Coulomb failure criterion, we calculate the failure criterion for a potential landslide at Barton-on-Sea, UK (Figure 1.14a). The soil parameters are assumed as cohesion $c = 8$ kN/m^2 and internal friction angle $\phi = 20°$. The outcome of the calculations is presented in Figure 1.14b using the MATLAB script in Code 1.5. Figure 1.14b shows that the failure envelope is a straight line demonstrating a linear correlation between normal stress (σ) and shear stress (τ).

Code 1.5 *MATLAB code for plotting the MC failure criterion for a potential slope failure assuming cohesion c = 8 kN/m² and internal friction angle φ = 20°. This code generates Figure 1.14b.*

```matlab
% This code plots the Mohr-Coulomb
% failure envelope for a soil
clc; clear; close all;
set(0,'DefaultAxesFontsize',16);
%
c=8; % soil cohesion (kN/m^2);
phi= 20; % internal friction angle (degree);
% 'tau' is shear stress
% 'zigma' is normal stress
zigma=-22:1:100; % normal stress (kN/m^2)
tau= c + zigma.* tan(phi.*pi./180);
% deformation is reported in millimeters
%
subplot('position',[0.2 0.2 0.45 0.45]);
plot(zigma,tau,'b', 'LineWidth',1.5);
xlim([-30 100]); ylim([0 50]);
xlabel('\sigma, Normal stress (kN/m^2)');
ylabel('\tau, Shear stress (kN/m^2)');
line([0 0],[-10 1000],'color','k','linewidth',1.5);
set(gca,'linewidth',1.5);
grid on;
%
% End of the code %
```

1.8 Problems for further study

Problem 1.1

By applying the seven steps discussed in this chapter for making mathematical models, develop a mathematical model for the deflection of a beam, $y(x)$, in the form of an ordinary differential equation.

Answer to Problem 1.1

$\frac{d^2 y(x)}{dx^2} = -\frac{M(x)}{EI}$, where $y(x)$ is the deflection at distance x from the origin, E is Young's modulus of elasticity, $M(x)$ is the bending moment at distance x, and I is the moment of inertia of the beam's cross section.

Problem 1.2

A dam reservoir, containing a water volume of 78×10^6 m³, has been contaminated by chemical pollutants with an initial concentration of 450 ppm. If the acceptable concentration of pollution in the reservoir is 75 ppm, how many days will it take for the pollution concentration to reach this acceptable level? Develop MATLAB code for this problem and plot the outcome.

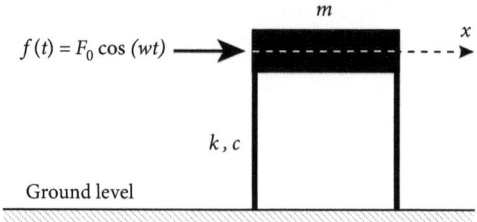

Figure 1.15 Sketch showing a single-storey building with stiffness coefficient k and damping coefficient c, under external load $f(t) = F_0 \cos(wt)$.

Problem 1.3

Consider a single-storey building with mass m, spring stiffness k, and damping coefficient $c = 0$ (an undamped structure) which is under external forcing $f(t) = F_0 \cos(wt)$, where F_0 is a constant value, t is time, and w is the frequency of the incident wave of wind or earthquake (Figure 1.15). Solve the oscillations of the building under this periodic external loading and discuss the results.

References

Bender, E.A. (2000). *An Introduction to Mathematical Modeling.* Dover Publications.
Das, B.M. (2019). *Advanced Soil Mechanics*, 5th edn. CRC Press.
Kreyszig, E. (2011). *Advanced Engineering Mathematics.* John Wiley & Sons.
Salsa, S. (2016). *Partial Differential Equations in Action: From Modelling to Theory.* Springer.

Chapter 2
Structural Engineering

2.1 Introduction

Structural analysis is one of the core subjects for several engineering disciplines. In civil engineering structures, such as buildings, bridges, and dams, analysis of structural members is of major importance for reliable and efficient design. Structural analysis focuses on modelling and evaluation of a structure's behaviour under the action of loads. It can help us identify critical load-bearing structural components and predict their response to loading. Typical structural components in civil engineering structures are tension members (e.g., bars and cables), beams and columns, disks, plates, and shells. Each category of structural members is specially designed to withstand specific types of loading. Their combination results in structures such as trusses, frames, and domes (e.g., Figure 2.1).

Structural analysis relies heavily on solid mechanics, the mechanical properties of engineering materials, and methods of engineering mathematics. In this chapter, after

A multistorey building damaged during the 1 January 2024 Noto Peninsula (Japan) M7.5 earthquake

Column

Beam

Foundation

Figure 2.1 Photo showing the structural elements of a multistorey building which collapsed during the 1 January 2024 Noto Peninsula earthquake in Japan.

A Practical Approach to Advanced Mathematical Modelling in Civil Engineering. Mohammad Heidarzadeh et al., Oxford University Press. © Mohammad Heidarzadeh, Theodosios K. Papathanasiou, Yurui Fan, Hamid Bahai (2025). DOI: 10.1093/9780191888656.003.0002

a brief introduction to solid mechanics and some basics on the mechanical properties of engineering materials, the governing equations and solution methodologies for several structural members are presented. The response of bars, beams, torsional bars, plates, and shells will be studied using analytical or numerical solution methodologies. Finally, the concept of stress functions and their application to problems of stress concentration are presented.

2.2 Mechanics of solids and structures

2.2.1 Stress tensor and conservation of momentum

Under the action of loads, a structure can translate and rotate (rigid body motions) but also deform. In static equilibrium, once the rigid body motions are appropriately constrained, the structure deforms and develops internal stresses, until a new equilibrium state is achieved. Rigid body motion in three-dimensional space includes three translations and three rotations. The load-bearing capacity of a structure depends on appropriate constraints on some or all of the rigid body motion degrees of freedom.

Mathematical modelling of the static or dynamic equilibrium of structures relies on the identification of physical quantities that are conserved. Conservation of **linear momentum**, **angular momentum**, and **energy** play crucial roles in structural analysis in the sense that they govern a structure's response (Love 1944). To derive the equations of equilibrium for a solid with density ρ, a differential volume with edge lengths dx, dy, dz is considered, as depicted in Figure 2.2. On each face of the volume, normal and shear stresses appear, representing the action of neighbouring particles in this free body diagram where the volume is isolated from the body of a solid.

Conservation of linear momentum (Newton's law) along the z axis (Figure 2.2) implies that forces due to stresses and body forces must be equal to the inertia force along the same direction, as represented by the following equation:

$$\sum F_z = \rho dx dy dz \frac{\partial^2 w}{\partial t^2} \tag{2.1}$$

In Eq (2.1), w represents the displacement along the z direction. On the face of the differential volume located in the xy plane, there is the action of the force $\sigma_{zz}(x, y, z)\, dxdy$, along direction z, due to the normal stress σ_{zz}. On the face parallel to the xy plane with vertical coordinates $(x, y, z + dz)$, there is the action of the force $\sigma_{zz}(x, y, z + dz)\, dxdy$. As the distance dz is very small, using Taylor's expansion and retaining only the linear term in dz, we have

$$\sigma_{zz}(x, y, z + dz)\, dxdy \approx \sigma_{zz}(x, y, z)\, dxdy + \frac{\partial \sigma_{zz}}{\partial z} dz dx dy \tag{2.2}$$

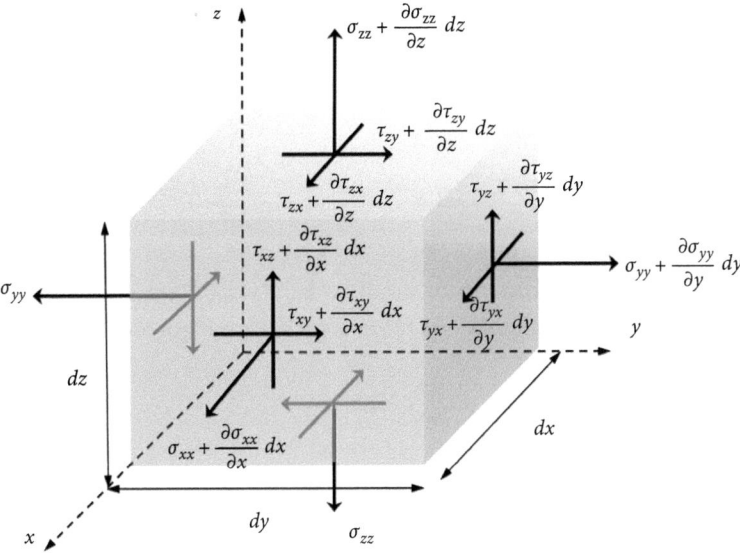

Figure 2.2 Free body diagram for a differential volume.

On the face in the yz plane, the shear force $\tau_{xz}(x, y, z)\, dydz$ acts along direction z (Figure 2.2). Similarly, on the face parallel to the yz plane located at $x + dx$, along the z direction, we can write

$$\tau_{xz}(x + dx, y, z)\, dydz \approx \tau_{xz}(x, y, z)\, dydz + \frac{\partial \tau_{xz}}{\partial x} dxdydz \qquad (2.3)$$

Finally, on the face in the zx plane, there is the action of the force $\tau_{yz}(x, y, z)\, dxdz$ along the z direction. At distance $y + dy$ the shear force acting along z is given by the following equation:

$$\tau_{yz}(x, y + dy, z)\, dzdx \approx \tau_{yz}(x, y, z)\, dzdx + \frac{\partial \tau_{yz}}{\partial y} dydzdx \qquad (2.4)$$

Accounting also for body forces, e.g., weight, Eq (2.1) in the z direction can be written as

$$\left(\sigma_{zz} + \frac{\partial \sigma_{zz}}{\partial z} dz\right) dxdy + \left(\tau_{xz} + \frac{\partial \tau_{xz}}{\partial x} dx\right) dydz + \left(\tau_{yz} + \frac{\partial \tau_{yz}}{\partial y} dy\right) dzdx - \sigma_{zz}dxdy - \tau_{xz}dydz$$

$$- \tau_{yz}dzdx + b_z dxdydz = \rho dxdydz\frac{\partial^2 w}{\partial t^2}$$

Working similarly for the x and y directions with displacements u and v, respectively, the three equations of equilibrium, expressing conservation of linear momentum, are derived as follows:

$$\frac{\partial \sigma_{xx}}{\partial x} + \frac{\partial \tau_{yx}}{\partial y} + \frac{\partial \tau_{zx}}{\partial z} + b_x = \rho \frac{\partial^2 u}{\partial t^2} \tag{2.5a}$$

$$\frac{\partial \tau_{xy}}{\partial x} + \frac{\partial \sigma_{yy}}{\partial y} + \frac{\partial \tau_{zy}}{\partial z} + b_y = \rho \frac{\partial^2 v}{\partial t^2} \tag{2.5b}$$

$$\frac{\partial \tau_{xz}}{\partial x} + \frac{\partial \tau_{yz}}{\partial y} + \frac{\partial \sigma_{zz}}{\partial z} + b_z = \rho \frac{\partial^2 w}{\partial t^2} \tag{2.5c}$$

Conservation of angular momentum needs to be analysed too. Regarding rotations around the z axis, equilibrium implies that if rotary inertia is ingored, it is

$$\sum M_z = 0 \tag{2.6}$$

The moments of forces around the z axis, with respect to the centre of the cube in Figure 2.2, are given by the following equation:

$$\left(\tau_{xy} + \frac{\partial \tau_{xy}}{\partial x} dx\right) dydz \frac{dx}{2} - \tau_{xy} dydz \frac{dx}{2} = \left(\tau_{yx} + \frac{\partial \tau_{yx}}{\partial y} dy\right) dzdx \frac{dy}{2} - \tau_{yx} dzdx \frac{dy}{2}$$

If terms with second-order differentials, i.e., $dx^2/2$ and $dy^2/2$, are ignored, since they are very small compared to the first-order terms, this becomes:

$$\tau_{xy} dydz \frac{dx}{2} = \tau_{yx} dzdx \frac{dy}{2}$$

Working similarly for the equilibrium of rotations around the x and y axes, the equations of angular momentum conservation are derived as:

$$\tau_{xy} = \tau_{yx} \tag{2.7a}$$

$$\tau_{yz} = \tau_{zy} \tag{2.7b}$$

$$\tau_{zx} = \tau_{xz} \tag{2.7c}$$

These equations imply symmetry of the shear stress components with respect to index reversal. Consequently, they imply the symmetry of the stress tensor defined as:

$$\boldsymbol{\sigma} = \begin{bmatrix} \sigma_{xx} & \tau_{xy} & \tau_{zx} \\ \tau_{xy} & \sigma_{yy} & \tau_{yz} \\ \tau_{zx} & \tau_{yz} & \sigma_{zz} \end{bmatrix} \tag{2.8}$$

By defining the body force vector $\boldsymbol{b} = \begin{bmatrix} b_x & b_y & b_z \end{bmatrix}^T$, along with the displacement vector $\boldsymbol{u} = \begin{bmatrix} u & v & w \end{bmatrix}^T$, the equations of equilibrium can be written in compact form as

$$div\,\boldsymbol{\sigma} + \boldsymbol{b} = \rho \frac{\partial^2 \boldsymbol{u}}{\partial t^2} \tag{2.9}$$

The stress tensor trace is equal to minus three times the pressure (p) at a given stress state, denoted by the following equation:

$$tr(\sigma) = \sigma_{xx} + \sigma_{yy} + \sigma_{zz} = -3p \tag{2.10}$$

The three stress tensor invariants (I_1, I_2, and I_3) are defined as follows:

$$I_1 = tr(\sigma) = \sigma_{xx} + \sigma_{yy} + \sigma_{zz} \tag{2.11a}$$

$$I_2 = \begin{vmatrix} \sigma_{yy} & \tau_{yz} \\ \tau_{yz} & \sigma_{zz} \end{vmatrix} + \begin{vmatrix} \sigma_{xx} & \tau_{zx} \\ \tau_{zx} & \sigma_{zz} \end{vmatrix} + \begin{vmatrix} \sigma_{xx} & \tau_{xy} \\ \tau_{xy} & \sigma_{yy} \end{vmatrix} = \sigma_{xx}\sigma_{yy} + \sigma_{yy}\sigma_{zz} + \sigma_{zz}\sigma_{xx} \tag{2.11b}$$
$$- \tau_{xy}^2 - \tau_{yz}^2 - \tau_{zx}^2$$

$$I_3 = \det(\sigma) = \sigma_{xx}\sigma_{yy}\sigma_{zz} + 2\tau_{xy}\tau_{yz}\tau_{zx} - \sigma_{zz}\tau_{xy}^2 - \sigma_{xx}\tau_{yz}^2 - \sigma_{yy}\tau_{zx}^2 \tag{2.11c}$$

The eigenvalues of the stress tensor are the following principal stresses (Fung et al. 2001):

$$\sigma_I = \frac{I_1}{3} + \frac{2}{3}\cos(\varphi)\sqrt{I_1^2 - 3I_2} \tag{2.12a}$$

$$\sigma_{II} = \frac{I_1}{3} + \frac{2}{3}\cos\left(\varphi - \frac{2\pi}{3}\right)\sqrt{I_1^2 - 3I_2} \tag{2.12b}$$

$$\sigma_{III} = \frac{I_1}{3} + \frac{2}{3}\cos\left(\varphi - \frac{4\pi}{3}\right)\sqrt{I_1^2 - 3I_2} \tag{2.12c}$$

where $\varphi = \frac{1}{3}\cos^{-1}\left(\frac{2I_1^3 - 9I_1 I_2 + 27I_3}{2(I_1^2 - 3I_2)^{3/2}}\right)$.

Principal stresses, being the eigenvalues of the stress tensor, correspond to principal directions that are the eigenvectors of the stress tensor. Each principal stress corresponds to one principal direction. Given a specific stress state inside a solid, principal directions are directions along which shear stresses disappear and the stress state can be expressed using only the three normal components σ_I, σ_{II}, and σ_{III}. A graphical method to calculate the principal stresses is Mohr's circle (Love 1944). Finally, using the mean stress $\sigma_m = tr(\sigma)/3$, the stress tensor can be decomposed into a **mean stress** and a **deviatoric** part as follows:

$$\begin{bmatrix} \sigma_{xx} & \tau_{xy} & \tau_{zx} \\ \tau_{xy} & \sigma_{yy} & \tau_{yz} \\ \tau_{zx} & \tau_{yz} & \sigma_{zz} \end{bmatrix} = \begin{bmatrix} \sigma_m & 0 & 0 \\ 0 & \sigma_m & 0 \\ 0 & 0 & \sigma_m \end{bmatrix} + \begin{bmatrix} \sigma_{xx} - \sigma_m & \tau_{xy} & \tau_{zx} \\ \tau_{xy} & \sigma_{yy} - \sigma_m & \tau_{yz} \\ \tau_{zx} & \tau_{yz} & \sigma_{zz} - \sigma_m \end{bmatrix} = \sigma_m + s$$

The mean stress part expresses changes in volume, while the deviatoric part s corresponds to volume distortion. The deviatoric part, being a stress tensor itself, has invariants $(J_1, J_2,$ and $J_3)$ which are:

$$J_1 = tr(s) = 0, \quad J_2 = \frac{1}{3}I_1^2 - I_2, \quad J_3 = \frac{2}{27}I_1^3 - \frac{1}{3}I_1I_2 + I_3$$

The quantity $\sigma_{VM} = \sqrt{3J_2}$ is called the von Mises stress and is a useful and popular parameter in the plasiticity theory of solids (Fung et al. 2001).

2.2.2 Kinematics, deformation, and strain tensor

Rigid body motions consist of three translational displacements and three rotations. Displacement along the x axis is referred to as surge, displacement along the y axis is termed sway, and vertical motion (motion along the z axis in Figure 2.3) is called heave. Rotation around the x axis is termed roll, around the y axis is named pitch, and around the z axis is called yaw (Figure 2.3). Once the rigid body motions are appropriately constrained, the action of loads leads to deformations.

Different types of forcing lead to different types of deformation. In the three-dimensional setting, the displacement of a particle at point $P(x, y, z)$ is expressed through the vector $\mathbf{u} = \begin{bmatrix} u & v & w \end{bmatrix}^T$, where $u = u(x, y, z, t)$ is the displacement along the x axis, $v = v(x, y, z, t)$ is the displacement along the y axis, and $w = w(x, y, z, t)$ is the displacement along the z axis. Relative differences in displacement between particles lead to deformation. Strains (ε) are nondimensional quantities constituting measures of deformation in an elastic solid. They represent displacements between particles in a solid relative to a reference state. There are two types of strain, **normal strains** and **shear strains**.

Normal strains express changes in length, relative to an initial state. Regarding a differential volume with edge lengths dx, dy, and dz, a material point at $P(x, y, z)$ is assumed to be displaced by $u(x, y, z, t)$ along the x axis. Similarly, a material point at $Q(x + dx, y, z)$

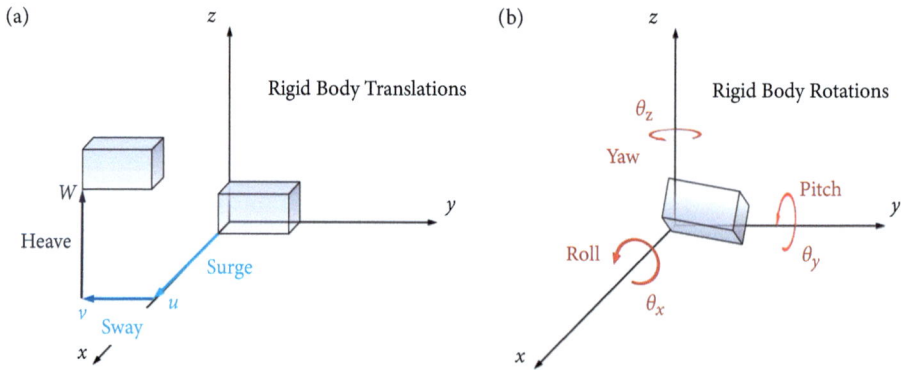

Figure 2.3 Rigid body motion consists of: (a) three translations, and (b) three rotations.

is displaced along the x axis by $u(x + dx, y, z, t)$. Since the distance dx is very small, using Taylor's expansion around point $P(x, y, z)$ and retaining only the linear term, we have

$$u(x + dx, y, z, t) = u(x, y, z, t) + \frac{\partial u}{\partial x} dx \tag{2.13}$$

The normal strains (ε_{xx}, ε_{yy}, and ε_{zz}) are depicted in Figure 2.4a. Working similarly for the displacement along the y and z directions, and given the definition of strain as change in length relative to a reference length, the normal strains are denoted as follows:

$$\varepsilon_{xx} = \frac{u(x + dx, y, z, t) - u(x, y, z, t)}{dx} = \frac{u(x, y, z, t) + \frac{\partial u}{\partial x} dx - u(x, y, z, t)}{dx} = \frac{\partial u}{\partial x} \tag{2.14a}$$

$$\varepsilon_{yy} = \frac{v(x, y + dy, z, t) - v(x, y, z, t)}{dy} = \frac{v(x, y, z, t) + \frac{\partial v}{\partial y} dy - v(x, y, z, t)}{dy} = \frac{\partial v}{\partial y} \tag{2.14b}$$

$$\varepsilon_{zz} = \frac{w(x, y, z + dz, t) - w(x, y, z, t)}{dz} = \frac{w(x, y, z, t) + \frac{\partial w}{\partial z} dz - w(x, y, z, t)}{dz} = \frac{\partial w}{\partial z} \tag{2.14c}$$

The shear strains (γ_{xy}, γ_{yz}, γ_{zx}, ε_{xy}, ε_{yz}, and ε_{zx}) express changes in shape. They are defined as changes in the angles of the differential cube depicted in Figure 2.2. With reference to Figure 2.4b, the engineering shear strain for small deformation angles is defined as

$$\gamma_{xy} = \alpha_{xy} + \beta_{xy} \approx \tan(\alpha_{xy}) + \tan(\beta_{xy}) = \frac{\frac{\partial v}{\partial x} dx}{dx} + \frac{\frac{\partial u}{\partial y} dy}{dy} = \frac{\partial v}{\partial x} + \frac{\partial u}{\partial y} \tag{2.15a}$$

$$\gamma_{yz} = \alpha_{yz} + \beta_{yz} \approx \tan(\alpha_{yz}) + \tan(\beta_{yz}) = \frac{\frac{\partial w}{\partial y} dy}{dy} + \frac{\frac{\partial v}{\partial z} dz}{dz} = \frac{\partial w}{\partial y} + \frac{\partial v}{\partial z} \tag{2.15b}$$

$$\gamma_{zx} = \alpha_{zx} + \beta_{zx} \approx \tan(\alpha_{zx}) + \tan(\beta_{zx}) = \frac{\frac{\partial u}{\partial z} dz}{dz} + \frac{\frac{\partial w}{\partial x} dx}{dx} = \frac{\partial u}{\partial z} + \frac{\partial w}{\partial x} \tag{2.15c}$$

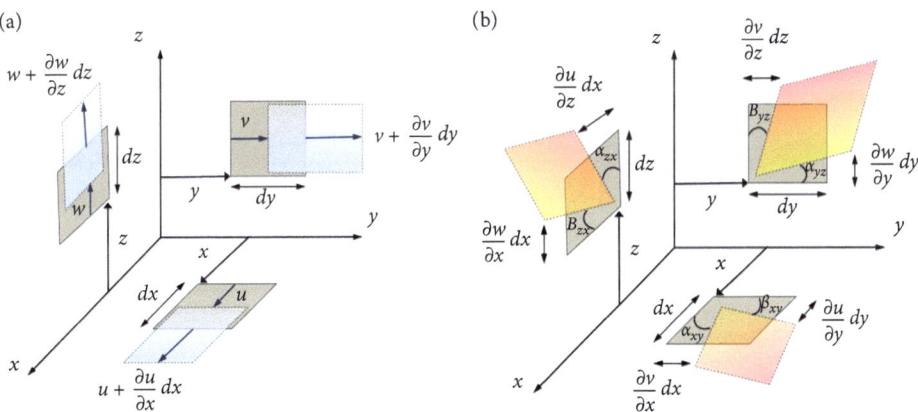

Figure 2.4 Definition of (a) normal and (b) shear strains in an elastic solid.

The shear strains ε_{xy}, ε_{yz}, and ε_{zx} are

$$\varepsilon_{xy} = \frac{1}{2}\gamma_{xy} = \frac{1}{2}\left(\frac{\partial v}{\partial x} + \frac{\partial u}{\partial y}\right) \tag{2.16a}$$

$$\varepsilon_{yz} = \frac{1}{2}\gamma_{yz} = \frac{1}{2}\left(\frac{\partial w}{\partial y} + \frac{\partial v}{\partial z}\right) \tag{2.16b}$$

$$\varepsilon_{zx} = \frac{1}{2}\gamma_{zx} = \frac{1}{2}\left(\frac{\partial u}{\partial z} + \frac{\partial w}{\partial x}\right) \tag{2.16c}$$

Consequently, the shear strains define the strain tensor as given by the following matrix:

$$\varepsilon = \begin{bmatrix} \varepsilon_{xx} & \varepsilon_{xy} & \varepsilon_{zx} \\ \varepsilon_{xy} & \varepsilon_{yy} & \varepsilon_{yz} \\ \varepsilon_{zx} & \varepsilon_{yz} & \varepsilon_{zz} \end{bmatrix} \tag{2.17}$$

Finally, changes in volume can be expressed through the **dilatation** (relative changes in volume) δ (Love 1944), which is defined as the trace of the strain tensor, denoted by the following equation:

$$\delta = tr(\varepsilon) = \varepsilon_{xx} + \varepsilon_{yy} + \varepsilon_{zz} = \frac{\partial u}{\partial x} + \frac{\partial v}{\partial y} + \frac{\partial w}{\partial z} \tag{2.18}$$

2.2.3 External loads and stress resultants

Under the action of external loads, an appropriately supported structure develops internal stresses and deforms. If the structure is not fully constrained, in the sense that some possible rigid body motion is not restrained, then it could translate or rotate like a mechanism when external loads are applied to it (see Section 2.2.1). Several cases of external loads are depicted in Figure 2.5. In general, loads can be distributed on the volume (e.g., weight), areas/lines, or can be concentrated. In the general three-dimensional case, concentrated loads include the following:

- externally applied forces F_x, F_y, and F_z;
- externally applied moments T_x, T_y, and T_z.

Although there are several types of loads, there are only two types of stresses developed internally in a solid, namely normal stresses and shear stresses. Of major importance for engineering calculations are the **stress resultants** on a cross section of a structure. Stress resultants include axial and shear forces, and bending and torsional moments. Their calculation is possible if the stresses acting on an area A are known. With reference to Figure 2.5, we have:

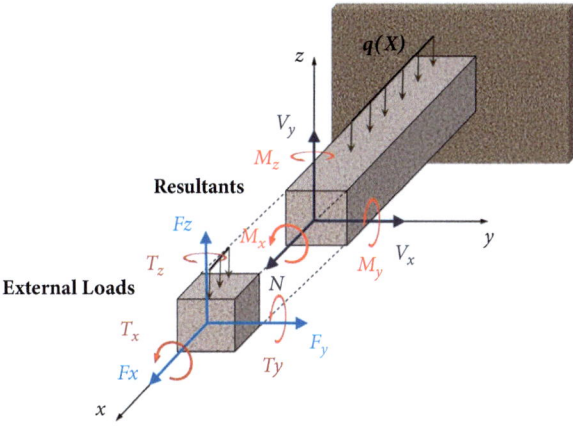

Figure 2.5 External loads and resultants inside a structural component.

- Axial force $N = \int_A \sigma_{xx} dA$
- Shear forces $V_x = \int_A \tau_{xy} dA, \quad V_y = \int_A \tau_{zx} dA$
- Bending moments $M_y = \int_A z\sigma_{xx} dA, \quad M_z = \int_A y\sigma_{xx} dA$
- Torsional moment $M_x = \int_A \left(y\tau_{zx} - z\tau_{xy} \right) dA$

Alternatively, the stress resultants in certain cases can be calculated using the equations of equilibrium. This is the case for isostatic (statically determinate) structures. Consequently, the above formulas can be used to calculate the stresses (Fung et al. 2001).

2.2.4 Deformation energy and work

When an elastic solid deforms, potential energy is stored in it due to the change of state. This energy is called strain energy. The strain energy density for a linear elastic solid is

$$dW = \frac{1}{2} \left(\sigma_{xx}\varepsilon_{xx} + \sigma_{yy}\varepsilon_{yy} + \sigma_{zz}\varepsilon_{zz} + \tau_{xy}\gamma_{xy} + \tau_{yz}\gamma_{yz} + \tau_{zx}\gamma_{zx} \right) \quad (2.19)$$

The strain energy inside the solid can be calculated as the volume integral of the strain energy density, i.e., $W = \int_V dW$ over the volume of the solid (Fung et al. 2001).

2.2.5 Material behaviour

Different solid materials feature differences in their behaviour during deformation. The basic categories of engineering materials (Callister 1993) are (Figure 2.6):

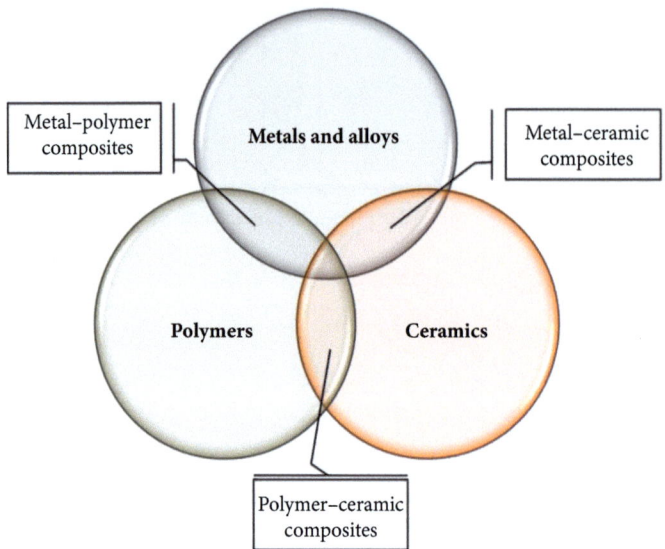

Figure 2.6 A classification of engineering materials.

- metals and alloys
- ceramics
- polymers
- composite materials.

Examples of metals and alloys used extensively in structural engineering are steel, aluminium, and cast iron. Alloys are materials resulting typically as a combination of metals or combination of metals with nonmetallic elements in small concentrations. Steel, an alloy of major importance for construction, is produced as a combination of iron and carbon. Ceramic materials are inorganic and can be classified as crystalline or noncrystalline. Some ceramic materials important in construction are silicon carbide (SiC), which is typically employed as a refractory material, and cement, which when mixed with aggregate produces mortar or concrete. Polymers are organic materials and can be classified as thermosetting, thermoplastic, or rubbers. Polymers in construction appear as adhesives, foams, paints, or sealants.

Composite materials are combinations of the categories above (Figure 2.6). Examples of composite materials include reinforced concrete, plywood, and fibre-reinforced polymers. Desirable properties of composite materials include increased strength-to-weight ratio and directionality in the material properties, e.g., increased strength along the direction of fibres.

Materials can be classified according to their internal structure into the following categories (Fung et al. 2001; Callister 1993):

(i) **Homogeneous or heterogeneous**: Homogeneous materials have the same properties (e.g., density or conductivity) at every point, while heterogeneous materials have material properties that depend on the specific point coordinates. Heterogeneous materials are functionally graded materials (FGM). In FGM, mechanical properties like Young's modulus are functions of the material point coordinates, i.e., $E = E(x, y, z)$.

(ii) **Isotropic or anisotropic:** Isotropic materials are those where certain material properties (e.g., flexibility or conductivity) do not depend on specific directions inside the solid body. Metals and alloys are typically isotropic materials. Anisotropic materials have moduli that change along different directions. A typical example of an anisotropic material is wood. In the case of wood, the strength (i.e., the ability to withstand tension in particular) along the grain is greater than the strength perpendicular to the direction of grain. Wood is an example of an orthotropic material. Anisotropic materials can be:

- **Transversely isotropic:** This happens when material properties are isotropic on a specific plane inside the material and there is symmetry with respect to an axis perpendicular to this plane. Some types of sedimentary rocks can be considered as transversely isotropic materials.
- **Orthotropic:** In this case, the material properties differ along three mutually orthogonal axes of rotational symmetry inside the material.
- **Fully anisotropic:** This occurs when the properties along each direction in the material are different.

(iii) **Ductile or brittle:** These concepts are typically used in a comparative framework classifying one material as being more ductile or more brittle than another material. Ductility is a measure of the material's ability to develop permanent deformations before rupture. A brittle material does not develop significant permanent deformations before fracture (Figure 2.7).

The response of deformable solids can be characterized as follows:

- **Linear elastic** behaviour occurs when stress and strain components maintain direct proportionality without resulting in permanent (plastic) deformations (Figure 2.7).
- **Nonlinear elastic** behaviour manifests when stress and strain components do not exhibit direct proportionality, yet no permanent (plastic) deformations occur.
- **Plastic** behaviour is when permanent deformations occur that do not vanish under the removal of the cause, e.g., stress. There are different types of plastic behaviour, and the differences are typically related to how imperfections inside the material move or multiply during the action of large-magnitude loads or excessive deformations.
- **Viscoelastic** behaviour happens when the material exhibits both elastic and viscous characteristics. In this setting, the elastic behaviour is rate dependent, and hereditary effects of the deformation influence the final state of the material under the action of loads.
- **Viscoplastic** behaviour occurs when the material exhibits rate-dependent inelastic behaviour.

A typical stress–strain diagram for a relatively brittle and a relatively ductile material under uniaxial loading is depicted in Figure 2.7. It is worth mentioning that two types of stress–

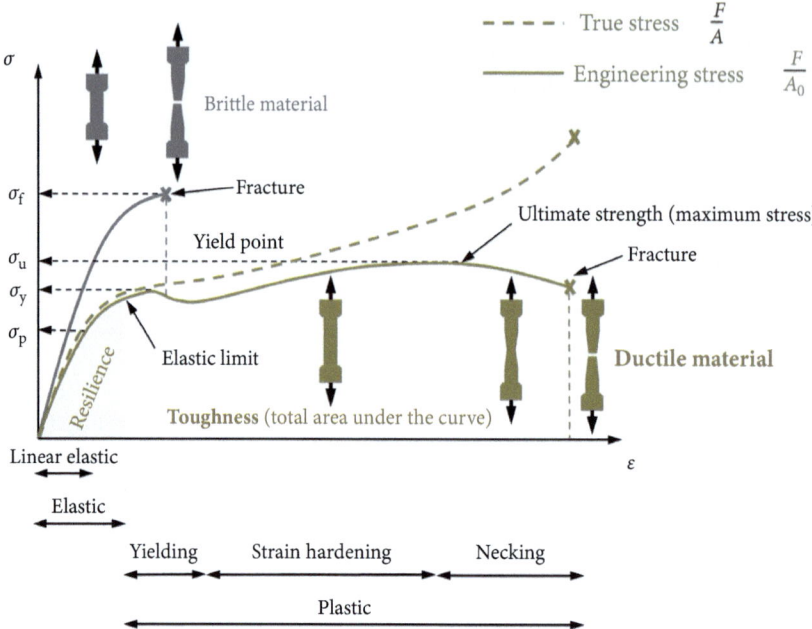

Figure 2.7 Stress–strain diagram for a brittle and a ductile material specimen under tension.

strain curve can be computed. They have to do with the specific stress–strain measures used. The stress in uniaxial tension tests is defined as the ratio of applied force (F) over the specimen cross-sectional area (A), i.e., F/A. Initially the area of the specimen is A_0. However, as the load increases, the specimen elongates, and the material cross section decreases to a new value A that depends on the magnitude of the applied load and the deformation state. The ratio F/A is the **true stress**. Typically, in engineering calculations, the ratio of the applied load over the area prior to deformation, F/A_0, is used. This ratio is termed the **engineering stress**.

Referring to Figure 2.7, for small values of stress and strain, both brittle and ductile materials exhibit linear elastic behaviour, followed by a nonlinear elastic regime. Up to this point, all deformations are reversible in the sense that if the loading is removed, the material returns to the original undeformed state. Once the stress increases beyond a certain point, the brittle material ruptures and fracture occurs. In the case of a ductile material, once the yield point is reached, permanent deformations occur and plastic behaviour emerges (Figure 2.7). In the case of several materials, e.g., structural steel, yielding is followed by a stress–strain curve that shows increasing stress with increasing strain. This is the strain-hardening regime where the slope of the stress–strain curve is positive. The behaviour changes when the ultimate strength point is reached. This is the maximum stress that the solid can withstand. It is followed by the necking band, where stress decreases with increasing strain. Finally, fracture occurs at the end point of this regime. Note that the behaviour in terms of the true stress is different, featuring a curve with positive slope up to the fracture point (Figure 2.7). This is due to the reduction of the cross-sectional area of the specimen that leads to higher values of the true stress measure. For actual engineering calculations, typically the engineering stress is employed.

With reference to the stress–strain diagram shown in Figure 2.7, some important features are as follows:

- **Proportionality limit** (σ_p) is the greatest stress that appears in an elastic material while the material features linear elastic behaviour.
- **Elastic limit** is the maximum stress that a material can sustain without developing permanent deformation.
- **Modulus of elasticity** (E) is the slope (tangent) of the stress–strain curve in the linear elastic region.
- **Yield strength** (σ_y) is a stress value indicating the transition from elastic to plastic deformation. It approximates the elastic limit and is typically defined as the stress level at which the material exhibits a small, specified permanent deformation value.
- **Ultimate strength** (σ_u) is the maximum stress that appears in the engineering stress–strain diagram. It corresponds to necking initiation in several materials.
- **Resilience** is the area between the stress–strain curve and the strain axis, up to the elastic limit. It represents the capability of the material to store strain energy until the maximum elastic deformation.
- **Toughness** indicates the ability of a material to absorb energy without failure (fracture). The toughness modulus is represented by the total area under the stress–strain diagram.

Of major importance for structural engineering is the linear elastic regime. In this regime the structure behaviour is predictable and any deformation is not permanent. That is, the structure retains its original state after the loading effects are removed. The linear elastic stress–strain relation in this regime is governed by Hooke's law. For an anisotropic solid, using Voigt notation, where stresses and strains are arranged in vectors, Hooke's law is expressed as follows:

$$
\begin{bmatrix} \sigma_{xx} \\ \sigma_{yy} \\ \sigma_{zz} \\ \tau_{yz} \\ \tau_{xz} \\ \tau_{xy} \end{bmatrix} = \begin{bmatrix} C_{11} & C_{12} & C_{13} & C_{14} & C_{15} & C_{16} \\ C_{12} & C_{22} & C_{23} & C_{24} & C_{25} & C_{26} \\ C_{13} & C_{23} & C_{33} & C_{34} & C_{35} & C_{36} \\ C_{14} & C_{24} & C_{34} & C_{44} & C_{45} & C_{46} \\ C_{15} & C_{25} & C_{35} & C_{36} & C_{55} & C_{56} \\ C_{16} & C_{26} & C_{36} & C_{46} & C_{56} & C_{66} \end{bmatrix} \begin{bmatrix} \varepsilon_{xx} \\ \varepsilon_{yy} \\ \varepsilon_{zz} \\ \gamma_{yz} \\ \gamma_{xz} \\ \gamma_{xy} \end{bmatrix} \tag{2.20}
$$

where the constants C_{ij} represent elastic moduli of the material. For an orthotropic material, it is denoted by the following equation:

$$
\begin{bmatrix} \sigma_{xx} \\ \sigma_{yy} \\ \sigma_{zz} \\ \tau_{yz} \\ \tau_{xz} \\ \tau_{xy} \end{bmatrix} = \begin{bmatrix} C_{11} & C_{12} & C_{13} & 0 & 0 & 0 \\ C_{12} & C_{22} & C_{23} & 0 & 0 & 0 \\ C_{13} & C_{23} & C_{33} & 0 & 0 & 0 \\ 0 & 0 & 0 & C_{44} & 0 & 0 \\ 0 & 0 & 0 & 0 & C_{55} & 0 \\ 0 & 0 & 0 & 0 & 0 & C_{66} \end{bmatrix} \begin{bmatrix} \varepsilon_{xx} \\ \varepsilon_{yy} \\ \varepsilon_{zz} \\ \gamma_{yz} \\ \gamma_{xz} \\ \gamma_{xy} \end{bmatrix}
\tag{2.21}
$$

For a transversely isotropic material, with x being the axis of symmetry, the equation becomes

$$
\begin{bmatrix} \sigma_{xx} \\ \sigma_{yy} \\ \sigma_{zz} \\ \tau_{yz} \\ \tau_{xz} \\ \tau_{xy} \end{bmatrix} = \begin{bmatrix} C_{11} & C_{12} & C_{13} & 0 & 0 & 0 \\ C_{12} & C_{22} & C_{23} & 0 & 0 & 0 \\ C_{13} & C_{23} & C_{33} & 0 & 0 & 0 \\ 0 & 0 & 0 & C_{44} & 0 & 0 \\ 0 & 0 & 0 & 0 & C_{55} & 0 \\ 0 & 0 & 0 & 0 & 0 & C_{66} \end{bmatrix} \begin{bmatrix} \varepsilon_{xx} \\ \varepsilon_{yy} \\ \varepsilon_{zz} \\ \gamma_{yz} \\ \gamma_{xz} \\ \gamma_{xy} \end{bmatrix}
\tag{2.22}
$$

Finally, for an isotropic material with Young's modulus E and Poisson ratio v, if the Lamé constants $\lambda = \frac{Ev}{(1+v)(1-2v)}$ and $\mu = G = \frac{E}{2(1+v)}$ are introduced, the equation takes the following form:

$$
\begin{bmatrix} \sigma_{xx} \\ \sigma_{yy} \\ \sigma_{zz} \\ \tau_{yz} \\ \tau_{xz} \\ \tau_{xy} \end{bmatrix} = \begin{bmatrix} 2\mu + \lambda & \lambda & \lambda & 0 & 0 & 0 \\ \lambda & 2\mu + \lambda & \lambda & 0 & 0 & 0 \\ \lambda & \lambda & 2\mu + \lambda & 0 & 0 & 0 \\ 0 & 0 & 0 & \mu & 0 & 0 \\ 0 & 0 & 0 & 0 & \mu & 0 \\ 0 & 0 & 0 & 0 & 0 & \mu \end{bmatrix} \begin{bmatrix} \varepsilon_{xx} \\ \varepsilon_{yy} \\ \varepsilon_{zz} \\ \gamma_{yz} \\ \gamma_{xz} \\ \gamma_{xy} \end{bmatrix}
\tag{2.23}
$$

Poisson's ratio (v) is a nondimensional number that quantifies the Poisson effect, which describes the tendency of a material to expand in directions perpendicular to the direction of compression and to contract in directions perpendicular to the direction of stretching. Typical engineering materials have Poisson's ratio values in the interval between $v = 0$ and $v = 1/2$. In the case where $v = 1/2$, the material is incompressible.

2.2.6 Thermal stresses

Strain develops in elastic bodies due to differences in temperature as well. This type of strain is termed thermal strain and is proportional to the temperature difference (ΔT) and the linear thermal expansion coefficient (a_L). In typical engineering materials, temperature differences can cause only normal strains and therefore are related to normal stresses.

Hooke's law for an isotropic material featuring thermal stresses along with mechanical stresses is written as follows:

$$
\begin{bmatrix} \sigma_{xx} \\ \sigma_{yy} \\ \sigma_{zz} \\ \tau_{yz} \\ \tau_{xz} \\ \tau_{xy} \end{bmatrix} = \begin{bmatrix} 2\mu + \lambda & \lambda & \lambda & 0 & 0 & 0 \\ \lambda & 2\mu + \lambda & \lambda & 0 & 0 & 0 \\ \lambda & \lambda & 2\mu + \lambda & 0 & 0 & 0 \\ 0 & 0 & 0 & \mu & 0 & 0 \\ 0 & 0 & 0 & 0 & \mu & 0 \\ 0 & 0 & 0 & 0 & 0 & \mu \end{bmatrix} \begin{bmatrix} \varepsilon_{xx} \\ \varepsilon_{yy} \\ \varepsilon_{zz} \\ \gamma_{yz} \\ \gamma_{xz} \\ \gamma_{xy} \end{bmatrix} - \begin{bmatrix} a_L \Delta T \\ a_L \Delta T \\ a_L \Delta T \\ 0 \\ 0 \\ 0 \end{bmatrix}
$$

$$(2.24)$$

2.3 Bars

Bars are typically slender prismatic structural components that can sustain axial loads. The geometric characteristics of bars enable their use in lightweight structures such as trusses (Figure 2.8). Bars are used as tension members in bracing systems, trusses, and other applications in civil engineering. Figure 2.8a depicts a truss bridge.

The kinematic field associated with the deformation of bars is the axial displacement along the principal axis of the member, i.e., u. Due to the small cross-sectional area, Poisson effects can be neglected in simple bar models and the analysis becomes one-dimensional, relating the axial displacement (u) to the axial resultant force (N), as shown in Figure 2.8b.

2.3.1 Governing equations

Bar (or rod) structural members can sustain axial loads, typically tensile, and in response they undergo axial deformations. Consequently, in the analysis of bars it is the axial force

(a)   (b)

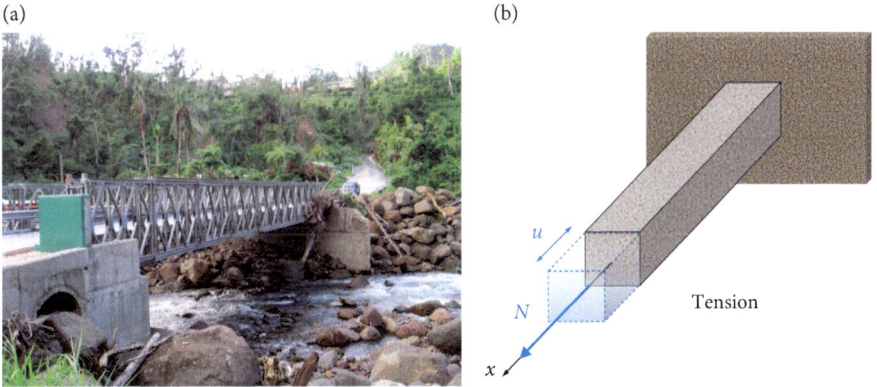

Figure 2.8 (a) A truss bridge in Dominica. (b) Axial deformation of a bar under the action of axial load.

(F_x) that is the cause of deformation. The corresponding internal resultant is typically denoted as $N = N(x, t)$ and depends on the axial spatial coordinate x and time t.

Application of Newton's second law of motion to a differential volume of a bar produces the following equation of motion:

$$N(x + dx, t) - N(x, t) + f(x, t) A dx = \rho A dx \frac{\partial^2 u}{\partial t^2} \tag{2.25}$$

where $f(x, t)$ is a volume-distributed load, ρ is the bar material density, and A is the cross-sectional area. Since the differential element with length dx is very small, the axial force at point $x + dx$ can be approximated (using a Taylor series expansion) as follows:

$$N(x + dx, t) \approx N(x, t) + \frac{\partial N}{\partial x} dx \tag{2.26}$$

The product $\rho A dx$ denotes the differential volume's mass (i.e., dm). Finally, the differential equation of motion for a bar in terms of the axial force and axial displacement is

$$\frac{\partial N}{\partial x} + f(x, t) A = \rho A \frac{\partial^2 u}{\partial t^2} \tag{2.27}$$

This is a single equation for two unknown fields. The final equation can be expressed in terms of the axial displacement field. Since the normal stress along the x axis is $\sigma_{xx} = N/A$, the previous equation is rewritten as

$$\frac{\partial (A \sigma_{xx})}{\partial x} + f(x, t) A = \rho A \frac{\partial^2 u}{\partial t^2} \tag{2.28}$$

At this point, assumptions on the material behaviour need to be made. Assuming an isotropic material, linear elastic behaviour, and neglecting the Poisson effect, Hooke's law implies $\sigma_{xx} = E \varepsilon_{xx}$. Using Hooke's law and the definition of strain ε_{xx}, the equation of motion for the axial deformation of the bar becomes

$$\frac{\partial}{\partial x} \left(EA \frac{\partial u}{\partial x} \right) + f(x, t) A = \rho A \frac{\partial^2 u}{\partial t^2}. \tag{2.29}$$

Equation (2.29) accounts for a spatially varying Young's modulus $E(x)$ and cross-sectional area $A(x)$. If these quantities are constant, Eq (2.29) takes the following form:

$$\frac{\partial^2 u}{\partial x^2} - \frac{1}{c^2} \frac{\partial^2 u}{\partial t^2} = \frac{f(x, t)}{E} \tag{2.30}$$

where $c = \sqrt{E/\rho}$ is the speed of longitudinal waves in the bar.

2.3.2 Static deformation of bars

Assuming that all forces acting on the bar do not depend on time, or their variation with time is very small, the inertia effects can be neglected, and the static deformation of the bar is governed by the following Ordinary Differential Equation (ODE):

$$\frac{d}{dx}\left(EA\frac{du}{dx}\right) + f(x)A = 0 \tag{2.31}$$

By integrating Eq (2.31) with respect to x, we obtain

$$EA\frac{du}{dx} + \int f(x)\,A dx = C_1 \tag{2.32}$$

where C_1 is an integration constant. Dividing by EA and integrating the equation again, we have

$$u(x) = -\int \frac{1}{EA}\int f(x)\,A dx dx + C_1 \int \frac{1}{EA}dx + C_2 \tag{2.33}$$

Using Hooke's law, the axial force can be written as $N(x) = \sigma_{xx}A = AE\varepsilon_{xx}$. Hence, the axial force can be calculated by Eq (2.32) as

$$N(x) = -\int f(x)\,A dx + C_1 \tag{2.34}$$

The general solution, Eq (2.33), features two integration constants, namely C_1 and C_2. This comes as no surprise since the governing equation is a second-order differential equation. The integration constants play an important role in the simulation of the bar deformation as they allow for the satisfaction of specific boundary conditions, which in turn model constraints of motion and applied loads in real-world configurations. Two types of boundary condition are useful and popular (see Table 2.1):

    (i) a fixed displacement location
    (ii) a specified external load.

These two conditions cannot appear simultaneously at the same end point of the bar but are complementary. The basic rules are:

- If the axial displacement is known at an end point of the bar, then the axial load is unknown. It corresponds to the reaction force at the constraint. It can be obtained using Eq (2.34), after the calculation of C_1 from $N(x_o)$, where $x = x_o$ is the location of the constraint.
- If the axial force is known at an end point of the bar, then the axial displacement is unknown. It is calculated by Eq (2.33) after the calculation of C_1 and C_2 by setting $x = L$, where L is the bar length. The **elongation** of the bar is the total change in the bar length. It can be calculated by setting $x = L$ in Eq (2.33).

Table 2.1 Boundary conditions for a bar.

End-point kinematic constraints and applied loads	Boundary conditions for bars
	Fixed end $\qquad\qquad u = 0$
	Prescribed axial force $\quad N = P \to EA\dfrac{du}{dx} = P \to \dfrac{du}{dx} = \dfrac{P}{EA}$
	Free end $\qquad\qquad N = 0 \to EA\dfrac{du}{dx} = 0 \to \dfrac{du}{dx} = 0$

Example calculations for the elongation of a bar with distributed and point loads, the elongation of a crane bridge cable, and thermal stresses in a bar are shown in Boxes 2.1, 2.2, and 2.3, respectively.

Box 2.1: Example (Elongation of a bar with distributed and point load)

A bar of length L and cross-sectional area A is considered. A uniformly distributed load $f(x) = f$ and a point force P at $x = L$ are acting on the bar, which deforms slowly until static equilibrium is achieved. The point $x = 0$ is fixed and the bar is linearly elastic with Young's modulus E. Find the elongation.

Solution

Using Eq (2.33) for the displacement with a constant distributed load f and constant cross section A, we have

$$u(x) = -\int \frac{1}{EA}\int fA\,dx\,dx + C_1\int \frac{1}{EA}\,dx + C_2 = -\frac{f}{2E}x^2 + \frac{C_1}{EA}x + C_2$$

Since the bar is fixed at the point $x = 0$,

$$u(0) = -\frac{f}{2E}0^2 + \frac{C_1}{EA}0 + C_2 = C_2 = 0$$

Now, using Eq (2.34) for the axial force, it is $N(x) = -\int fA\,dx + C_1 = -fAx + C_1$. But the axial force at $x = L$ is P, resulting in the following equation:

$$N(L) = -fAL + C_1 \text{ or } C_1 = P + fAL.$$

Finally, the displacement along the bar is

$$u(x) = -\frac{f}{2E}x^2 + \frac{(P+fAL)}{EA}x = \frac{Px}{EA} + \frac{fL}{E}x - \frac{f}{2E}x^2$$

Using this equation, and substituting $x = L$, the elongation becomes: $u(L) = \frac{PL}{EA} + \frac{fL^2}{2E}$.

Here, note that the total elongation is composed of two parts. The term $\frac{PL}{EA}$ is associated with the action of the concentrated load P at the end point of the bar. This is the elongation for a bar where there are no distributed loads. The second term $\frac{fL^2}{2E}$ is the contribution of the uniformly distributed load f to the total elongation.

By assuming values such as point load $P = 10$ kN, distributed load $f = 10$ MN/m^3, cross-sectional area $A = 0.001$ m^2, and elastic modulus $E = 210$ GPa, the MATLAB script in Code 2.1 calculates and plots the elongation of the bar as a function of the bar length (L). The outcome is plotted in Figure 2.9. The MATLAB code and other scripts presented in this chapter can be downloaded from the following website: http://www.oup.com/AdvancedMathematicalModelling

Code 2.1. *MATLAB code to plot the elongation of a bar with distributed and point loads as a function of the length of the bar (L). This code generates Figure 2.9. This MATLAB code (and other scripts in this chapter) can be downloaded from the following website: http://www.oup.com/AdvancedMathematicalModelling*

```
% calculation of a bar structural member elongation
clear all; clc; close all;

set(0,'defaultaxesfontsize',16);

E=210*10^9; % Young's modulus in Pa
A=0.001;    % Cross-section area in m^2
P=10*10^3; % Point load in N
f=10000*10^3;  % distributed axial load in N/m^3

L=0:0.01:5; % bar length in m
uL=P*L/(E*A)+f*L.^2/(2*E); % elongation in m

figure(1)
plot(L,uL,'b-','linewidth',1.5)
ylabel('elongation u_L (m)','FontWeight','bold','FontSize',22);
xlabel('L (m)','FontWeight','bold','FontSize',22);

annotation(figure(1),'line',[0.208333 0.1958333],...
    [0.803535637 0.782937365]);
annotation(figure(1),'line',[0.2085937 0.1960937],...
    [0.783557235421166 0.762958963282937]);
```

Code 2.1. *Continued*

```
annotation(figure(1),'line',[0.208984375 0.1964843],...
    [0.762768898488119 0.74217062634989]);
annotation(figure(1),'line',[0.2091145 0.196614],...
    [0.740900647948161 0.720302375809932]);
annotation(figure(1),'rectangle',...
    [0.210375 0.74622030 0.2651458 0.04103671],...
    'FaceColor',[0.8 0.8 0.8]);
annotation(figure(1),'line',[0.209505208 0.209505208],...
    [0.810825053995679 0.717872570194383],'LineWidth',3);
annotation(figure(1),'arrow',[0.4278645833 0.465624999],...
    [0.7672786177 0.7678185745],'Color',[1 0 0],'LineWidth',1,...
    'HeadWidth',8,'HeadStyle','plain','HeadLength',8);
annotation(figure(1),'arrow',[0.379427083 0.41718749999],...
    [0.7661987041 0.766738660],'Color',[1 0 0],'LineWidth',1,...
    'HeadWidth',8,'HeadStyle','plain','HeadLength',8);
annotation(figure(1),'arrow',[0.3304687499 0.3682291666],...
    [0.7661987041 0.7667386609],'Color',[1 0 0],'LineWidth',1,...
    'HeadWidth',8,'HeadStyle','plain','HeadLength',8);
annotation(figure(1),'arrow',[0.2242187499 0.2619791666],...
    [0.7661987041 0.7667386609],'Color',[1 0 0],'LineWidth',1,...
    'HeadWidth',8,'HeadStyle','plain','HeadLength',8);
annotation(figure(1),'arrow',[0.2770833333 0.3148437499],...
    [0.766738660 0.767278617],'Color',[1 0 0],'LineWidth',1,...
    'HeadWidth',8,'HeadStyle','plain','HeadLength',8);
annotation(figure(1),'textbox',...
    [0.3523437499 0.802995679 0.0304687505 0.06749460171],...
    'Color',[1 0 0],'String',{'f'},'FontSize',24,'EdgeColor',[1 1 1]);
annotation(figure(1),'doublearrow',[0.211458333 0.475],...
    [0.694384449 0.6922246220]);
annotation(figure(1),'textbox',...
    [0.3372395833 0.6626069 0.028385416 0.06749460171],...
    'Color',[0.14901960 0.14901960 0.149019607],...
    'String',{'L'},'FontWeight','bold','FontSize',24,'FitBoxToText',...
    'off','EdgeColor',[1 1 1],'BackgroundColor',[1 1 1]);
annotation(figure(1),'arrow',[0.476692708 0.54596354],...
    [0.76870842 0.7686285097],...
    'Color',[0.0745098039 0.6235294117 1],...
    'LineWidth',2,'HeadStyle','plain');
annotation(figure(1),'arrow',[0.47708333 0.5098958],...
    [0.692574514038876 0.692494600431965],'Color',[0 0 1]);
annotation(figure(1),'textbox',...
    [0.5244791666 0.77869762 0.02500000 0.06749460],...
    'Color',[0.074509803 0.62352941 1],...
    'String',{'P'},'FontWeight','bold','FontSize',24, ...
    'FitBoxToText','off','EdgeColor',[1 1 1]);
annotation(figure(1),'textbox',...
    [0.513281249 0.657207342 0.02500000 0.067494601],...
    'Color',[0 0 1],'String','u_L','FontWeight','bold','FontSize',24,...
    'FitBoxToText','off','EdgeColor',[1 1 1]);
```

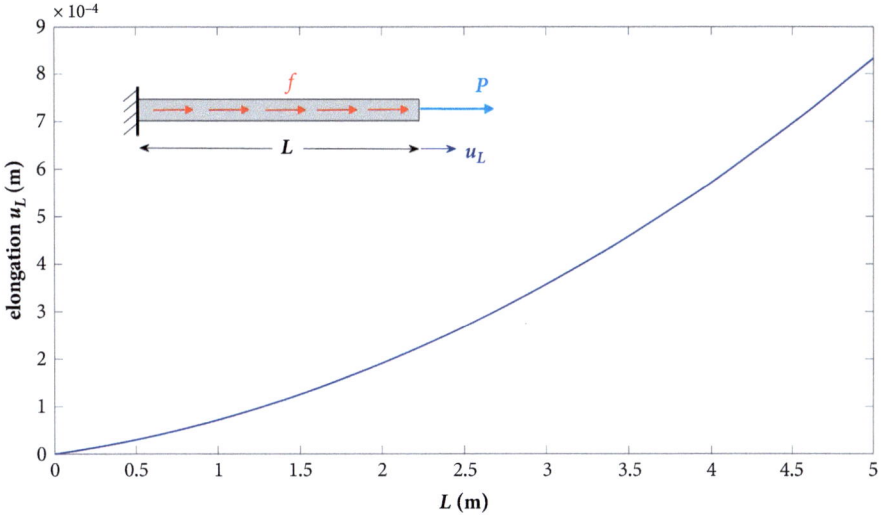

Figure 2.9 A graph showing the elongation of a bar with distributed ($10\ MN/m^3$) and point ($10\ kN$) loads as a function of the length of the bar (L). The properties of the bar are: cross-sectional area $A = 0.001\ m^2$ and elastic modulus $E = 210\ GPa$. This figure is generated using Code 2.1.

Box 2.2: Example (Elongation of a crane bridge cable)

A crane bridge cable of length $L = 20\ m$ supports a weight of $M = 400$ kg (Figure 2.10). The material properties of the cable are: density $\rho = 7800\ kg/m^3$, Young's modulus 200 GPa. The radius of the cable cross section is $R = 1\ cm$. The cable is assumed to be fixed at the upper part of the crane. What is the cable elongation and what is the force acting on the upper part of the crane due to the action of the cable and the weight?

Solution

The cross section of the cable can be calculated using the disk area formula as

$$A = \pi R^2 = 3.14 \times 0.01^2 = 3.14 \times 10^{-4} m^2.$$

The distributed load due to the cable self-weight is

$$f = \frac{cable\ weight}{cable\ volume} = \frac{\rho A L g}{A L} = \rho g = 7800 \times 9.81 = 76518\ N/m^3$$

In fact, the distributed load due to cable's self-weight equals the specific weight of the cable material, which is ρg. The weight acting at point $x = L$ is $P = Mg = 500 \times 9.81 = 3924\ N$.

Box 2.2 *Continued*

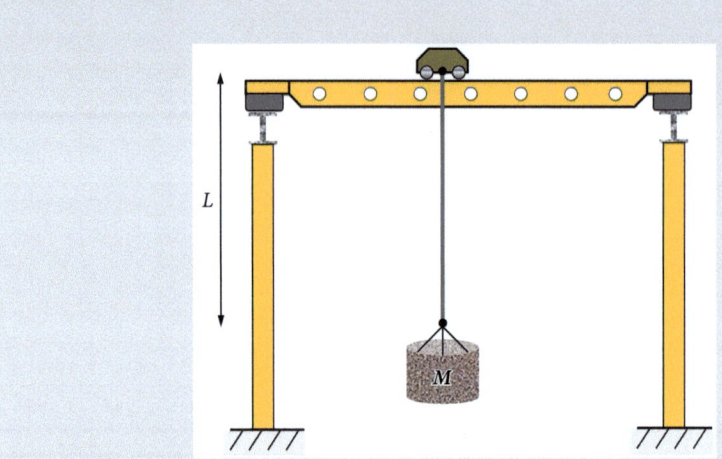

Figure 2.10 A sketch showing a crane bridge with static loading.

The elongation is therefore calculated as follows:

$$u(L) = \frac{PL}{EA} + \frac{fL^2}{2E} = \frac{3924 \times 20}{200 \times 10^5 \times 3.14} + \frac{76518 \times 20^2}{400 \times 10^9} = 0.0012 + 7.65 \times 10^{-5} \; m$$

In this case, the contribution of the cable self-weight to the total elongation is very small and the total elongation is approximately 1.2 *mm*.

The total force acting on the crane bridge is

$$N(0) = -\rho g A0 + \rho g AL + P = 7.6518 \times 3.14 \times 20 + 3924 = 480.53 + 3924 = 4404.53 \; N.$$

Box 2.3: Example (Thermal stresses in a bar)

Calculate the thermal stresses in a bar of length L and cross-sectional area A (both ends fixed) due to a temperature change of magnitude $\Delta T = 30° \; C$. The bar material is isotropic with Young's modulus E and volumetric thermal expansion coefficient a_V. Under the same conditions, calculate the elongation of the bar with one end free.

Solution

The temperature difference is $\Delta T = 30°C = 30 \; K$. Furthermore, the linear expansion coefficient is $a_L = \frac{1}{3}a_V$. If both ends are fixed, thermal expansion is restrained. Consequently, the thermal stress is given by the following equation:

$$\sigma_{xx} = \sigma_T = EAa_L\Delta T = EA\frac{1}{3}a_V30 = 10EAa_V$$

If one end is free, there is no thermal stress and the elongation can be calculated using the thermal strain:

$$\varepsilon_{xx} = \frac{du}{dx} = \varepsilon_T = a_L \Delta T = 10a_V$$

or, integrating, $u(x) = 10a_V x + C_1$.

Assuming that the bar end at $x = 0$ is fixed, then $u(0) = C_1 = 0$. The elongation is therefore $u(L) = 10a_V L$.

2.4 Torsion

Torsion effects refer to the twisting of a structural member and are typically induced by the action of a torque or loads acting eccentrically. In the analysis of structural members for civil engineering structures, these effects are particularly important in cases of lateral torsional buckling of columns and lateral application of point loads (Figure 2.11).

2.4.1 Governing equations

Torsion is induced by the action of torque (torsional moment) T_x on a shaft. Consequently, each section of the structural member rotates with respect to the other sections around the x axis, as can be seen in Figure 2.11b. This rotation is denoted as $\theta_x = \theta(x, t)$ and the internal torsional moment as $M_x = M(x, t)$ since, in the case of torsional vibrations for instance, it depends on both the axial spatial coordinate x and time t.

In the following, we consider only cylindrical cross sections under torsion. The analysis is quite similar to that of bars under the action of axial loads. The basic difference is that

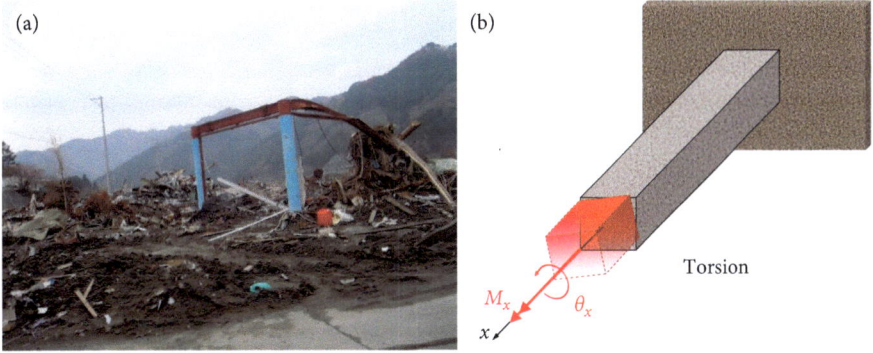

Figure 2.11 (a) Torsional deformation of a beam during the March 2011 M9 Japan earthquake and tsunami. (b) Sketch showing torsional deformation.

conservation of angular momentum will be used. Conservation of angular momentum for a differential volume of the member produces the following equation of motion:

$$M(x + dx, t) - M(x, t) + \mu(x, t) A dx = I_R \frac{\partial^2 \theta}{\partial t^2} \tag{2.35}$$

where $I_R = \rho dx \frac{\pi R^4}{2}$ denotes the mass moment of inertia of the differential volume with cylindrical cross section and radius R, and $\mu(x, t)$ is a distributed torsional moment along the structural member volume. Since the differential element with length dx is very small, using Taylor's series, the axial force at point $x + dx$ can be approximated as

$$M(x + dx, t) \approx M(x, t) + \frac{\partial M}{\partial x} dx \tag{2.36}$$

Finally, the differential equation of motion for a bar in terms of the axial force and axial displacement is produced, which takes the following form:

$$\frac{\partial M}{\partial x} + \mu(x, t) A = \rho \frac{\pi R^4}{2} \frac{\partial^2 \theta}{\partial t^2} \tag{2.37}$$

This is a single equation for two unknown fields. The final equation can be expressed in terms of the rotation angle (θ). Torsion induces only shear stresses which, due to the axial symmetry of the bar, depends on the distance r from the cross-section centre, i.e., $\tau = \tau(r, x, t)$. The moment due to the action of shear stress at a differential annular area located between radius r and $r + dr$ is denoted as follows:

$$dM = rdF = r\tau dA \tag{2.38}$$

The differential area can be calculated as $dA = \pi(r + dr)^2 - \pi r^2 = 2\pi r dr$ by neglecting the second-order differentials. At this point, assumptions on the material behaviour need to be made. Assuming an isotropic material, linear elastic behaviour, and neglecting the Poisson effect, Hooke's law implies $\tau = G\gamma$, where G is the shear modulus. For small deformation angles, the strain can be approximated as

$$\tan(\gamma) \approx \gamma = r \frac{\partial \theta}{\partial x} \tag{2.39}$$

Consequently, the torsional moment becomes

$$M = \int_A dM = \int_A r\tau dA = G \frac{\partial \theta}{\partial x} \int_A r^2 dA = 2\pi G \frac{\partial \theta}{\partial x} \int_0^R r^3 dr = \frac{\pi R^4 G}{2} \frac{\partial \theta}{\partial x} \tag{2.40}$$

Finally, the following second-order partial differential equation occurs:

$$\frac{\partial}{\partial x}\left(GJ\frac{\partial\theta}{\partial x}\right) + \mu(x,t)A = \rho\frac{R^2}{2}A\frac{\partial^2\theta}{\partial t^2} \tag{2.41}$$

where $J = \frac{\pi R^4}{2} = \frac{R^2}{2}A$ is the polar second moment of area (Timosenko 1955).

Equation (2.41) accounts for a spatially varying shear modulus $G(x)$ and cross-sectional area radius $R(x)$ along the axial dimension. If these quantities attain constant values, Eq (2.41) takes the following form:

$$\frac{\partial^2\theta}{\partial x^2} - \frac{1}{c^2}\frac{\partial^2\theta}{\partial t^2} = \frac{2\mu(x,t)}{R^2G} \tag{2.42}$$

where $c = \sqrt{G/\rho}$ is the speed of shear waves in the cylindrical shaft.

2.4.2 Static torsional deformation of cylinders

Assuming that all forces acting on the bar do not depend on time, or that their variation with time is very small, inertia effects can be neglected and the static deformation of the bar is governed by the following ODE:

$$\frac{\partial}{\partial x}\left(GJ\frac{\partial\theta}{\partial x}\right) + \mu(x)A = 0 \tag{2.43}$$

By integrating Eq (2.43) with respect to x, we have

$$GJ\frac{\partial\theta}{\partial x} + \int\mu(x)A dx = C_1 \tag{2.44}$$

where C_1 is an integration constant. By dividing by GJ and integrating again, we have

$$\theta(x) = -\int\frac{1}{GJ}\int\mu(x)A dx\, dx + \frac{C_1}{GJ}x + C_2 \tag{2.45}$$

Using Eq (2.40), the moment can be written as follows:

$$M(x) = GJ\frac{\partial\theta}{\partial x} = -\int\mu(x)A dx + C_1 \tag{2.46}$$

The general solution of Eq (2.45) features two integration constants, namely C_1 and C_2, similar to the equation for axial deformation of bars, Eq (2.33). The integration constants play an important role as they allow for the satisfaction of specific boundary conditions, which in turn model constraints of motion and applied loads in real-world configurations.

Two types of boundary conditions are popular: a fixed rotation point, and a specified external load. These two conditions cannot appear simultaneously at the same end point but are complementary. The following two rules need to be considered:

- If the rotation angle is known at an end point, the moment is unknown; it corresponds to the reaction moment at the constraint. It can be obtained using Eq (2.46), after the calculation of C_1, as $M(x_o)$, where $x = x_o$ is the location of the constraint.
- If the moment is known at an end point of the bar, the rotation angle is unknown. It is calculated by Eq (2.45) after the calculation of C_1 and C_2 by setting $x = L$, where L is the length of the structural member.

An example calculation for the torsion in a bar with uniformly distributed and point loads is shown in Box 2.4.

Box 2.4: Example (Torsion with uniformly distributed and point load)

A structural member of length L and circular cross-sectional area of radius R is considered. A uniformly distributed load $\mu(x) = \mu$ and a moment T at $x = L$ are acting on the member, which deforms slowly until static equilibrium is achieved. The point $x = 0$ is fixed, and the structural member is linearly elastic with shear modulus G. Find the total rotation angle.

Solution

Using Eq (2.45) for the rotation angle with a constant distributed load μ and constant $J = \frac{\pi R^4}{2}$ polar second moment of area, we have

$$\theta(x) = -\int \frac{1}{GJ} \int \mu\, A dx\, dx + C_1 x + C_2 = -\frac{\mu A}{2GJ}x^2 + \frac{C_1}{GJ}x + C_2$$

Since at point $x = 0$ there is no rotation, it can be written as

$$u(0) = -\frac{\mu A}{2GJ}0^2 + \frac{C_1}{GJ}0 + C_2 = C_2 = 0.$$

Now using Eq (2.46) for the moment, it is $M(x) = -\int \mu\, A dx + C_1 = -\mu A x + C_1$. But the moment at $x = L$ is T, i.e., $M(L) = T$, which results in $M(L) = -\mu AL + C_1$ or $C_1 = T + \mu AL$.

Finally, the rotation angle along the member is calculated as

$$\theta(x) = -\frac{\mu A}{2GJ}x^2 + \frac{(T + \mu AL)}{GJ}x = \frac{Tx}{GJ} + \frac{\mu AL}{GJ}x - \frac{\mu A}{2GJ}x^2$$

We know that, for this member, $A/J = 2/R^2$. Consequently, by replacing $x = L$, the rotation angle becomes $\theta(L) = \frac{TL}{GJ} + \frac{\mu L^2}{GR^2}$.

Note that the total angle is composed of two parts. The term $\frac{TL}{GJ}$ is associated with the action of the concentrated moment T at the end point of the member. The second term $(\frac{\mu L^2}{GR^2})$ is the contribution of the uniformly distributed load (μ) to the total angle.

The MATLAB script in Code 2.2 calculates and plots the rotation angle (θ) of a structural member of length $L = 3\ m$ and three cases of circular cross-sectional area (radius $R_1 = 0.1$ m, $R_2 = 0.15\ m$, and $R_3 = 0.2\ m$) as a function of distance (x). The member is under uniformly distributed load $\mu = 10\ MN/m^2$ and a moment $T = 10\ kN \cdot m$ at $x = 3\ m$. Assume $G = 190\ GPa$. The results are plotted in Figure 2.12. To download Code 2.2, visit the following website: http://www.oup.com/AdvancedMathematicalModelling

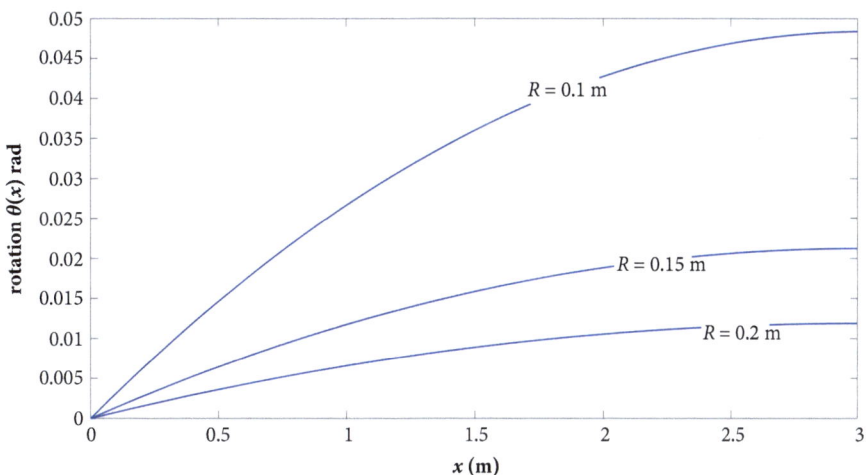

Figure 2.12 A graph showing the rotation angle (θ) of a structural member of length $L = 3\ m$ and a circular cross-sectional area as a function of distance (x), for different cross-section radius values. The member is under a uniformly distributed load $\mu = 10\ MN/m^2$ and a moment $T = 10\ kN \cdot m$ at $x = 3\ m$. Assume $G = 190\ GPa$. This figure is produced using Code 2.2.

Code 2.2. *MATLAB code to calculate and plot the rotation angle (θ) of a structural member with length L = 3 m and a circular cross-sectional area as a function of distance (x), for different cross-section radius values. The member is under a uniformly distributed load μ = 10 MN/m² and a moment T = 10 kN · m at x = 3 m. Assume G = 190 GPa. This code generates Figure 2.12. This MATLAB code (and other scripts in this chapter) can be downloaded from the following website:* http://www.oup.com/AdvancedMathematicalModelling

```matlab
% calculation of rotation angle under torsion
clear all; clc; close all;

set(0,'defaultaxesfontsize',16);

G=190*10^9; % Young's modulus in Pa
R1=0.1;    % cross-section radius in m: case 1
R2=0.15;    % cross-section radius in m: case 2
R3=0.2;    % cross-section radius in m: case 3
T=10*10^3; % Moment a x=L in Nm
miu=10*10^6;  % distributed moment load in N/m^2
L=3;       % Length in m

A1=pi*R1^2; % Area 1
A2=pi*R2^2; % Area 2
A3=pi*R3^2; % Area 3

J1=pi*R1^4/2; % Polar second moment of area 1
J2=pi*R2^4/2; % Polar second moment of area 2
J3=pi*R3^4/2; % Polar second moment of area 3

x=0:0.01:3; % axial distance from fixed end
th1=T*x/(G*J1)+miu*A1*L/(G*J1)*x-miu*A1/(2*G*J1)*x.^2;
th2=T*x/(G*J2)+miu*A2*L/(G*J2)*x-miu*A2/(2*G*J2)*x.^2;
th3=T*x/(G*J3)+miu*A3*L/(G*J3)*x-miu*A3/(2*G*J3)*x.^2;

figure(1)
plot(x,th1,'b-',x,th2,'b-',x,th3,'b','linewidth',1.5)
ylabel('rotation \ theta(x) rad','FontWeight','bold','FontSize',22);
xlabel('x (m)','FontWeight','bold','FontSize',22);
annotation(figure(1),'textbox',...
    [0.57395833333333 0.750619869431956 0.0687500000000022 0.0545356381274196],...
    'String',{'R=0.1 m'},'FontSize',18,'FitBoxToText','off','EdgeColor',...
    [1 1 1],'BackgroundColor',[1 1 1]);
annotation(figure(1),'textbox',...
    [0.658072916666661 0.406667385630659 0.077343750000005 0.0545356381274195],...
    'String','R=0.15 m','FontSize',18,'FitBoxToText','off','EdgeColor',...
    [1 1 1],'BackgroundColor',[1 1 1]);
annotation(figure(1),'textbox',...
    [0.746093749999994 0.272758098373639 0.0687500000000022 0.0545356381274196],...
    'String','R=0.2 m','FontSize',18,'FitBoxToText','off','EdgeColor',...
    [1 1 1],'BackgroundColor',[1 1 1]);
```

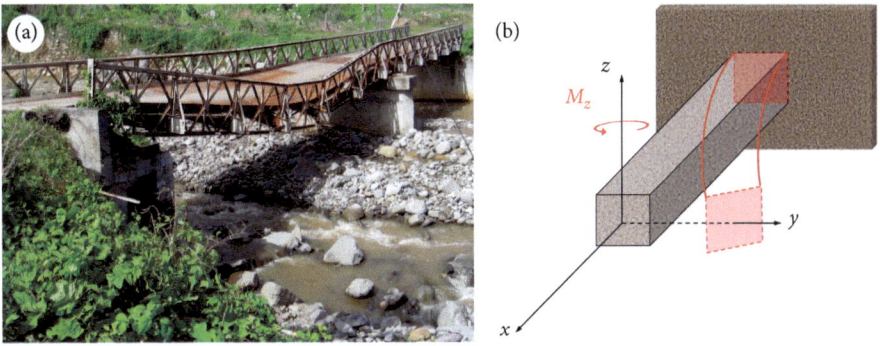

Figure 2.13 (a) Vertical bending of a bridge beam in Dominica, which was damaged following the 2017 Hurricane Maria in this island country. (b) A sketch showing lateral bending of a beam.

2.5 Beams

Beams are typically slender, prismatic structural members that can sustain transverse loads and bending moments (Figure 2.13). In response they undergo transverse flexural deformations and rotations. Therefore, to study the flexure of a beam with span along the x axis, the forces F_y and F_z need to be considered, along with the moments T_y and T_z. If small deformations are considered, bending along the y direction does not depend on bending along the z direction and vice versa. Consequently, it suffices to study a beam with span along the x axis under the action of F_z and T_y. This implies bending upwards or downwards. The other case (action of F_z and T_y with shear resultant V_y and bending moment M_z) implies bending to the right or left, as shown in Figure 2.13b.

2.5.1 Governing equations

Due to the action of a transverse vertical load $q(x)$ and/or the force F_z and moment T_y, a beam will deform along the z axis and develop internal resultants in the form of a shear force V_z and bending moment M_y. Since in the following we only study each point along the beam span x, the deflection will be equal to the displacement along z, $w(x, t)$, a function that also accounts for dynamic response.

Utilizing Newton's law of motion to uphold the principle of linear momentum conservation, the free body diagram, representing a differential volume of the beam, can be expressed as

$$V(x + dx, t) - V(x, t) + q(x, t) A dx = dm \frac{\partial^2 w}{\partial t^2} \tag{2.47}$$

Since the differential element with length dx is very small, using Taylor's series (see Appendix A) the shear force at point $x + dx$ can be approximated as

$$V(x + dx, t) \approx V(x, t) + \frac{\partial V}{\partial x} dx \tag{2.48}$$

The differential mass is $dm = \rho \cdot dV = \rho \cdot A \cdot dx$. Finally, the differential equation for the translational motion of the beam segment in terms of the shear force and the deflection w is

$$\frac{\partial V}{\partial x} + q(x, t) = \rho A \frac{\partial^2 w}{\partial t^2} \tag{2.49}$$

Conservation of angular momentum must be studied as well since the beam segment also undergoes rotations. It is written as follows:

$$M(x + dx, t) - V dx + q(x, t) dx \frac{dx}{2} - M(x, t) = I_R \frac{\partial^2 \theta_y}{\partial t^2} \tag{2.50}$$

where I_R denotes the rotary inertia of the beam section. Using Taylor's expansion for $M(x + dx, t)$ and eliminating the term $q(x, t) dx \frac{dx}{2}$ along with all other differentials of second and higher orders, it takes the following form:

$$\frac{\partial M}{\partial x} - V = I_R \frac{\partial^2 \theta_y}{\partial t^2} \tag{2.51}$$

The differential equations (2.49) and (2.51) must be solved for the deflection w and rotation θ_y.

2.5.2 Euler–Bernoulli beams

Examining the deflection of a beam segment with length L along the z direction, it is evident that the lower part in the deformed configuration will be elongated, while the upper part of the beam will be subjected to compression if the forcing is downwards. There will be a specific position along the thickness of the beam that corresponds to a fibre that is neither being elongated nor is shrinking. Thus, the length of this particular fibre (line) after deformation will be equal to the undeformed beam length L. This fibre is referred to as the **neutral axis** or **neutral line**.

The Euler–Bernoulli kinematic assumptions regarding the deformation of the beam are:

- The cross sections of the beam after deformation remain plane.
- The cross sections of the beam under deformation remain perpendicular to the neutral axis.

The strain ε_{xx} of a fibre at distance z from the neutral line is expressed as

$$\varepsilon_{xx} = \frac{\Delta L}{L} = \frac{R\varphi - (R + z)\varphi}{R\varphi} = -\frac{z}{R} = -z\kappa \tag{2.52}$$

where R and κ denote the radius of curvature and curvature, respectively, for the beam segment. The action of pure bending moments will produce only normal stresses on the beam cross section. The bending moment is therefore counterbalanced according to the following equation:

$$M = \int_A dM = \int_A z\,dF = \int_A z\sigma_{xx}\,dA \tag{2.53}$$

Assuming linear elastic material behaviour and neglecting Poisson's effect, Hooke's law in the form $\sigma_{xx} = E\varepsilon_{xx}$ can be applied. Therefore, Eq (2.53) now takes the following form:

$$M = -\int_A E\kappa z^2\,dA = -E\kappa \int_A z^2\,dA = -EI_{yy}\kappa \tag{2.54}$$

where, $\int_A z^2\,dA = I_{yy}$ is the second moment of the area of the cross section with respect to the y axis. The product EI_{yy} is termed **flexural rigidity** and is a measure of the beam's resistance to bending.

The Euler–Bernoulli bending theory assumptions imply that the rotary inertia effects are negligible and hence, according to the equilibrium Eq (2.51), the shear force is equal to the first derivative of the bending moment:

$$\frac{\partial M}{\partial x} - V = 0 \tag{2.55}$$

By differentiating Eq (2.55) and using Eq (2.54), we have

$$\frac{\partial V}{\partial x} = \frac{\partial^2 M}{\partial x^2} = \frac{\partial^2}{\partial x^2}\left(EI_{yy}\kappa\right) \tag{2.56}$$

By substituting in the linear momentum conservation equation, Eq (2.49), and using the linearized curvature-deflection expression $\kappa = d^2w/dx^2$ (Timoshenko 1955), we have

$$\rho A \frac{\partial^2 w}{\partial t^2} + \frac{\partial^2}{\partial x^2}\left(EI_{yy}\frac{\partial^2 w}{\partial x^2}\right) = q\,(x,t) \tag{2.57}$$

Equation (2.57) is the Partial Differential Equation (PDE) that governs the flexural vibrations of an Euler–Bernoulli beam with spatially varying material and geometric properties. In the case where EI_{yy} does not depend on the spatial variable x, Eq (2.57) can be rewritten as

$$\frac{\partial^4 w}{\partial x^4} + \frac{1}{c^2}\frac{\partial^2 w}{\partial t^2} = \frac{q\,(x,t)}{EI_{yy}} \tag{2.58}$$

where $c = \sqrt{\frac{EI_{yy}}{\rho A}}$ is the speed of flexural waves in the beam.

Note that, according to Eq (2.31), the resistance of a bar to tension is characterized by the product EA. That is, it increases with increasing Young's modulus and increasing cross-sectional area. When bending is considered, the resistance (expressed through flexural rigidity EI_{yy} or EI_{zz}) increases with increasing Young's modulus, while now it does not depend on the cross-sectional area A itself but on the second moment of area, I_{yy} or I_{zz}. The general rule is: for a given cross-sectional area, as the distance of the material points from the neutral line increases, so does the resistance to bending. It is this observation that led to the use of I-, T-, or L-type cross sections for beams or columns in construction to achieve optimum resistance to bending with minimum material usage.

2.5.3 Static deformation of prismatic beams

Assuming that all forces acting on the bar do not depend on time, or their variation with time is very small, the inertia effects can be neglected, and the static deformation of the bar is governed by the following ODE:

$$\frac{d^2}{dx^2}\left(E I_{yy}\frac{d^2 w}{dx^2}\right) = q(x) \tag{2.59}$$

Integrating Eq (2.59) with respect to x twice, it becomes

$$E I_{yy}\frac{d^2 w}{dx^2} = \int\int q(x)\, dx\, dx + C_1 x + C_2 \tag{2.60}$$

where C_1 and C_2 are integration constants. Dividing by EI_{yy} and integrating it again twice, we have

$$w(x) = \int\int \frac{1}{E I_{yy}}\int\int q(x)\, dx^4 + \int\int \frac{C_1 x}{EI_{yy}}\, dx\, dx + \int\int \frac{C_2}{EI_{yy}}\, dx\, dx + C_3 x + C_4 \tag{2.61}$$

If constant flexural rigidity is assumed, Eq (2.61) simplifies to the following form:

$$w(x) = \frac{1}{EI_{yy}}\int\int\int\int q(x)\, dx^4 + \frac{C_1 x^3}{6EI_{yy}} + \frac{C_2 x^2}{2EI_{yy}} + C_3 x + C_4 \tag{2.62}$$

The bending moment $M(x) = -EI_{yy}\frac{d^2 w}{dx^2}$ can be computed using Eq (2.60):

$$M(x) = -\int\int q(x)\, dxdx - C_1 x - C_2 \tag{2.63}$$

The shear force, being equal to the derivative of the bending moment, is given by the following equation:

$$V(x) = -\int q(x)\, dx - C_1 \tag{2.64}$$

When the case of a constantly distributed load is considered, e.g., the weight distribution of a uniform beam, Eq (2.62) simplifies further after four integrations to take the following form:

$$w(x) = \frac{qx^4}{24EI_{yy}} + \frac{C_1 x^3}{6EI_{yy}} + \frac{C_2 x^2}{2EI_{yy}} + C_3 x + C_4 \tag{2.65}$$

Four integration constants appear in the equation for the deflection, i.e., Eq (2.62) for variable flexural rigidity or Eq (2.65) for constant flexural rigidity. This fact is compatible with the theory of ODEs since the general solution of a fourth-order equation should feature four integration constants. For specific BVPs these constants can be determined using the boundary conditions. Several cases occur when Euler–Bernoulli beams are considered, depending on the externally applied constraints of motion and loading scenarios. The typical cases are summarized in Table 2.2.

Example calculations for a cantilever beam with uniform load and a simply supported beam with uniform load are shown in Boxes 2.5 and 2.6, respectively.

Table 2.2 Boundary conditions for the Euler–Bernoulli beam.

End-point kinematic constraints and applied loads	Boundary conditions for Euler–Bernoulli beam	
	Fixed end	$w = 0$
		$\theta = 0 \rightarrow \frac{dw}{dx} = 0$
	Hinge	$w = 0$
		$M = 0 \rightarrow \frac{d^2 w}{dx^2} = 0$
	Roller	$w = 0$
		$M = 0 \rightarrow \frac{d^2 w}{d^2} = 0$
	Prescribed bending moment and shear force	$M = M_0 \rightarrow \frac{d^2 w}{dx^2} = -\frac{M_0}{E I_{yy}}$
		$V = V_0 \rightarrow \frac{d}{dx}\left(E I_{yy} \frac{d^2 w}{dx^2}\right) = -V_0$
	Free end	$M = 0 \rightarrow \frac{d^2 w}{dx^2} = 0$
		$V = 0 \rightarrow \frac{d^3 w}{dx^3} = 0$

Box 2.5: Example (A cantilever beam with uniform load)

A cantilever beam of length L and second moment of area I_{yy} is fixed at point $x = 0$ and is free at point $x = L$. The material's Young's modulus is E. A uniformly distributed load q is acting on the beam (Figure 2.14). Calculate the deflection, slope, bending moment, and shear force distribution.

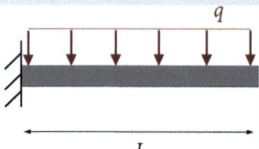

Figure 2.14 A cantilever beam. One end of the beam is fixed, and its other end is free.

Solution

According to Eq (2.65), the deflection is given as

$$w(x) = \frac{qx^4}{24EI_{yy}} + \frac{C_1 x^3}{6EI_{yy}} + \frac{C_2 x^2}{2EI_{yy}} + C_3 x + C_4$$

The slope is the first derivative of the displacement:

$$\theta(x) = \frac{qx^3}{6EI_{yy}} + \frac{C_1 x^2}{3EI_{yy}} + \frac{C_2 x}{EI_{yy}} + C_3$$

Since at the fixed end we have $w(0) = \theta(0) = 0$, the integration constants C_3 and C_4 must vanish, and therefore $C_3 = C_4 = 0$.

The shear force is equal to zero at the free end, and therefore we have

$$V(0) = -L - C_1 = 0 \text{ or } C_1 = -qL.$$

The bending moment is equal to zero at the free end ($x = L$). Therefore, it is expressed as

$$M(L) = -\frac{q}{2}L^2 - C_1 L - C_2 = 0 \text{ or } C_2 = -\frac{q}{2}L^2 + qLL = \frac{q}{2}L^2$$

Finally, the results are:

$$w(x) = \frac{qx^4}{24EI_{yy}} - \frac{qLx^3}{6EI_{yy}} + \frac{qL^2 x^2}{4EI_{yy}} = \frac{qx^2}{24EI_{yy}}\left(x^2 - 4Lx + 6L^2\right)$$

$$\theta(x) = \frac{qx^3}{6EI_{yy}} - \frac{qLx^2}{2EI_{yy}} + \frac{qL^2 x}{2EI_{yy}} = \frac{qx}{6EI_{yy}}\left(x^2 - 3Lx + 3L^2\right)$$

$$M(x) = -\frac{q}{2}\left(x^2 - 2Lx + L^2\right)$$

$$V(x) = q(L - x)$$

Box 2.6: Example (A simply supported beam with uniform load)

Solve the previous problem for a simply supported beam with uniform load as shown in Figure 2.15.

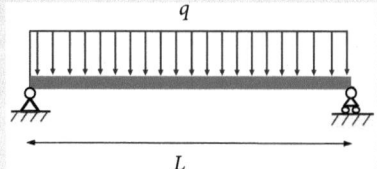

Figure 2.15 A simply supported beam. One end of the beam is supported by a hinge and the other by a roller.

Solution

In the case of a simply supported beam, the boundary conditions are $w(0) = M(0) = w(L) = M(L) = 0$.

According to Eq (2.65), the deflection is given by the following equation:

$$w(x) = \frac{qx^4}{24EI_{yy}} + \frac{C_1 x^3}{6EI_{yy}} + \frac{C_2 x^2}{2EI_{yy}} + C_3 x + C_4$$

Since $w(0) = 0$, we have $C_4 = 0$. The bending moment at $x = 0$ vanishes. Therefore, we have

$$M(0) = -\frac{q}{2} \cdot 0^2 - C_1 \cdot 0 - C_2 = 0 \text{ or } C_2 = 0.$$

The bending moment at $x = L$ is zero as well, which implies

$$M(L) = -\frac{q}{2} L^2 - C_1 L = 0 \text{ or } C_1 = -\frac{q}{2} L$$

Since the deflection at $x = L$ is zero, we can write

$$w(L) = \frac{qL^4}{24EI_{yy}} - \frac{qL^4}{12EI_{yy}} + C_3 L = 0 \text{ or } C_3 = \frac{qL^3}{24EI_{yy}}$$

Finally, the results are:

$$w(x) = \frac{qx^4}{24EI_{yy}} - \frac{qLx^3}{12EI_{yy}} + \frac{qL^3}{24EI_{yy}}x = \frac{qx}{24EI_{yy}}\left(x^3 - 2Lx^2 + L^3\right)$$

$$\theta(x) = \frac{qx^3}{6EI_{yy}} - \frac{qLx^2}{4EI_{yy}} + \frac{qL^3}{24EI_{yy}} = \frac{q}{24EI_{yy}}\left(4x^3 - 6Lx^2 + L^3\right)$$

$$M(x) = \frac{q}{2}x(L - x)$$

$$V(x) = q\left(\frac{L}{2} - x\right)$$

Box 2.6 *Continued*

Code 2.3 calculates and plots the variations of deflection (w), slope (θ), bending moment (M), and shear force (V) distribution in a simply supported beam of length L = 5 m, having a rectangular cross section with each side measuring 0.3 m, under uniform load q = −100 kN/m as a function of distance (x). Assume that the Young's modulus of the beam material is E = 50 GPa. The second moment or area for a rectangular cross section with side h is I_{yy} = $h^4/12$. The outcome is shown in Figure 2.16.

Code 2.3. *MATLAB code to calculate and plot variations of deflection (w), slope (θ), bending moment (M), and shear force (V) distribution in a simply supported beam of length L = 5 m, having a rectangular cross section with each side measuring 0.3 m, under uniform load q = −100 kN/m as a function of distance (x). Assume that the Young's modulus of the beam material is E = 50 GPa. This code generates Figure 2.16. To download this code, visit the following website: http://www.oup. com/AdvancedMathematicalModelling*

```
% calculate and plot deflection, slope, bending moment and shear force in a
% simply supported beam
clear all;clc;close all;

set(0,'defaultaxesfontsize',16);

E=50*10^9; % Young's modulus in Pa
h=0.3;     % cross-section side in m
I=h^4/12;  % second moment of area
L=5;       % beam span in m
q=-100*1063;   % distributed load N/m
x=0:0.01:5; % axial distance from hinge

w=q*x/(24*E*I).*(x.^3-2*L*x.^2+L^3);
sl=q/(24*E*I).*(4*x.^3-6*L*x.^2+L^3);
M=q/2*x.*(L-x);
V=q*(L/2-x);

figure(1)
subplot(2,2,1)
plot(x,w,'b-',LineWidth=1.5)
xlabel('x (m)','FontWeight','bold','FontSize',22);
ylabel('w (m)','FontWeight','bold','FontSize',22);
subplot(2,2,2)
plot(x,sl,'k-',LineWidth=1.5)
xlabel('x (m)','FontWeight','bold','FontSize',22);
ylabel('\ theta (rad)','FontWeight','bold','FontSize',22);
subplot(2,2,3)
plot(x,M,'r-',LineWidth=1.5)
xlabel('x (m)','FontWeight','bold','FontSize',22);
ylabel('M (Nm)','FontWeight','bold','FontSize',22);
subplot(2,2,4)
plot(x,V,'g-',LineWidth=1.5)
xlabel('x (m)','FontWeight','bold','FontSize',22);
ylabel('V (N)','FontWeight','bold','FontSize',22);
```

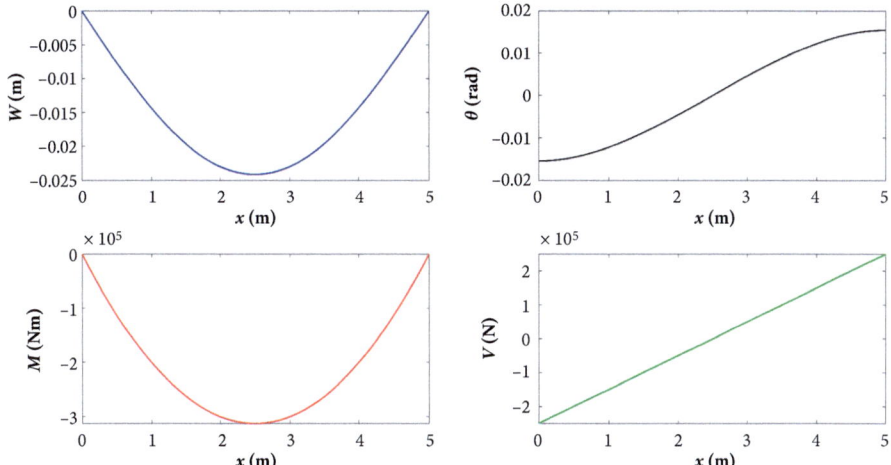

Figure 2.16 A graph showing variations of deflection (w), slope (θ), bending moment (M), and shear force (v) distribution in a simply supported beam of length $L = 5\ m$, having a rectangular cross section with each side measuring $0.3\ m$, and under uniform load $q = -100\ kN/m$ as a function of distance (x). The Young's modulus of the beam material is assumed to be $E = 50\ GPa$. This figure is generated using Code 2.3.

2.5.4 Beams on elastic foundations

Analysis of beams can be performed in the case where the beam rests on elastic foundations as well. This scenario is common in soil–structure interaction, pavement engineering, and railway engineering (Figure 2.17). The foundation is assumed to produce a reaction force ($F_{foundation}$) that depends on the deflection values and opposes the deflection itself, resembling the action of springs (Hetenyi 1946).

If the reaction of the foundation is proportional to the beam deflection, i.e., $F_{foundation} = -kw$, where k is the spring stiffness coefficient and w the distributed reaction load from the foundation (known as Winkler reaction), then the concept of a **Winkler foundation** applies (Winkler 1867). The proportionality constant k expresses the stiffness of the foundation. Small values of k represent relatively soft soils, while large values of k denote relatively stiff soils. The conservation of linear momentum in the presence of a Winkler foundation becomes

$$\frac{\partial V}{\partial x} - kw + q\left(x, t\right) = \rho A \frac{\partial^2 w}{\partial t^2} \tag{2.66}$$

The conservation of angular momentum is not affected since the presence of the Winkler reaction produces the higher-order term $kwdx^2/2$ in the rotation equilibrium equation for the differential volume of the beam and therefore can be ignored.

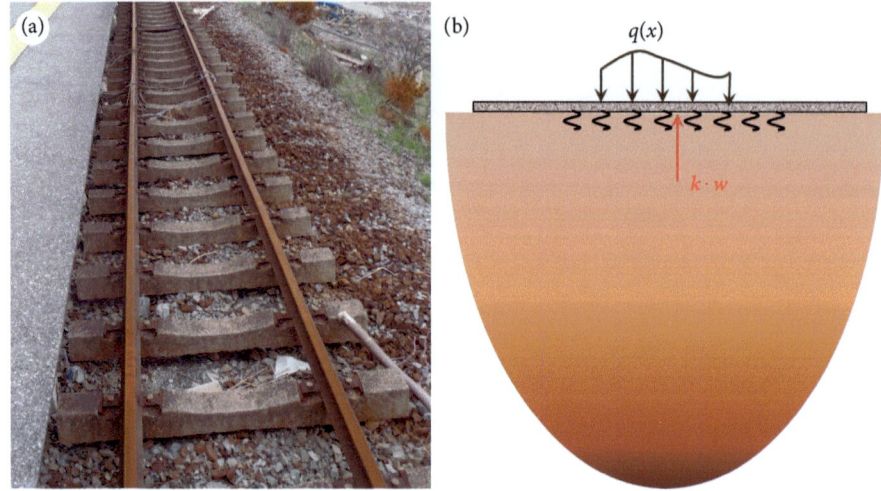

Figure 2.17 (a) A railway track can be studied as a beam resting on an elastic foundation. (b) Sketch of the concept of the Winkler elastic foundation.

Using the Euler–Bernoulli approximation, which is $V = -\frac{d}{dx}\left(EI_{yy}\frac{d^2w}{dx^2}\right)$, the equilibrium in Eq (2.66) becomes

$$\rho A\frac{\partial^2 w}{\partial t^2} + \frac{d^2}{dx^2}\left(EI_{yy}\frac{d^2w}{dx^2}\right) + kw = q\,(x,t) \tag{2.67}$$

Assuming a uniform cross section and an isotropic, homogeneous material, the static equilibrium of an Euler–Bernoulli beam on a Winkler foundation is governed by the following fourth-order ODE:

$$EI_{yy}\frac{d^4w}{dx^4} + kw = q\,(x) \tag{2.68}$$

The general solution for this equation is

$$w\,(x) = e^{\beta x}\left[C_1\cos\left(\beta x\right) + C_2\sin\left(\beta x\right)\right] + e^{-\beta x}\left[C_3\cos\left(\beta x\right) + C_4\sin\left(\beta x\right)\right] + f_q \tag{2.69}$$

where $\beta = \left(\frac{k}{4EI_{yy}}\right)^{1/4}$ and f_q is a partial solution that depends on the specific form of the distributed load q.

In the case of an elastic beam with infinite width, the model of the beam on the Winkler-type foundation has the exact same form with $EI_{yy} = \frac{EH^3}{12(1-\nu^2)}$, where H is the beam height and ν the beam material Poisson's ratio. This case can be useful for pavement engineering applications.

An example calculation for the deflection of a beam with infinite length is shown in Box 2.7.

Box 2.7: Example (Deflection of a beam with infinite length)

Find the deflection of a beam with infinite length and flexural rigidity EI_{yy}, resting on a Winkler foundation with constant k and loaded by a concentrated load P. Plot the solution using MATLAB for the case where $EI_{yy} = 1000\ Nm^2$, $P = -1\ N$, and for two different values of k, $k_1 = 1\ N/m^2$ and $k_2 = 2\ N/m^2$.

Solution

The origin is placed at the location of the load. Consequently, due to symmetry, the slope at $x = 0$ must be zero and only the part of the beam along the positive horizontal axis $x > 0$ needs to be considered. Since the deflection must vanish for $x \to +\infty$, the general solution takes the following form:

$$w(x) = e^{-\beta x}\left[C_3 \cos(\beta x) + C_4 \sin(\beta x)\right]$$

Then,

$$\theta(x) = \frac{dw}{dx} = -\lambda e^{-\beta x}\left[C_3 \cos(\beta x) + C_4 \sin(\beta x)\right]$$
$$+ e^{-\beta x}\left[-C_3 \lambda \sin(\beta x) + C_4 \lambda \cos(\beta x)\right]$$

Since $\theta(0) = 0$, we have $-\beta C_3 + \beta C_4 = 0$, or $C_3 = C_4 = C$. The value of the remaining constant C can be calculated from the overall equilibrium of vertical forces. It should be

$$\int_{-\infty}^{+\infty} kw\,dx = 2\int_0^{\infty} kw\,dx = P$$

The integral can be calculated using integration by parts as follows:

$$2\int_0^{\infty} kw\,dx = 2Ck\int_0^{\infty} e^{-\beta x}\left[\cos(\beta x) + \sin(\beta x)\right]dx = \frac{2kC}{\beta}$$

Then, $C = \frac{P\beta}{2k}$ and the solution becomes

$$w(x) = \frac{P\beta}{2k}e^{-\beta x}\left[\cos(\beta x) + \sin(\beta x)\right] \quad \text{for } x \geq 0$$
$$w(x) = \frac{P\beta}{2k}e^{\lambda x}\left[\cos(\beta x) - \sin(\beta x)\right] \quad \text{for } x < 0$$

The slope, bending moment, and shear force can be calculated using the first, second, and third derivatives of the deflection, respectively, since $\theta(x) = \frac{dw}{dx}$, $M(x) = -EI_{yy}\frac{d^2w}{dx^2}$, and $V(x) = \frac{dM}{dx}$.

Box 2.7 *Continued*

The MATLAB script in Code 2.4 calculates and plots the deflection of the beam, with its results shown in Figure 2.18.

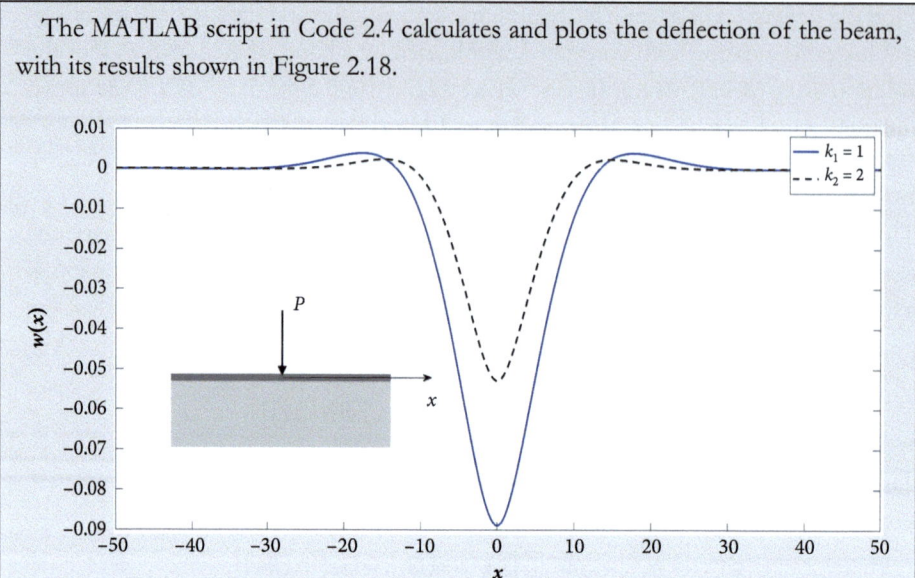

Figure 2.18 Deflection of beam on a Winkler foundation under the action of unit load, for two different values of the Winkler parameter. The properties of the beam and the load are $EI_{yy} = 1000\ Nm^2$ and $P = -1\ N$. Two different values of k are considered: $k_1 = 1\ N/m^2$ and $k_2 = 2\ N/m^2$. This figure is generated using the MATLAB script in Code 2.4.

Code 2.4. *MATLAB code to plot the deflection of a beam on a Winkler foundation under the action of a point load for two different values of the Winkler parameter. The assumptions are $EI_{yy} = 1000\ Nm^2$ and $P = -1\ N$. Two different values of k are considered: $k_1 = 1\ N/m^2$ and $k_2 = 2\ N/m^2$. This code generates Figure 2.18. To download this code, visit the following website: http://www.oup.com/AdvancedMathematicalModelling*

```
% deflection of beam on Winkler foundation
clear all; clc; close all;

set(0,'defaultaxesfontsize',16);

EI=1000; % flexural rigidity
k1=1; % Winkler constant 1
k2=2;    % Winkler constant 2
beta1=(k1/(EI))^(1/4); beta2=(k2/(EI))^(1/4);

P=-1;    % unit load at x=0
L=50;    % beam half-length simulating infinite beam

x_n=-L:0.01:-0.01; % negative axis coordinates
x_p=0:0.01:L;      % positive axis coordinates
x=[x_n x_p];
% solution for k1
```

```
w1_n=P*beta1/(2*k1)*exp(beta1*x_n).*(cos(beta1*x_n)-sin(beta1*x_n));
w1_p=P*beta1/(2*k1)*exp(-beta1*x_p).*(cos(beta1*x_p)+sin(beta1*x_p));
w1=[w1_n w1_p];
% solution for k2
w2_n=P*beta2/(2*k2)*exp(beta2*x_n).*(cos(beta2*x_n)-sin(beta2*x_n));
w2_p=P*beta2/(2*k2)*exp(-beta2*x_p).*(cos(beta2*x_p)+sin(beta2*x_p));
w2=[w2_n w2_p];

figure(1)
plot(x,w1,'b-',x,w2,'k--','LineWidth',2)
ylabel('w(x)',FontWeight='bold',FontSize=22);
xlabel('x',FontWeight='bold',FontSize=22);
legend('k_1=1','k_2=2','fontsize',18)
annotation(figure(1),'rectangle',...
    [0.185895833333333 0.277537796976242 0.223479166666667 0.133909287257017],...
    'Color',[0.8 0.8 0.8],'FaceColor',[0.8 0.8 0.8]);
annotation(figure(1),'rectangle',...
    [0.1859375 0.412526997840172 0.2234375 0.0129589632829359],...
    'Color',[0.501960784313725 0.501960784313725 0.501960784313725],...
    'FaceColor',[0.501960784313725 0.501960784313725 0.501960784313725]);
annotation(figure(1),'arrow',[0.3 0.3],[0.554345572354211 0.423596112311015],...
    'LineWidth',2,'HeadStyle','plain');
annotation(figure(1),'arrow',[0.300390625 0.451432291666667],...
    [0.418006479481641 0.419006479481641]);
annotation(figure(1),'textbox',...
    [0.307031249999999 0.535177104647162 0.0359375007015963 0.0631749471886884],...
    'String',{'P'},'FontSize',22,'EdgeColor',[1 1 1]);
annotation(figure(1),'textbox',...
    [0.442708333333333 0.345652266634205 0.0328125006270906 0.0631749471886884],...
    'String',{'x'},'FontSize',22,'EdgeColor',[1 1 1]);
```

2.6 Buckling of columns

Columns are structural members, similar in characteristics to beams, that sustain vertical loads. If a column is very slender it may buckle under the action of relatively small compressive loads (Figure 2.19a). In the following, a simple model for the buckling of simply supported columns, under the Euler–Bernoulli assumptions for bending, is studied. This model is the first buckling model introduced by Euler (Timoshenko 1955) and helps civil engineers to easily identify the critical buckling load of a column. A column under buckling can be seen in Figure 2.19a. Selecting a location along the longitudinal axis of the column, i.e., the vertical axis z, the equilibrium of a part defined by a cross-section cut at level z can be studied using the free body diagram (Figure 2.19b). The deflection at that point is denoted as $v(z)$.

The column has internal resultants that enforce equilibrium, and these reactions are the axial force (P) and the bending moment (M). The free body diagram equilibrium has the following form:

$$M = Pv \tag{2.70}$$

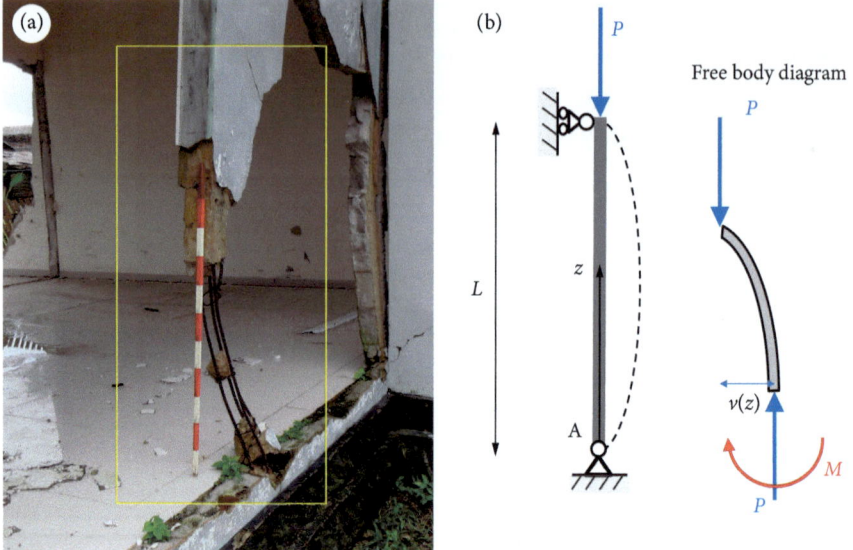

Figure 2.19 (a) A buckled column is shown during the 2018 Anak Krakatau tsunami in Sunda Strait, Indonesia. (b) A sketch showing the Euler–Bernoulli concept for buckling of columns and the respective free body diagram.

where v is the deflection. Using the relation between bending moment (M) and deflection (v), a second-order ODE is produced for the column's buckling:

$$EI\frac{d^2v}{dz^2} + Pv = 0 \tag{2.71}$$

where EI is the flexural rigidity. The general solution of this second-order differential equation takes the following form:

$$v(z) = C_1 \cos(\alpha z) + C_2 \sin(\alpha z) \tag{2.72}$$

where $\alpha = \sqrt{P/EI}$. Since there is no deflection at $z = 0$ and $z = L$, we must have $v(0) = 0$ and $v(L) = 0$. The first condition leads to $v(0) = C_1 \cos(0) + C_2 \sin(0) = C_1 = 0$. The second condition implies that $C_2 \sin(\alpha L) = 0$, and therefore if $C_2 \neq 0$ it is $\alpha L = n\pi$. The buckling loads are then written as

$$P_n = \frac{n^2\pi^2 EI}{L^2} \tag{2.73}$$

for $n = 1, 2, 3, \ldots$ The case $n = 1$ corresponds to the critical or Euler buckling load $P_{cr} = \frac{\pi^2 EI}{L^2}$. The buckling states of the column will then have the form $v_n(z) = C_2 \sin(n\pi z/L)$ and are plotted in Figure 2.20 for $n = 1, 2$, $C_2 = 0.01$, and $L = 4$ m using the MATLAB script in Code 2.5.

Code 2.5. *MATLAB code to calculate and plot the buckling modes of a column with length $L = 4$ m, which is supported by a hinge at one end and a roller at the other end. This code*

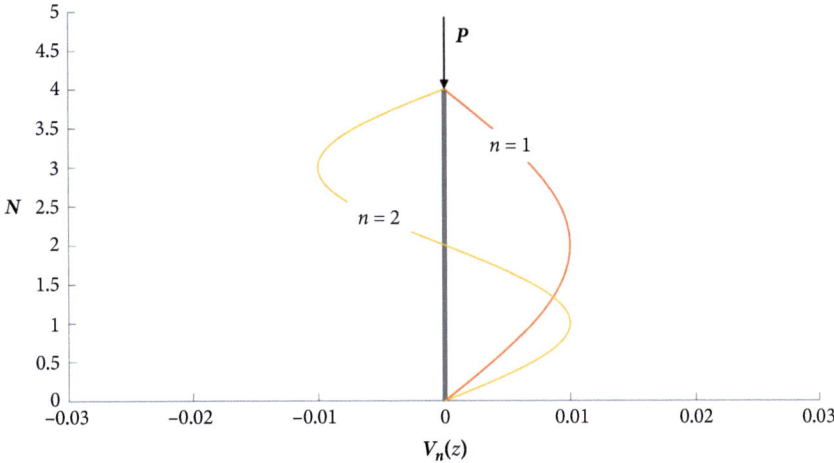

Figure 2.20 The buckling modes of a column with length $L = 4$ m, which is supported by a hinge at one end and by a roller at the other end. This figure is generated using the MATLAB script in Code 2.5.

produces Figure 2.20. To download this code, visit the following website: http://www.oup.com/ AdvancedMathematicalModelling

```matlab
% Buckling modes of a column
clear all;clc;close all;

set(0,'defaultaxesfontsize',16);

L=4; % column height in m
z=0:0.01:4; % z co-ordinate
N=2;
for n=1:N
v(n,:)=0.01*sin(n*pi*z/L);
end

figure(1)
hold on
plot(zeros(size(z)),z,LineWidth=6,Color=[0.5 0.5 0.5])
for n=1:N
plot(v(n,:),z,LineWidth=2); xlim([-0.03 0.03]); ylim([0 5])
end
hold off
ylabel('z',FontWeight='bold',FontSize=22);
xlabel('v_n(z)',FontWeight='bold',FontSize=22);
annotation(figure(1),'arrow',[0.51770833 0.5178385],...
    [0.913606911 0.763498920],'LineWidth',2,'HeadStyle','plain');
annotation(figure(1),'textbox',...
    [0.528125 0.84889200 0.039322917 0.0674946017],...
    'String',{'P'},'FontWeight','bold','FontSize',24,'EdgeColor',[1 1 1]);
annotation(figure(1),'textbox',...
    [0.562239 0.61185097 0.07161458 0.067494601],...
```

Code 2.5. *Continued*

```
   'String',{'n = 1'},'FontSize',24,'EdgeColor',[1 1 1],...
   'BackgroundColor',[1 1 1]);
annotation(figure(1),'textbox',...
   [0.4197916666 0.46552267 0.0716145 0.067494601],...
   'String',{'n = 2'},'FontSize',24,'EdgeColor',[1 1 1],...
   'BackgroundColor',[1 1 1]);
```

2.7 Plane stress, plane strain, and stress concentration

In certain cases, structural members have faults, like cracks, pores, or slits. Furthermore, at the locations of joints, holes are drilled when bolted connections are used. These locations are referred to as stress concentration locations (Sanford 2003). This is because the stresses there become significantly larger than their typical value as calculated using stress resultants. Locations of stress concentration are often the critical points in structural members. They are the locations where failures occur. To analyse stress concentration, more advanced methods need to be employed. The basic equations used are the equations of linear elasticity.

2.7.1 Equations of linear elasticity

In Sections 2.2.1 and 2.2.2 we studied the equations of equilibrium and the definition of strains in an elastic solid in three spatial dimensions. Two important cases, where two-dimensional analysis applies to the solutions of problems in linear elasticity, are plane stress and plane strain. In the case of plane stress parallel to the $x - y$ plane, we have $\sigma_{zz} = \sigma_{zx} = \sigma_{zy} = 0$. Hooke's law for an isotropic material in the case of plane stress has the following form:

$$\begin{bmatrix} \sigma_{xx} \\ \sigma_{yy} \\ \tau_{xy} \end{bmatrix} = \frac{E}{1 - v^2} \begin{bmatrix} 1 & v & 0 \\ v & 1 & 0 \\ 0 & 0 & \frac{1-v}{2} \end{bmatrix} \begin{bmatrix} \varepsilon_{xx} \\ \varepsilon_{yy} \\ \gamma_{xy} \end{bmatrix} \tag{2.74}$$

In the case of plane strain parallel to the $x - y$ plane the displacement along the z coordinate vanishes and, furthermore, we have $\varepsilon_{zz} = \varepsilon_{zx} = \varepsilon_{zy} = 0$. In the case of plane strain, the following expression holds: $\sigma_{zz} = v(\sigma_{xx} + \sigma_{zy})$. Hooke's law for an isotropic material in the case of plane strain is expressed as follows:

$$\begin{bmatrix} \sigma_{xx} \\ \sigma_{yy} \\ \tau_{xy} \end{bmatrix} = \frac{E}{(1 + v)(1 - 2v)} \begin{bmatrix} 1 - v & v & 0 \\ v & 1 - v & 0 \\ 0 & 0 & \frac{1}{2} - v \end{bmatrix} \begin{bmatrix} \varepsilon_{xx} \\ \varepsilon_{yy} \\ \gamma_{xy} \end{bmatrix} \tag{2.75}$$

The case of plane stress applies to solids of small thickness where all loads are in the plane of the solid's horizontal span (perpendicular to the thickness). The case of plain strain is suitable for elongated solids, where the loads are uniform along the elongated dimension.

2.7.2 Airy stress function

When plane stress and plane strain problems are considered, the stress field can be calculated in several cases using the Airy stress function (Airy 1863). The Airy stress function is a scalar function $\Phi(x, y)$ such that

$$\sigma_{xx} = \frac{\partial^2 \Phi}{\partial y^2}, \ \sigma_{yy} = \frac{\partial^2 \Phi}{\partial x^2}, \ \tau_{xy} = -\frac{\partial^2 \Phi}{\partial x \partial y} \tag{2.76}$$

Considering the equilibrium equations, Eq (2.5), in the absence of body forces and differentiating the first equation with respect to x, differentiating the second equation with respect to y, and adding them together, we obtain the following:

$$\frac{\partial^2 \sigma_{xx}}{\partial x^2} + \frac{\partial^2 \sigma_{yy}}{\partial y^2} + 2\frac{\partial^2 \tau_{xy}}{\partial x \partial y} = 0 \tag{2.77}$$

Notice that, according to its definition, the Airy function satisfies this last equation automatically. Using the definition of normal strains, we can write

$$\frac{\partial^2 \varepsilon_{xx}}{\partial y^2} + \frac{\partial^2 \varepsilon_{yy}}{\partial x^2} = \frac{\partial^3 u}{\partial x \partial y^2} + \frac{\partial^3 v}{\partial y \partial x^2} \tag{2.78}$$

From the definition of shear strains, we have

$$\frac{\partial^2 \varepsilon_{xy}}{\partial x \partial y} = \frac{1}{2}\left(\frac{\partial^3 u}{\partial x \partial y^2} + \frac{\partial^3 v}{\partial y \partial x^2} \right) \tag{2.79}$$

and therefore,

$$\frac{\partial^2 \varepsilon_{xx}}{\partial y^2} + \frac{\partial^2 \varepsilon_{yy}}{\partial x^2} = 2\frac{\partial^2 \varepsilon_{xy}}{\partial x \partial y} \tag{2.80}$$

This is one of Saint-Venant's compatibility equations for strains (Love 1944). Using Hooke's law, Eq (2.80) can be expressed in terms of stresses. In the case of plane stress, for example, we have

$$\frac{\partial^2 \left(\sigma_{xx} - v\sigma_{yy} \right)}{\partial x^2} + \frac{\partial^2 \left(\sigma_{yy} - v\sigma_{xx} \right)}{\partial y^2} - 2\left(1 + v \right)\frac{\partial^2 \tau_{xy}}{\partial x \partial y} = 0 \tag{2.81}$$

Using Eq (2.77) to eliminate the shear stresses from Eq (2.81), we can have the following:

$$\frac{\partial^2 \sigma_{xx}}{\partial x^2} + \frac{\partial^2 \sigma_{yy}}{\partial x^2} + \frac{\partial^2 \sigma_{xx}}{\partial y^2} + \frac{\partial^2 \sigma_{yy}}{\partial y^2} = 0 \tag{2.82}$$

Utilizing the definition of stresses through the Airy function, Eq (2.82) can be rewritten as

$$\frac{\partial^4 \Phi}{\partial x^4} + 2\frac{\partial^4 \Phi}{\partial x^2 \partial y^2} + \frac{\partial^4 \Phi}{\partial y^4} = 0 \tag{2.83}$$

This means the Airy stress function satisfies the biharmonic equation. The same result occurs when plane strain conditions are assumed. Furthermore, the result holds in the case where body forces are present and are produced by a potential $V(x, y)$ as follows:

$$b_x = -\frac{\partial V}{\partial x}, \quad b_y = -\frac{\partial V}{\partial y} \tag{2.85}$$

In this case, the stresses are defined from the Airy stress function as follows:

$$\sigma_{xx} = \frac{\partial^2 \Phi}{\partial y^2} + V, \quad \sigma_{yy} = \frac{\partial^2 \Phi}{\partial x^2} + V, \quad \tau_{xy} = -\frac{\partial^2 \Phi}{\partial x \partial y}. \tag{2.86}$$

The Airy stress function can also be expressed in polar coordinates (r, θ), such that $x = r\cos\theta$ and $y = r\sin\theta$. For plane stress and plane strain problems in polar coordinates, the stress components become

$$\sigma_{rr} = \frac{1}{r}\frac{\partial \Phi}{\partial r} + \frac{1}{r^2}\frac{\partial^2 \Phi}{\partial \theta^2}, \quad \sigma_{\theta\theta} = \frac{\partial^2 \Phi}{\partial r^2}, \quad \tau_{r\theta} = -\frac{1}{r^2}\frac{\partial \Phi}{\partial \theta} - \frac{1}{r}\frac{\partial^2 \Phi}{\partial r \partial \theta}. \tag{2.87}$$

The biharmonic equation, satisfied by the Airy stress function, in polar coordinates becomes

$$\frac{1}{r}\frac{\partial}{\partial r}\left[r\frac{\partial}{\partial r}\left(\frac{1}{r}\frac{\partial}{\partial r}\left[r\frac{\partial \Phi}{\partial r}\right]\right)\right] + \frac{2}{r^2}\frac{\partial^4 \Phi}{\partial r^2 \partial \theta^2} - \frac{2}{r^3}\frac{\partial^3 \Phi}{\partial r \partial \theta^2} + \frac{4}{r^4}\frac{\partial^2 \Phi}{\partial \theta^2} + \frac{1}{r^4}\frac{\partial^4 \Phi}{\partial \theta^4} = 0 \tag{2.88}$$

The general solution of the biharmonic Eq (2.88) in polar coordinates was derived by Michell (1899) and has the following form:

$$\begin{aligned}
\Phi(r, \theta) &= A_0 r^2 + B_0 \ln(r) + C_0 r^2 \ln(r) \\
&+ \left[D_0 + D_1 r^2 + D_2 \ln(r) + D_3 r^2 \ln(r)\right]\theta \\
&+ \left[E_0 r + E_1 r^{-1} + E_2 r^3 + E_3 r \ln(r) + E_4 r\theta\right]\cos(\theta) \\
&+ \left[F_0 r + F_1 r^{-1} + F_2 r^3 + F_3 r \ln(r) + F_4 r\theta\right]\sin(\theta) \\
&+ \sum_{n=2}^{\infty} \left(A_{1n} r^n + A_{2n} r^{-n} + A_{3n} r^{n+2} + A_{4n} r^{-n+2}\right)\cos(n\theta) \\
&+ \sum_{n=2}^{\infty} \left(B_{1n} r^n + B_{2n} r^{-n} + B_{3n} r^{n+2} + B_{4n} r^{-n+2}\right)\sin(n\theta)
\end{aligned} \tag{2.89}$$

Given the general solution for the Airy stress function, the stresses can be calculated using Eq (2.87). The solutions to specific problems can be calculated by considering specific terms in Eq (2.89) and applying the given boundary conditions of the problem in hand to determine the constants appearing in each term.

Example calculations for a circular cylindrical tube with internal pressure and the stress concentration around a circular hole are shown in Boxes 2.8 and 2.9, respectively.

Box 2.8: Example (Circular cylindrical tube with internal pressure: Lamé solution)

Consider a pipe with internal radius $r = a$ and external radius $r = b$ as shown in Figure 2.21. Internally the pipe is under pressure p_i, while the external pressure is p_o. Find the radial stress σ_{rr} and hoop stress $\sigma_{\theta\theta}$ at a pipe cross section.

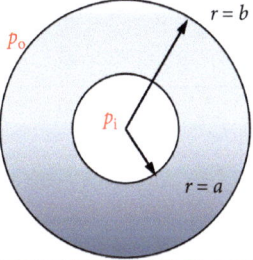

Figure 2.21 Pipe cross section with internal and external pressures applied.

Solution

This is a problem with axial symmetry and was first solved by the French mathematician Gabriel Lamé. Due to the axial symmetry, the stress function does not depend on θ and takes the following form:

$$\Phi(r, \theta) = A_0 r^2 + B_0 \ln(r) + C_0 r^2 \ln(r)$$

Then, $\sigma_{rr} = \frac{1}{r}\frac{\partial \Phi}{\partial r} = 2A_0 + \frac{B_0}{r^2} + 2C_0 \ln(r) + C_0$. The function $\ln(r)$, if kept in the solution, will give rise to unphysical displacements (multivalued; Fung et al. 2001). Therefore, the radial stress will finally have the following form:

$$\sigma_{rr} = 2A_0 + \frac{B_0}{r^2} + C_0 = A + \frac{B}{r^2}$$

The constants A and B can be calculated from the boundary conditions:

$$\sigma_{rr} = -p_i \text{ at } r = a$$

$$\sigma_{rr} = -p_o \text{ at } r = b$$

Box 2.8 *Continued*

Finally, the stresses are calculated as follows:

$$\sigma_{rr} = -p_i \frac{\left(\frac{b}{r}\right)^2 - 1}{\left(\frac{b}{a}\right)^2 - 1} - p_o \frac{1 - \left(\frac{a}{r}\right)^2}{1 - \left(\frac{a}{b}\right)^2}$$

$$\sigma_{\theta\theta} = p_i \frac{\left(\frac{b}{r}\right)^2 + 1}{\left(\frac{b}{a}\right)^2 - 1} - p_o \frac{1 + \left(\frac{a}{r}\right)^2}{1 - \left(\frac{a}{b}\right)^2}$$

We note that in this problem we have $\tau_{r\theta} = 0$ due to the axial symmetry.

The MATLAB script in Code 2.6 is used to plot the radial and hoop stresses in a pipe with $a = 0.2\ m$ and $b = 0.3\ m$, where $p_i = 100\ MPa$ and $p_o = 0$. The results are shown in Figure 2.22.

Code 2.6. *MATLAB code to calculate and plot the radial and hoop stresses in a pipe, using the Lamé solution. The assumptions are: $a = 0.2\ m$, $b = 0.3\ m$, $p_i = 100\ MPa$, and $p_o = 0$. This code generates Figure 2.22.*

```matlab
% calculate and plot stresses in a pipe under pressure
clear all;clc;close all;

set(0,'defaultaxesfontsize',16);

a=0.2; % internal radius in m
b=0.3; % external radius in m
p_i=100*10^6; % internal pressure in Pa
p_o=0; % external pressure in Pa

r=a:0.001:b;

s_radial=-p_i*((b./r).^2-1)./((b/a)^2-1)...
        -p_o*(1-(a./r).^2)./(1-(a/b)^2);
s_hoop=p_i*((b./r).^2+1)./((b/a)^2-1)...
        -p_o*(1+(a./r).^2)./(1-(a/b)^2);

figure(1)
plot(r,s_radial/10^6,'k',r,s_hoop/10^6,'b-',LineWidth=1.5)
xlabel('r (m)','FontWeight','bold','FontSize',22);
ylabel('stress (MPa)','FontWeight','bold','FontSize',22);
annotation(figure(1),'textbox',...
    [0.620833333333332 0.224701942093684 0.0395833333333344 0.08693304710711773],...
    'String',{'\sigma_{rr}'},'FontSize',24,'FitBoxToText','off',...
    'EdgeColor',[1 1 1],'BackgroundColor',[1 1 1]);
annotation(figure(1),'textbox',...
    [0.620052083333331 0.643168464771867 0.0395833333333343 0.08693304710711769],...
    'Color',[0 0 1],'String','\sigma_{\theta\theta}','FontSize',24,...
    'FitBoxToText','off','EdgeColor',[1 1 1],'BackgroundColor',[1 1 1]);
```

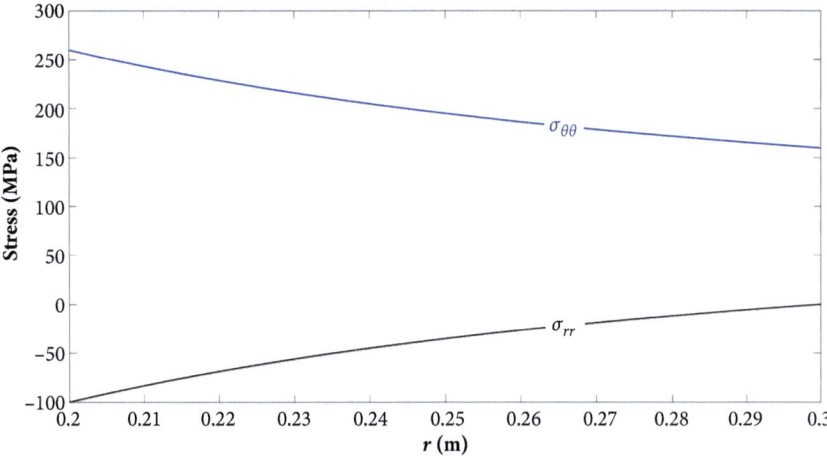

Figure 2.22 Plot of the radial and hoop stresses in a pipe, using the Lamé solution. The assumptions are: $a = 0.2\ m$, $b = 0.3\ m$, $p_i = 100\ MPa$, and $p_o = 0$. This figure is produced using Code 2.6.

Box 2.9: Example (Stress concentration around a circular hole: Kirsch solution)

The following problem was solved by Kirsch (1898). Consider a thin elastic plate with a circular hole of radius R. For the (x, y) coordinate system, the origin coincides with the centre of the circular hole. The elastic plate extends to infinity where a tensile normal stress (σ_∞) along the x axis is applied (Figure 2.23). Assuming plane stress conditions, find the stress field at the vicinity of the circular hole.

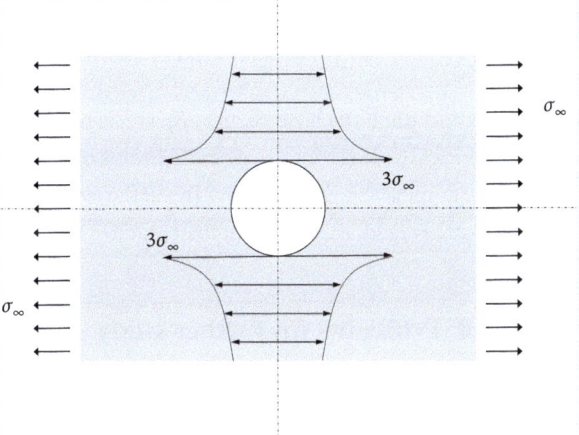

Figure 2.23 Infinite plate under plane stress conditions with a circular hole.

Box 2.9 *Continued*

Solution

For a hole of radius R, Kirsch presented the following stress field:

$$\sigma_{rr} = \frac{\sigma_\infty}{2}\left(1 - \frac{R^2}{r^2}\right) + \frac{\sigma_\infty}{2}\left(1 - 4\frac{R^2}{r^2} + 3\frac{R^4}{r^4}\right)\cos 2\theta$$

$$\sigma_{\theta\theta} = \frac{\sigma_\infty}{2}\left(1 + \frac{R^2}{r^2}\right) - \frac{\sigma_\infty}{2}\left(1 + 3\frac{R^4}{r^4}\right)\cos, 2\theta$$

$$\sigma_{r\theta} = -\frac{\sigma_\infty}{2}\left(1 + 2\frac{R^2}{r^2} - 3\frac{R^4}{r^4}\right)\sin 2\theta$$

For $r = R$, the axial stress σ_{rr} and the shear stress $\sigma_{r\theta}$ vanish because the hole surface is traction free. The circumferential component $\sigma_{\theta\theta}$ is $\sigma_{\theta\theta} = \sigma_\infty - 2\sigma_\infty \cos 2\theta$. This quantity is maximized for $\cos 2\theta = -1$, that is, $\theta = \pi/2$ and $\theta = 3\pi/2$. In these cases, the circumferential stress at $r = R$ attains the value $\sigma_{max} = 3\sigma_\infty$. The stress concentration factor K_t is defined as follows:

$$K_t = \frac{\sigma_{max}}{\sigma_\infty}$$

And in this case, it is $K_t = 3$.

The stress concentration factor for the more general case of an elliptical hole with major semi-axis a and minor semi-axis b has been derived by Inglis (1913) and has the following form:

$$K_t = \frac{\sigma_{max}}{\sigma_\infty} = \left(1 + 2\frac{a}{b}\right) \tag{2.90}$$

In the case of a circle, $a = b$ and the Kirsch prediction $K_t = 3$ is retrieved. It is interesting to note that as $b \to 0$, the elliptical hole resembles a crack. In this case $K_t \to \infty$ with stresses at the vicinity of the crack attaining very high values. In such cases, another quantity that indicates the severity of the stress field, the stress intensity factor, is useful (Sanford 2003).

2.8 Problems for further study

Problem 2.1

Calculate the elongation of an elastic bar with length $L = 2$, having its left-hand end fixed, a distributed constant load f, and a point load of magnitude P acting on the right-hand end. The cross-sectional area A is tapered and has the form $A(x) = 3 - x$, where x is the distance from the fixed end. The Young's modulus is E.

Problem 2.2

Calculate the dilatation of a solid if its displacement field is $u(x, y, z) = Ax + By + Cz$, $v(x, y, z) = Bx + Cy + Az$, and $w(x, y, z) = Cx + Ay + Bz$, where A, B, and C are constants.

Answer to Problem 2.2

$\delta = A + B + C.$

Problem 2.3

Calculate the principal stresses for the following stress state using Eqs (2.12a), (2.12b), and (**2.12c**). Compare these values with the eigenvalues of σ (see Appendix A for the calculation of eigenvalues):

$$\sigma = \begin{bmatrix} 2 & 1 & 0 \\ 1 & -1 & 1 \\ 0 & 1 & 2 \end{bmatrix}$$

Problem 2.4

Find the deflection of a beam with infinite length and flexural rigidity EI_{yy}, resting on a Winkler foundation with constant k and loaded by the point loads P and $-P$ applied at $x = \varepsilon$ and $x = -\varepsilon$, respectively. Plot the solution using MATLAB for the case where $EI_{yy} = 1000\ Nm^2$, $P = -1\ N$, and two different values of k: $k_1 = 1\ N/m^2$ and $k_2 = 2\ N/m^2$.

Problem 2.5

By appropriately reducing terms from the stress function used for the pipe with internal and external pressure, $\Phi(r, \theta) = A_0 r^2 + B_0 \ln(r) + C_0 r^2 \ln(r)$, find the radial stress σ_{rr} and hoop stress $\sigma_{\theta\theta}$ in a circular disk of radius α under external pressure p.

Answer to Problem 2.5

$\sigma_{rr} = \sigma_{\theta\theta} = p.$

References

Airy, G.B. (1863). On the strains in the interior of beams, *Philosophical Transactions of the Royal Society*, 153, 49–79.

Callister, W.D. Jr (1993). *Materials Science and Engineering: An Introduction*, 3rd edn. John Wiley & Sons, Inc.

Fung, Y.C., Tong, P., and Chen, X. (2001). *Classical and Computational Solid Mechanics*. World Scientific Publishing Company.

Goodier, J.N. and Hoff N.J. (eds.) (1960). *Structural Mechanics*. Pergamon Press.

Hetenyi, M. (1946). *Beams on Elastic Foundation: Theory with Applications in the Fields of Civil and Mechanical Engineering*. University of Michigan Press.

Inglis, C.E. (1913). Stresses in a plate due to the presence of cracks and sharp corners. *Transactions of the Institute of Naval Architects*, 55(219–241), 193–198.

Kirsch, G. (1898). Die Theorie der Elastizität und die Bedürfnisse der Festigkeitslehre. *Zeitschrift des Vereines deutscher Ingenieure*, 42, 797–807.

Love, A.E.H. (1944). *A Treatise on the Mathematical Theory of Elasticity*, 4th edn. Dover Publications.

Michell, J.H. (1899). On the direct determination of stress in an elastic solid, with application to the theory of plates. *Proceedings of the London Mathematical Society*, 31(1), 100–124.

Sanford, R.J. (2003). *Principles of Fracture Mechanics*. Pearson Education, Inc

Timoshenko, S. (1955). *Strength of Materials*, 3rd edn. Von Nostrand.

Winkler, E. (1867). *Die Lehre von der Elastizität und Festigkeit*. Dominicus.

Chapter 3
Geotechnical Engineering

3.1 Introduction

Geotechnical engineering is a pioneering and firmly established subdiscipline within civil engineering. It is dedicated to addressing a wide array of engineering challenges, focusing on understanding the behaviour of soil and rock under various loading conditions. Moreover, it evaluates the suitability of soil and foundations to support applied loads from buildings and structures, while also studying the movement of water through porous media, encompassing phenomena like seepage, groundwater flow, capillarity, infiltration, and saturation.

Among the most popular problems in geotechnical engineering is water seepage through the soil. Figure 3.1 shows two earth dams (also known as embankment or earth-fill dams) which are widely used worldwide for storing water for domestic or industrial uses as well as for irrigation and flood control purposes. The large amount of stored water behind the dam (Figure 3.1) puts high pressures on the dam body and its foundation, and will move through these media in the form of water seepage. Seepage refers to the gradual movement of water along zigzag paths amidst the particles of porous soil media. Due to water seepage, the hydrostatic water pressure gradually increases in the soil, posing a potential threat to the stability and integrity of the earth dam and its foundation.

Predicting seepage paths and calculating variations in hydrostatic pressure due to seepage are crucial for designing earth dams. In Figure 3.2, the contours of isopotential (representing areas with the same hydrostatic pressure or water head) and flow paths for a typical earth dam (Figure 3.2a) and a concrete dam (Figure 3.2b) are displayed, indicating their orthogonal nature.

This chapter covers not only water seepage through soil but also various other geotechnical engineering problems, including soil consolidation, settlement, and groundwater flow through the soil, all of which are addressed and solved using mathematical approaches. The chapter presents a number of numerical examples with solutions and accompanying MATLAB code to demonstrate the application of knowledge to real-world problems in geotechnical engineering. Additionally, it provides various practice problems for further reader engagement.

A Practical Approach to Advanced Mathematical Modelling in Civil Engineering. Mohammad Heidarzadeh et al.,
Oxford University Press. © Mohammad Heidarzadeh, Theodosios K. Papathanasiou, Yurui Fan, Hamid Bahai (2025).
DOI: 10.1093/9780191888656.003.0003

Figure 3.1 Photos of two earth-fill dams, the Whaley Bridge and Torside dams, both in the United Kingdom. The dam bodies and the reservoirs are marked.

3.2 Seepage and Darcy's law

3.2.1 Governing equation of seepage

For the earth dam shown in Figure 3.3, consider a two-dimensional soil element with dimensions of dx and dy in the x and y directions, respectively. Let the permeability coefficient (k) of the soil in the x and y directions be k_x and k_y, respectively. From soil

(a) Earth dam

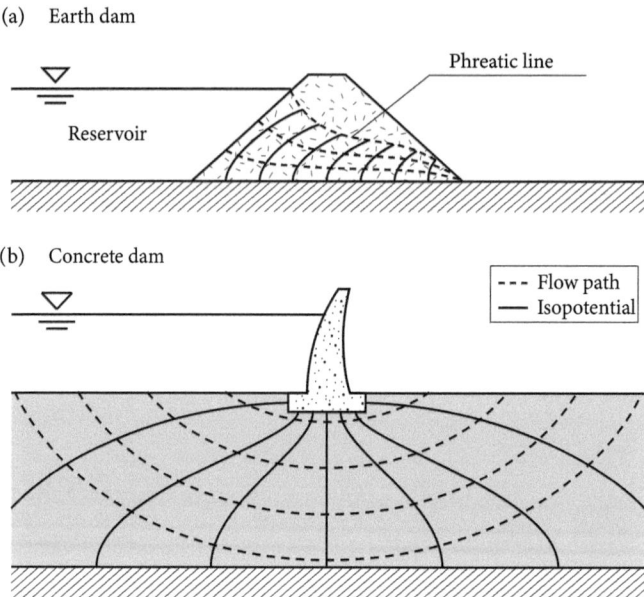

(b) Concrete dam

Figure 3.2 Sketches demonstrating water seepage through the dam body (a) and its foundation (b) in earth and concrete dams, respectively. The solid and dashed lines are the contours of isopotential and flow path, respectively.

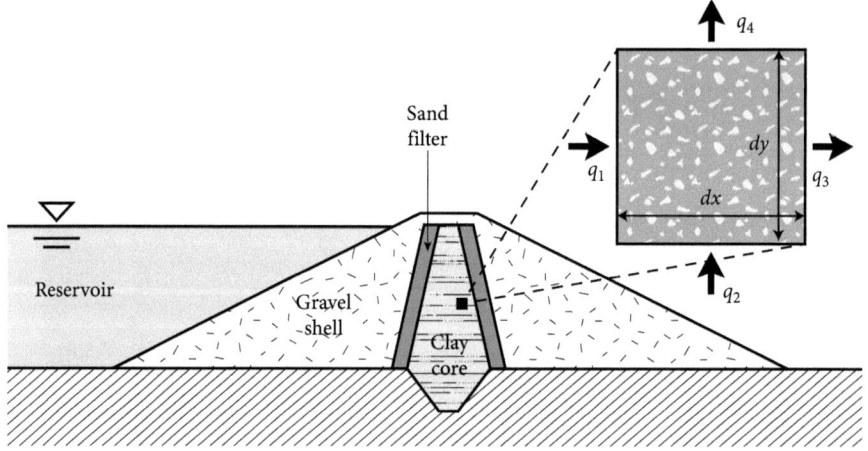

Figure 3.3 Sketch showing a two-dimensional soil element in an earth-fill dam.

mechanics, we know that the velocity of water flow through the soil (v) is given by Darcy's law as follows (Das 2019):

$$v = k \cdot i \tag{3.1}$$

in which, i is the hydraulic gradient defined as

$$i = \frac{\Delta h}{L} \tag{3.2}$$

where h is the piezometric water head at any point in the soil, Δh is the piezometric head difference between two points in the soil located at a distance L from each other. From basic fluid mechanics, the amount of flow (q) in a unit width of a soil section with water depth and velocity of d and v, respectively, is given by

$$q = v \cdot d \tag{3.3}$$

Assuming an incompressible medium and a steady flow, the amount of flow entering the soil element in Figure 3.3 (q_{in}) is equal to the flow leaving the cell (q_{out}):

$$q_{in} = q_{out} \tag{3.4}$$

According to Figure 3.3, the total amount of flow entering the element is $q_{in} = q_1 + q_2$, while the total existing flow is $q_{out} = q_3 + q_4$. Applying Eq (3.3), we have $q_1 = v_x \cdot dy$, $q_2 = v_y \cdot dx$, $q_3 = \left(v_x + \frac{\partial v_x}{\partial x} dx\right) \cdot dy$, and $q_4 = \left(v_y + \frac{\partial v_y}{\partial y} dy\right) \cdot dx$. Therefore, Eq (3.4) becomes

$$v_x \cdot dy + v_y \cdot dx = \left(v_x + \frac{\partial v_x}{\partial x} dx\right) \cdot dy + \left(v_y + \frac{\partial v_y}{\partial y} dy\right) \cdot dx \tag{3.5}$$

After rearranging the terms, it transforms into the following form:

$$\frac{\partial v_x}{\partial x} dx \cdot dy + \frac{\partial v_y}{\partial y} dy \cdot dx = 0 \tag{3.6}$$

Equation (3.6) can be further simplified to

$$\frac{\partial v_x}{\partial x} + \frac{\partial v_y}{\partial y} = 0 \tag{3.7}$$

Utilizing Darcy's law, Eq (3.1), we express the water velocity components in the x and y directions as

$$v_x = k_x \cdot i_x \tag{3.8}$$

$$v_y = k_y \cdot i_y \tag{3.9}$$

in which $i_x = \frac{\partial h}{\partial x}$ and $i_y = \frac{\partial h}{\partial y}$. Upon substituting Eqs (3.8) and (3.9) into Eq (3.7), we derive the following equation:

$$k_x \frac{\partial^2 h}{\partial x^2} + k_y \frac{\partial^2 h}{\partial y^2} = 0 \tag{3.10}$$

For isotropic soils, the permeability coefficients in the x and y directions are equal, represented as $k_x = k_y$. This expression aids in further simplifying Eq (3.10) to the following form:

$$\frac{\partial^2 h}{\partial x^2} + \frac{\partial^2 h}{\partial y^2} = 0 \tag{3.11}$$

Equation (3.11) is in Laplacian form and serves as the governing equation that describes water seepage through the soil. Solving Eq (3.11) allows for the determination of the piezometric water head (h) at each point and the associated flow path.

3.2.2 Solving the differential equation of seepage

Here we solve the seepage through the foundation of the earth-fill dam shown in Figure 3.4. The computational domain is the dam foundation, which is shown separately in Figure 3.4a with its boundary conditions identified and marked in Figure 3.4b. The primary objective is to solve Eq (3.11). The separation of variables method is applied, which gives the following general form for the solution:

$$h(x, y) = X_{(x)} \cdot Y_{(y)} \tag{3.12}$$

Here, $h(x, y)$ is the piezometric water head at any arbitrary location in the computational domain identified by x and y coordinates. $X_{(x)}$ and $Y_{(y)}$ are arbitrary and are exclusively dependent on x and y, respectively. Upon differentiating $h(x, y)$ with respect to x and y,

(a)   Earth dam

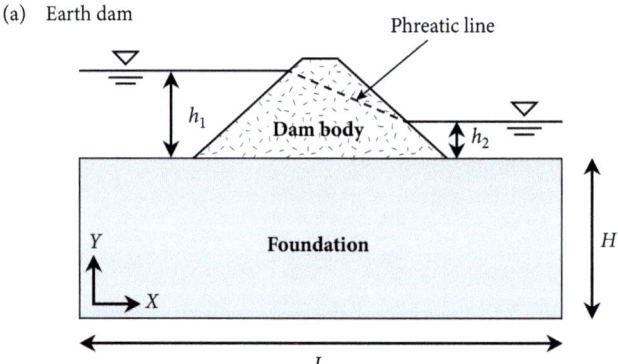

(b)   Boundary conditions of foundation

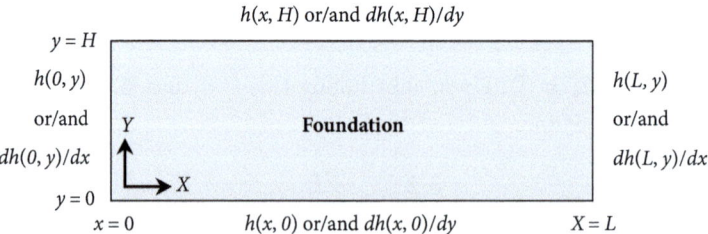

Figure 3.4 (a) Sketch showing an earth-fill dam with its body and foundation. (b) The foundation of the dam as the computational domain with its boundary conditions identified and marked.

the resulting derivatives are: $\frac{\partial h}{\partial x} = X' \cdot Y$, $\frac{\partial h}{\partial y} = X \cdot Y'$, $\frac{\partial^2 h}{\partial x^2} = X'' \cdot Y$, and $\frac{\partial^2 h}{\partial y^2} = X \cdot Y''$. These terms are substituted in Eq (3.11), resulting in the following equation:

$$X'' \cdot Y + X \cdot Y'' = 0 \tag{3.13}$$

Dividing both sides of Eq (3.13) by $X \cdot Y$,

$$\frac{X''}{X} + \frac{Y''}{Y} = 0 \tag{3.14}$$

Rearranging, we get

$$\frac{X''}{X} = -\frac{Y''}{Y} \tag{3.15}$$

In Eq (3.15), since the left-hand side depends exclusively on x and the right-hand side is solely a function of y, both sides must equal a constant value. This constant could be zero, negative, or positive:

$$\frac{X''}{X} = -\frac{Y''}{Y} = \begin{cases} \lambda^2 \\ 0 \\ -\lambda^2 \end{cases} \tag{3.16}$$

Equation (3.16) requires solving for all three cases involving constant values of 0, λ^2, and $-\lambda^2$ to determine which one yields a feasible solution, considering the specific boundary conditions of the problem. Assuming, for instance, that the value λ^2 represents the correct answer, Eq (3.16) takes the following form:

$$\frac{X''}{X} = -\frac{Y''}{Y} = \lambda^2 \tag{3.17}$$

Equation (3.17) results in two ODEs as follows:

$$X'' - \lambda^2 X = 0 \tag{3.18}$$

$$Y'' + \lambda^2 Y = 0 \tag{3.19}$$

We know that the general solutions for these ODEs are as follows:

$$X(x) = A \sinh(\lambda x) + B \cosh(\lambda x) \tag{3.20}$$

$$Y(y) = C \sin(\lambda y) + D \cos(\lambda y) \tag{3.21}$$

where A, B, C, and D denote constant values. Upon substituting the functions $X(x)$ and $Y(y)$ into Eq (3.12), the solution for the seepage problem, $h(x, y)$, takes the following form:

$$h(x, y) = \{A \sinh(\lambda x) + B \cosh(\lambda x)\} \times \{C \sin(\lambda y) + D \cos(\lambda y)\} \tag{3.22}$$

Table 3.1 Different types of boundary conditions used in addressing seepage problems

Type of boundary	Description	Mathematical expression[*]
Permeable	Water can move freely in and out of the boundary; for example, sand or gravel layers	$h = 0$
Impermeable	Water cannot move in or out of the boundary; for example, an impermeable rock layer	$\dfrac{\partial h}{\partial n} = 0$
Water head	There is a static water head with a water height of h_0 at the boundary; for example, a dam reservoir	$h = h_0$

[*]h is the piezometric water head.

It is worth noting that the constant value λ was chosen arbitrarily. Therefore, any other number such as $\lambda_1, \lambda_2, \lambda_3$, and so on can serve as a valid solution. We understand that in differential equations, if P_1, P_2, P_3, and so forth are valid answers, then any linear combination of them also constitutes a solution. Thus, one potential solution of the differential equation can be the sum of all these solutions. Employing these fundamental concepts, an answer to the seepage problem, inspired by Eq (3.22), is presented as follows:

$$h(x,y) = \sum_{n=0}^{\infty} \{A_n \sinh(\lambda_n x) + B_n \cosh(\lambda_n x)\} \times \{C_n \sin(\lambda_n y) + D_n \cos(\lambda_n y)\} \quad (3.23)$$

The coefficients λ_n, A_n, B_n, C_n, and D_n are determined by applying the specific boundary conditions of the problem. Table 3.1 outlines various types of boundary conditions.

Equation (3.23) was derived under the assumption that the correct solution to Eq (3.16) is λ^2. Nevertheless, it is essential to note that in certain situations, this assumption may not hold true, particularly depending on the specific boundary conditions of the problem. By assuming that the correct solution to Eq (3.16) is $-\lambda^2$, and by following the same procedure described earlier, Eqs (3.17)–(3.23), the form of the solution for the seepage problem, $h(x, y)$, will be as follows:

$$h(x,y) = \sum_{n=0}^{\infty} \{A_n \sin(\lambda_n x) + B_n \cos(\lambda_n x)\} \times \{C_n \sinh(\lambda_n y) + D_n \cosh(\lambda_n y)\} \quad (3.24)$$

3.2.3 Numerical example: Seepage through a clay layer

Consider the embankment dam shown in Figure 3.5. The reservoir water elevation is 150 m. Downstream of the dam, an underground clay layer with dimensions of $a = 1000$ m, $b = 1500$ m (in the horizontal plane) is considered which is bound from the dam side with a constant hydrostatic water head of $h_0 = 150$ m. At the other three sides, the boundary conditions are sand layers. The clay layer is positioned horizontally at a certain depth. Consequently, its configuration and boundary layers are not visible from the bird's-eye view depicted in Figure 3.5. Since the three boundaries of the clay layer are made of sand, the piezometric water head is zero at these boundaries ($h = 0$) according

Figure 3.5 An embankment dam with its downstream area as the computational domain. The underground layer is assumed to be of clay materials with three boundary layers of sand materials. Photo modified from Mahab Consulting Engineers (1998).

to Table 3.1. This occurs because water pressure dissipates rapidly within sand layers, resulting in an inability to generate water pressure (i.e., $h = 0$). On the boundary with the dam body, a constant water head of 150 m is permanently established, i.e., $h_0 = 150\,m$. A summary of the computational domain and the boundary layers is shown in Figure 3.6. It is assumed that the computational domain is a rectangle.

It is of engineering value to determine the distribution of the piezometric water head within the clay layer as it changes from 150 m at the location of the dam ($x = 1000m$; see Figure 3.6) to zero at a distance of 1 km downstream of the dam ($x = 0$; Figure 3.6). For instance, if the downstream piezometric head is larger than expected, remedial counter-measures need to be applied such as construction of downstream earth-fills (also known as seepage berms) and installation of relief wells. Heidarzadeh et al. (2015) presented examples of such measures in dam projects.

For solving the seepage problem outlined in Figures 3.5 and 3.6, we begin with one of Eqs (3.23) or (3.24) and impose the specific boundary conditions of this problem. It can be proved that the correct answer to Eq (3.16) is λ^2 and thus we apply the solution in the form of Eq (3.23) rather than Eq (3.24).

At the southern boundary $\left(y = 0\right)$, the water head is zero: $h\left(x, y\right) = 0$. Therefore, by substituting these values into Eq (3.23), we have

$$0 = \sum_{n=0}^{\infty} \{A_n \sinh\left(\lambda_n x\right) + B_n \cosh\left(\lambda_n x\right)\} \times \{D_n\} \tag{3.25}$$

For Eq (3.25) to hold true, we require $D_n = 0$.

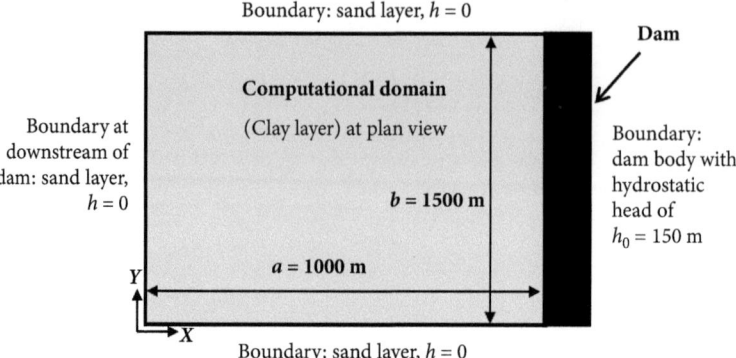

Figure 3.6 Sketch showing the computational domain and the boundary layers for the problem shown in Figure 3.5. It is assumed that the computational domain is a rectangle.

At the northern boundary $(y = b)$, the water head is zero: $h(x, y) = 0$. We also know that $D_n = 0$. Therefore, Eq (3.23) can be written as follows:

$$0 = \sum_{n=0}^{\infty} \{A_n \sinh(\lambda_n x) + B_n \cosh(\lambda_n x)\} \times \{C_n \sin(b\lambda_n)\} \tag{3.26}$$

As $D_n = 0$ from the previous stage, it is clear that $C_n \neq 0$ because otherwise $h(x, y)$ will always be zero. Knowing this, Eq (3.26) gives the following outcome:

$$\sin(b\lambda_n) = 0 \tag{3.27}$$

Thus, $b\lambda_n = n\pi$. Hence, $\lambda_n = \frac{n\pi}{b}$. So far, the solution takes the following form:

$$h(x, y) = \sum_{n=0}^{\infty} \left\{ A_n \sinh\left(\frac{n\pi}{b}x\right) + B_n \cosh\left(\frac{n\pi}{b}x\right) \right\} \times C_n \sin\left(\frac{n\pi}{b}y\right) \tag{3.28}$$

By multiplying the coefficient C_n by A_n and B_n, new coefficients are produced: $a_n = C_n A_n$ and $b_n = C_n B_n$. Therefore, the equation becomes

$$h(x, y) = \sum_{n=0}^{\infty} \sin\left(\frac{n\pi}{b}y\right) \times \left\{ a_n \sinh\left(\frac{n\pi}{b}x\right) + b_n \cosh\left(\frac{n\pi}{b}x\right) \right\} \tag{3.29}$$

We have applied two boundary conditions so far and thus two other conditions remain unused. At the left-hand boundary $(x = 0$; downstream of the dam, Figure 3.6), the head is zero: $h(x, y) = 0$. By applying this condition in Eq (3.29), we have

$$0 = \sum_{n=0}^{\infty} \sin\left(\frac{n\pi}{b}y\right) \times \{b_n\} \tag{3.30}$$

Equation (3.30) requires that $b_n = 0$.

The last boundary condition occurs at the dam body when $x = a$, where $h(x, y) = h_0$ (Figure 3.6). By substituting this data into Eq (3.29) and taking into account the earlier establishment of $b_n = 0$, we derive the following:

$$h_0 = \sum_{n=0}^{\infty} \sin\left(\frac{n\pi}{b}y\right) \times \left\{ a_n \sinh\left(\frac{n\pi}{b}a\right) \right\} \tag{3.31}$$

As the term $a_n \sinh\left(\frac{n\pi}{b}a\right)$ is a constant, a new coefficient is introduced, defined as $E_n = a_n \sinh\left(\frac{n\pi}{b}a\right)$. Consequently, Eq (3.31) becomes

$$h_0 = \sum_{n=0}^{\infty} E_n \sin\left(\frac{n\pi}{b}y\right) \tag{3.32}$$

Equation (3.32) represents a Fourier series problem. By applying the Fourier formulas, E_n is determined as follows:

$$E_n = \frac{2}{b} \int_0^b h_0 \sin\left(\frac{n\pi}{b}y\right) dy = \frac{2h_0}{n\pi} \left[1 - (-1)^n\right] \tag{3.33}$$

Given that $a_n = \frac{E_n}{\sinh\left(\frac{n\pi}{b}a\right)}$ as derived earlier, we consequently have

$$a_n = \frac{2h_0 \left[1 - (-1)^n\right]}{n\pi \, \sinh\left(\frac{n\pi}{b}a\right)} \tag{3.34}$$

Finally, by substituting Eq (3.34) into Eq (3.29), the final solution is obtained which has the following form:

$$h(x, y) = \sum_{n=0}^{\infty} \frac{2h_0 \left[1 - (-1)^n\right]}{n\pi \, \sinh\left(\frac{n\pi}{b}a\right)} \sinh\left(\frac{n\pi}{b}x\right) \sin\left(\frac{n\pi}{b}y\right) \tag{3.35}$$

Equation (3.35) gives the piezometric water head at any location within the computational domain with coordinate (x, y). This final solution can be plotted using mathematical packages such as MATLAB. Here, Code 3.1 can be used to plot Eq (3.35). The outcome is shown in Figure 3.7. This MATLAB code and other scripts in this chapter can be downloaded at the following website: http://www.oup.com/AdvancedMathematicalModelling

The solution depicted in Figure 3.7 illustrates that the water head initiates at 150 m near the dam axis and gradually decreases in all directions until reaching zero at the boundaries to the west, north, and south within the computational domain (Figure 3.7). We note that the water head is theoretically expected to be zero at sand boundaries (west, north, and south boundaries) as these type of boundary do not permit the buildup of water pressure.

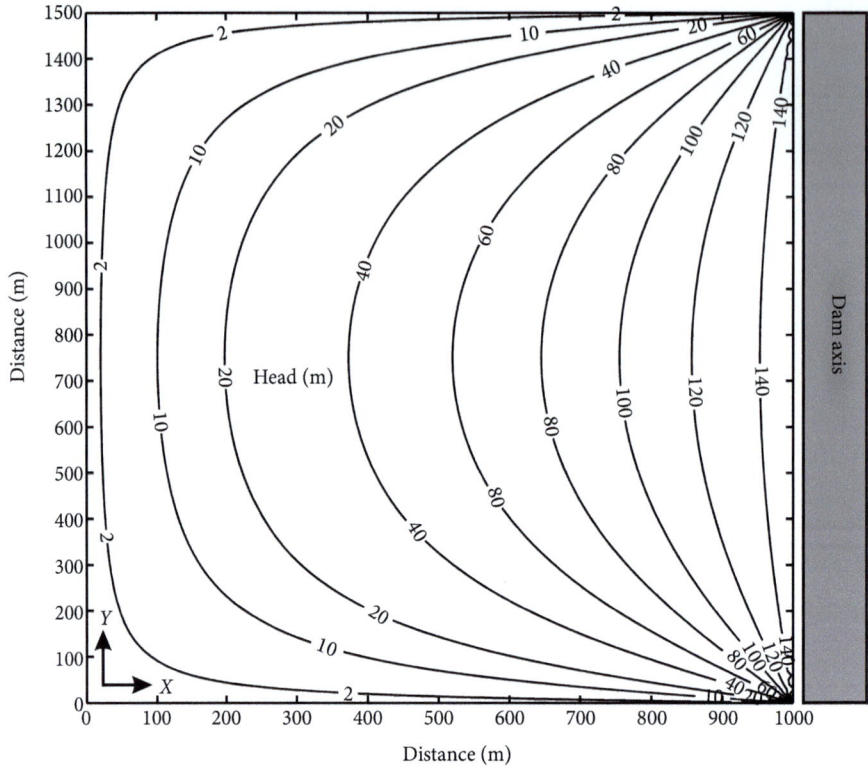

Figure 3.7 Plot of the solution of the seepage problem shown in Figures 3.5 and 3.6. This figure is generated using the MATLAB script in Code 3.1.

Code 3.1 *The MATLAB script used for plotting the solution of the problem shown in Figures 3.5 and 3.6. This code produces Figure 3.7. This MATLAB code (and the other MATLAB scripts in this chapter) can be downloaded from the following website: http://www.oup.com/ AdvancedMathematicalModelling*

```
% This code solves seepage in dam foundation
clc; clear; close all;
set(0,'DefaultAxesFontsize',17);
%
a=1000.0; b=1500.0; % Dimensions of the clay layer
h0=150.0; % Hydrostatic water head in the reservoir
dx=10; dy=10; % Spatial increment
N=100; % number of terms in Fourier series expansion
%
k=1; m=1; % Counters
%
for x=0:dx:a
    for y=0:dy:b
        s=0.0;
        for n=1:N
            temp=2.*h0.*(1-(-1).^n)./(pi.*n.*sinh(n.*pi.*a./b));
            s=s+temp.*sinh(n.*pi.*x./b).*sin(n.*pi.*y./b);
        end
        h(k,m)=s;
```

```
        k=k+1;
    end
    k=1;   m=m+1;
end
%
figure(3);
subplot('position',[0.1 0.1 0.45 0.95]);
[c,g]=contour(h,13); hold on;
clabel (c,g,'fontsize',17); caxis([0 150]);
colorbar ('location','northoutside');
set(gca,'linewidth',2.5);
set(gca,'XTick',[1:10:101]); set(gca,'YTick',[1:10:151]);
set(gca,'Xticklabel',[0:100:1000]); set(gca,'Yticklabel',[0:100:1500]);
set(gca,'linewidth',2.0);
% End of the code %
```

3.2.4 Numerical example: Seepage through a dam foundation

An earth-fill dam is holding 10 m-high water in its reservoir while the downstream water level is 2 m (Figure 3.8). The foundation of the dam is 50 m deep and 135 m long. Given the hydrostatic water head at the upstream (10 m) and downstream (2 m) sides of the dam, the piezometric water head in the foundation will be changing from 10 m on the upstream side to 2 m on the downstream side. In this example, the phreatic line is assumed to be a straight line, although in the real world it is a curve. We apply the Laplace differential equation, Eqs (3.23) or (3.24), to obtain the distribution of the piezometric head in the foundation. The sketch in Figure 3.9 shows the computational domain with the boundary conditions. By considering Eq (3.16) and accounting for the boundary conditions depicted in Figure 3.9, it can be readily demonstrated that neither 0 nor λ^2 can serve as solutions to Eq (3.16) for this specific problem. Therefore, the solution takes the form presented in Eq (3.24).

We now impose the boundary conditions on Eq (3.24). At the left boundary (i.e., $x = 0$) an impermeable layer is present, signifying $\frac{\partial h}{\partial x} = 0$. Upon differentiating Eq (3.24) relative

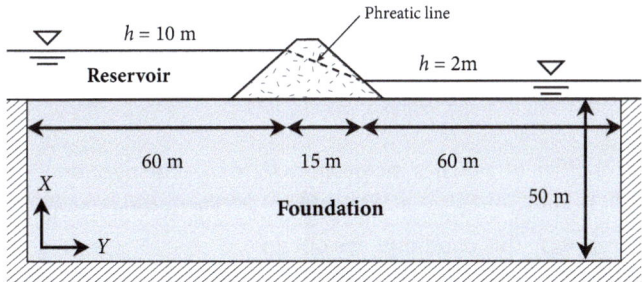

Figure 3.8 An embankment dam and its foundation. The contours of the isopotential lines through the foundation are desired. The phreatic line is assumed to be a straight line in this example; it is a curve in the real world.

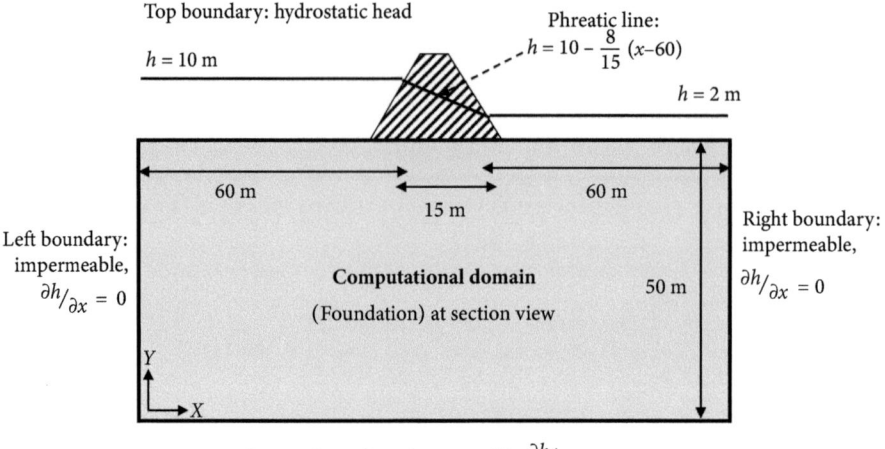

Figure 3.9 Sketch showing the computational domain and the boundary layers for the problem shown in Figure 3.8 (seepage through the foundation of an earth-fill dam).

to x and imposing this condition we get

$$\frac{\partial h\,(x,y)}{\partial x} = \sum_{n=0}^{\infty} \{A_n \lambda_n \cos\left(\lambda_n x\right) - B_n \lambda_n \sin\left(\lambda_n x\right)\} \times \left\{C_n \sinh\left(\lambda_n y\right) + D_n \cosh\left(\lambda_n y\right)\right\}$$

(3.36)

After applying the boundary condition, the equation becomes

$$0 = \sum_{n=0}^{\infty} \{A_n \lambda_n\} \times \left\{C_n \sinh\left(\lambda_n y\right) + D_n \cosh\left(\lambda_n y\right)\right\}$$ (3.37)

Equation (3.37) requires that $A_n = 0$. Similarly, applying the bottom boundary condition (i.e., $y = 0$, $\frac{\partial h}{\partial y} = 0$), leads to $C_n = 0$. Consequently, the solution for $h\,(x,y)$ takes the following form:

$$h\,(x,y) = \sum_{n=0}^{\infty} K_n \cos\left(\lambda_n x\right) \times \cosh\left(\lambda_n y\right)$$ (3.38)

Here, K_n is a new coefficient which is the product $B_n \times D_n$. The right boundary condition is an impermeable boundary (i.e., $x = 135$; $\frac{\partial h}{\partial x} = 0$). By differentiating Eq (3.38) with respect to x and imposing this condition, we obtain

$$\frac{\partial h\,(x,y)}{\partial x} = \sum_{n=0}^{\infty} -K_n \lambda_n \sin\left(\lambda_n x\right) \times \cosh\left(\lambda_n y\right)$$ (3.39)

After applying the boundary condition, the equation becomes

$$0 = \sum_{n=0}^{\infty} -K_n \lambda_n \sin(135\lambda_n) \times \cosh(\lambda_n y) \tag{3.40}$$

The preceding equation gives $\lambda_n = \frac{n\pi}{135}$. Consequently, the solution takes the following form:

$$h(x, y) = \sum_{n=0}^{\infty} K_n \cos\left(\frac{n\pi}{135}x\right) \times \cosh\left(\frac{n\pi}{135}y\right) \tag{3.41}$$

At this stage, only one boundary condition (i.e., the top condition) remains, the application of which will yield the coefficient K_n. Rewriting Eq (3.41) in the following form is advantageous to leverage the Fourier expansion formulas:

$$h(x, y) = K_0 + \sum_{n=1}^{\infty} K_n \cos\left(\frac{n\pi}{135}x\right) \times \cosh\left(\frac{n\pi}{135}y\right) \tag{3.42}$$

The top boundary condition at $y = 50$, which delineates the variations of the hydrostatic head across the dam width, can be summarized as

$$h(x, 50) = \begin{cases} 10 & \text{if } 0 < x < 60 \\ 10 - \frac{8}{15}(x - 60) & \text{if } 60 < x < 75 \\ 2 & \text{if } 75 < x < 135 \end{cases} \tag{3.43}$$

Applying the top boundary condition implies

$$K_0 + \sum_{n=1}^{\infty} K_n \cosh\left(\frac{n\pi}{135}50\right) \times \cos\left(\frac{n\pi}{135}x\right) = \begin{cases} 10 & \text{if } 0 < x < 60 \\ 10 - \frac{8}{15}(x - 60) & \text{if } 60 < x < 75 \\ 2 & \text{if } 75 < x < 135 \end{cases} \tag{3.44}$$

In the above equation, the term $K_n \cosh\left(\frac{n\pi}{135}50\right)$ is a constant. Hence, we assume it as $E_n = K_n \cosh\left(\frac{n\pi}{135}50\right)$ and consider $E_0 = K_0$. Therefore, Eq (3.44) can be rewritten as

$$E_0 + \sum_{n=1}^{\infty} E_n \times \cos\left(\frac{n\pi}{135}x\right) = \begin{cases} 10 & \text{if } 0 < x < 60 \\ 10 - \frac{8}{15}(x - 60) & \text{if } 60 < x < 75 \\ 2 & \text{if } 75 < x < 135 \end{cases} \tag{3.45}$$

The right side of Eq (3.45) can be expanded using a Fourier series of the form $a_0 + \sum_{n=1}^{\infty} a_n \times \cos\left(\frac{n\pi}{135}x\right)$. In fact, $E_0 = a_0$ and $E_n = a_n$. Applying the Fourier series formula, it can be shown that $a_0 = 6$ and a_n is as follows:

$$a_n = \frac{144}{n^2\pi^2}\left[\cos\left(\frac{4}{9}n\pi\right) - \cos\left(\frac{5}{9}n\pi\right)\right] \tag{3.46}$$

Therefore, the two coefficients of K_0 and K_n in Eq (3.44) are determined as

$$K_0 = E_0 = a_0 = 6 \tag{3.47}$$

and

$$K_n = \frac{E_n}{\cosh\left(\frac{n\pi}{135}50\right)} = \frac{a_n}{\cosh\left(\frac{n\pi}{135}50\right)} = \frac{144}{n^2\pi^2\cosh\left(\frac{n\pi}{135}50\right)}\left[\cos\left(\frac{4}{9}n\pi\right) - \cos\left(\frac{5}{9}n\pi\right)\right] \tag{3.48}$$

Finally, the solution for $h(x,y)$ becomes

$$h(x,y) = 6 + \sum_{n=1}^{\infty}\frac{144}{n^2\pi^2\cosh\left(\frac{n\pi}{135}50\right)}\left[\cos\left(\frac{4}{9}n\pi\right) - \cos\left(\frac{5}{9}n\pi\right)\right]$$
$$\cos\left(\frac{n\pi}{135}x\right) \times \cosh\left(\frac{n\pi}{135}y\right) \tag{3.49}$$

The final solution, Eq (3.49), is plotted in the computational domain as shown in Figure 3.10 using Code 3.2. It can be seen that the water piezometric head in the foundation varies from a maximum value of 10 m at the upstream side of the dam to a minimum value of 2 m at the downstream side. This MATLAB code is available at: http://www.oup.com/AdvancedMathematicalModelling.

Code 3.2 *The MATLAB script used for plotting the solution of the problem shown in Figures 3.8 and 3.9, which is the flow through the foundation of an earth-fill dam. This code, which generates Figure 3.10, can be downloaded from the following website: http://www.oup.com/AdvancedMathematicalModelling.*

```
% This code solve seepage through the
% foundation of an earth dam
clear; close all;clc;
set(0,'DefaultAxesFontsize',17);
%
a=135; % length of foundation
b=50; % height of foundation
N=100; % number of terms in Fourier Series
dx=1; % in meter; space increment in X-direction
dy=1; % in meter; space increment in Y-direction
k=1; m=1; % counters
%
    for x=0:dx:a
    for y=0:dy:b
        %
```

```
        s=0.0;
        for n=1:N
        temp=144.*(cos(4.*n.*pi./9)-cos(5.*n.*pi./9))./...
            (cosh(50.*n.*pi./135).*(n.*pi).^2);
        s=s+temp.*cos(n.*pi.*x./135).*cosh(n.*pi.*y./135);
        end
        %
        h(k,m)=6+s;
        k=k+1;
    end
    k=1;   m=m+1;
    end
    %
figure(1)
subplot('position',[0.1 0.1 0.7 0.45]);
[c,g]=contour(h,60);
colorbar; clabel(c,g);
%
set(gca,'XTick',[1:10:136]); set(gca,'YTick',[1:10:51]);
set(gca,'Xticklabel',[0:10:130]);
set(gca,'Yticklabel',[0:10:50]);
set(gca,'linewidth',2.5);
% End of the code %
```

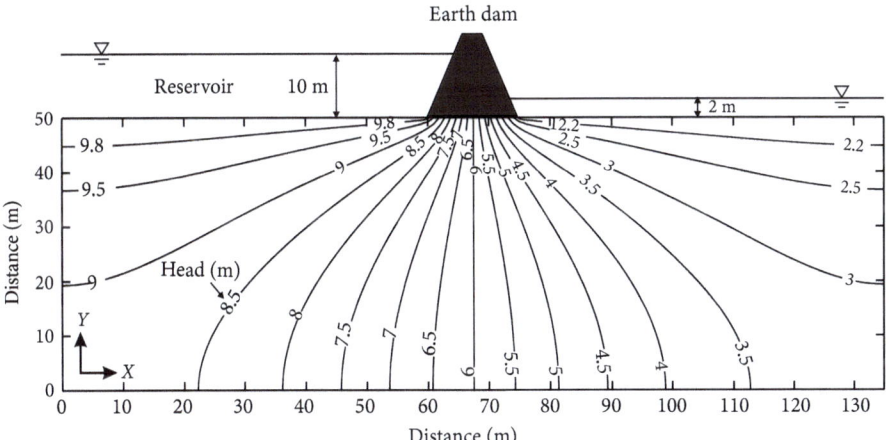

Figure 3.10 Plot of the solution of the seepage flow through the foundation of an earth-fill dam as shown in Figures 3.8 and 3.9. This figure is generated using the MATLAB script in Code 3.2.

3.3 Soil consolidation and settlement

Consolidation of soil is a time-related phenomenon during which water is squeezed out of the soil due to external loads. As a result, the soil shrinks and undergoes settlement. All of the pressure is carried by the water at the beginning of the load application because the water column transfers the load instantaneously from top to bottom of the water column, while soil particles require compression in order to carry the load. Therefore,

Figure 3.11 Soil consolidation due to embankment construction. In this photo, construction machines are working towards the construction of the Karkheh dam (Iran) in the 1990s. This photo is modified from Mahab Consulting Engineers (1998).

at the beginning, all of the load is carried by water and thus the pore water pressure is the maximum. This sudden increase in pore pressure is known as the 'excess pore pressure'. As a result of the sudden increase in pore pressure, water is drained out of the soil causing soil compression which helps the soil to carry the external load. In this time-dependent process, the soil experiences settlement as water drains from the medium, resulting in a gradual decrease in excess pore pressure until it eventually reaches zero, a phenomenon that may occur over an extended period of time, for instance spanning a few years.

A major external loading to the foundation and consequent consolidation settlement comes from construction of embankments such as earth-fill dams (Figure 3.11) or road embankments on soft soils. Such permanent long-term loading on the foundation results in a sudden increase of pore pressure and settlement over a long period of time. For construction purposes, it is important to closely monitor the status of excess pore pressure in the foundations of structures.

3.3.1 Governing equation of one-dimensional soil consolidation

Consider the multilayered foundation in Figure 3.12 with a two-dimensional soil element identified with dimensions of dx and dz. The clay layer in the middle is bounded by two sand layers from top and bottom. The external load ΔP is applied to the foundation from the ground level. The purpose here is to study the consolidation process in the clay layer whose thickness is $2d$ (Figure 3.12). As a result of the load ΔP, pore pressure increases immediately in both the clay and sand layers, but it dissipates rapidly in the sand layer because of its high permeability. For the clay layer, the pore pressure dissipates slowly since clay has a low permeability.

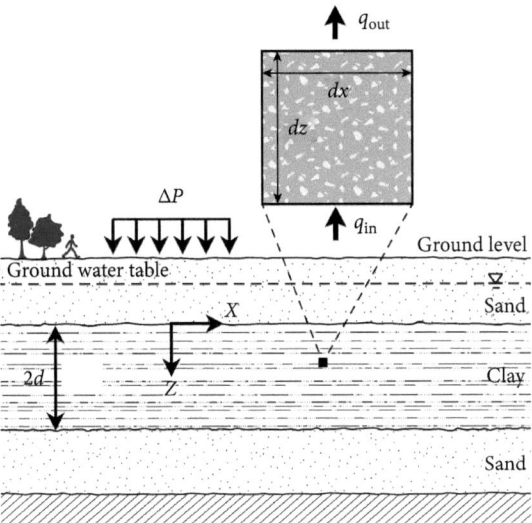

Figure 3.12 Sketch showing a two-dimensional soil element in a multilayered foundation under external loading ΔP.

For the soil cell depicted in Figure 3.12, the vertical soil volume change of the cell $\left(\frac{\partial V}{\partial t}\right)$ can be computed as the difference between the volume of water exiting the cell (q_{out}) and the volume entering the cell in the vertical direction (q_{in}). It is noted that we consider here only the settlement in the vertical direction, z. Considering water velocity in the z direction as v_z, we have $q_{in} = v_z \cdot dx$; and $q_{out} = \left(v_z + \frac{\partial v_z}{\partial z} dz\right) \cdot dx$. Therefore, the equation for $\frac{\partial V}{\partial t}$ is formulated as follows:

$$\frac{\partial V}{\partial t} = \left(v_z + \frac{\partial v_z}{\partial z} dz\right) dx - v_z \cdot dx \tag{3.50}$$

This equation is simplified to

$$\frac{\partial V}{\partial t} = \frac{\partial v_z}{\partial z} dz\, dx \tag{3.51}$$

The velocity term (v_z) in the above equation can be replaced by the water head (h) using Darcy's law $(v_z = k \cdot i_z, i_z = \frac{\partial h}{\partial z}$; see Eq (3.1)). The water head (h) at each point of the soil is given by $h = \frac{u}{\gamma_w}$, where u is the pore pressure and γ_w the specific weight of water, which is approximately 10 kN/m³. Therefore, $v_z = \frac{k}{\gamma_w} \frac{\partial u}{\partial z}$. Consequently, Eq (3.51) can be rewritten in the pore pressure term (u) as follows:

$$\frac{\partial V}{\partial t} = \frac{\partial \left(\frac{k}{\gamma_w} \frac{\partial u}{\partial z}\right)}{dz} dz\, dx \tag{3.52}$$

This can be rearranged to

$$\frac{\partial V}{\partial t} = \frac{k}{\gamma_w} \frac{\partial^2 u}{\partial z^2} dz\, dx \tag{3.53}$$

It is clear that the volume change of the cell $\left(\frac{\partial V}{\partial t}\right)$ corresponds to the change of the volume of voids (V_v) within the cell during the consolidation process. Consequently, we have

$$\frac{\partial V}{\partial t} = \frac{\partial V_v}{\partial t} \tag{3.54}$$

From soil mechanics (Das 2019), we understand that the total volume of the soil medium (V) is the sum of the volume of voids (V_v) and the volume of soil solids (V_s):

$$V = V_v + V_s \tag{3.55}$$

The void ratio (e) is defined as the ratio of the volume of voids (V_v) to the volume of soil solids $(V_s;$ Das 2019):

$$e = \frac{V_v}{V_s} \tag{3.56}$$

Here we rewrite Eq (3.54) using Eqs (3.55) and (3.56), resulting in the following expression:

$$\frac{\partial (V_v + V_s)}{\partial t} = \frac{\partial V_v}{\partial t} \tag{3.57}$$

This can be rearranged to

$$\frac{\partial (eV_s + V_s)}{\partial t} = \frac{\partial V_v}{\partial t} \tag{3.58}$$

and to

$$e \frac{\partial V_s}{\partial t} + V_s \frac{\partial e}{\partial t} + \frac{\partial V_s}{\partial t} = \frac{\partial V_v}{\partial t} \tag{3.59}$$

Assuming that the solid parts of the medium are not compressed, we have $\frac{\partial V_s}{\partial t} = 0$. Therefore, considering Eqs (3.59) and (3.54), we derive the following equation:

$$V_s \frac{\partial e}{\partial t} = \frac{\partial V_v}{\partial t} = \frac{\partial V}{\partial t} \tag{3.60}$$

It is evident from Eqs (3.55) and (3.56) that $V_s = \frac{V}{1+e_0}$, where e_0 represents the initial void ratio of the soil. The total volume (V) is $V = dx \cdot dz$ as illustrated in Figure 3.12, assuming a unit width for the soil cell. With these considerations, we can combine Eqs (3.53) and (3.60) to form the following expression:

$$\frac{dx\,dz}{1+e_0}\frac{\partial e}{\partial t} = \frac{k}{\gamma_w}\frac{\partial^2 u}{\partial z^2}dz\,dx \tag{3.61}$$

which can be rearranged to

$$\frac{k}{\gamma_w}\frac{\partial^2 u}{\partial z^2} = \frac{1}{1+e_0}\frac{\partial e}{\partial t} \tag{3.62}$$

Variations of the void ratio (∂e) depend on the variations of the loading to the medium (ΔP; Smith 2014): $\partial e = -a_v \cdot \Delta P$, where a_v is a constant called the 'coefficient of compressibility'. The total stress at each point in the soil (σ) is the sum of the effective stress (σ') and the pore water pressure (u), expressed as: $\sigma = \sigma' + u$. Throughout the consolidation process, the total stress remains relatively constant over time ($\frac{\partial \sigma}{\partial t} = 0$), while both the effective stress and pore pressure exhibit changes. At the beginning ($t = 0$), the pore pressure reaches its maximum, coinciding with zero effective stress ($\sigma' = 0$, $\sigma = u$). As time progresses, the effective stress gradually increases while the pore pressure decreases. When differentiating the total stress with respect to time, we get

$$0 = \frac{\partial \sigma'}{\partial t} + \frac{\partial u}{\partial t} \tag{3.63}$$

Assuming that the effective stress (σ') is proportional to the external loading (ΔP), we can say $\Delta P \sim \partial \sigma'$. Therefore, Eq (3.63) can be written as $\frac{\partial u}{\partial t} = -\frac{\Delta P}{\partial t}$. This gives the following equation for ∂e (remembering that $\partial e = -a_v \cdot \Delta P$ from above):

$$\frac{\partial e}{\partial t} = a_v \frac{\partial u}{\partial t} \tag{3.64}$$

Consequently, Eq (3.62) takes the following form:

$$\frac{k}{\gamma_w}\frac{\partial^2 u}{\partial z^2} = \frac{a_v}{1+e_0}\frac{\partial u}{\partial t} \tag{3.65}$$

The term $\frac{a_v}{1+e_0}$ is denoted m_v and is the 'coefficient of volume compressibility' (Das 2019). Therefore,

$$\frac{k}{\gamma_w}\frac{\partial^2 u}{\partial z^2} = m_v \frac{\partial u}{\partial t} \tag{3.66}$$

Here, we introduce a new term, $C_v = \frac{k}{m_v\,\gamma_w}$, called the 'coefficient of consolidation' (Das 2019). Consequently, the previous equation becomes

$$\frac{\partial u}{\partial t} = C_v \frac{\partial^2 u}{\partial z^2} \tag{3.67}$$

Equation (3.67) is the one-dimensional consolidation equation. By solving this equation, the amount of pore pressure at any point in the clay layer (z) and at any time (t) will be determined.

3.3.2 Solution of one-dimensional soil consolidation

To solve Eq (3.67), we assume the solution to be in the form of the production of two functions $Z(z)$ and $T(t)$ where $Z(z)$ is only a function of location (z) and $T(t)$ is a sole function of time (t), as follows:

$$u(z,t) = Z(z) \times T(t) \tag{3.68}$$

The equation for $u(z,t)$ is differentiated with respect to z and t, which gives $\frac{\partial u}{\partial z} = Z'T$, $\frac{\partial^2 u}{\partial z^2} = Z''T$, $\frac{\partial u}{\partial t} = ZT'$. Consequently, Eq (3.67) becomes

$$ZT' = C_v Z''T \tag{3.69}$$

Dividing both sides of the above equation by ZT,

$$\frac{T'}{T} = C_v \frac{Z''}{Z} \tag{3.70}$$

According to Eq (3.70), both sides of the equation must represent a constant, which could be either a negative constant (e.g., $-\lambda^2$), or a positive one (e.g., λ^2). The constant may be zero too.

Here, let's assume that $-\lambda^2$ is a valid solution for Eq (3.70). Therefore,

$$\frac{T'}{T} - C_v \frac{Z''}{Z} = -\lambda^2 \tag{3.71}$$

This equation yields two differential equations of first and second orders as follows:

$$T + \lambda_n{}^2 T = 0 \tag{3.72}$$

$$Z'' + \frac{\lambda_n{}^2}{C_v} Z = 0 \tag{3.73}$$

It is assumed that $K_n{}^2 = \frac{\lambda_n^2}{C_v}$. The solutions for Eqs (3.72) and (3.73) are $T(t) = C_1 e^{-\lambda_n^2 t}$ and $Z(z) = C_2 \sin(K_n z) + C_3 \cos(K_n z)$. Therefore, the solution for $u(z,t)$ takes the following form:

$$u(z,t) = \sum_{n=0}^{\infty} e^{-\lambda_n^2 t} [A \sin(K_n z) + B \cos(K_n z)] \tag{3.74}$$

Note that a new set of coefficients (i.e., A and B) are introduced by multiplying C_1 by C_2 and C_3. The coefficients A and B can be obtained by imposing the boundary conditions.

Depending on the type of the boundary conditions of a specific consolidation problem, the solution to Eq (3.70) can be either λ^2 or $-\lambda^2$. If the positive value (i.e., λ^2) is valid for a specific consolidation problem, it can be shown in the same way that the solution to the Eq (3.70) will be of the following form:

$$u(z, t) = \sum_{n=0}^{\infty} e^{\lambda_n^2 t} [A \sinh(K_n z) + B \cosh(K_n z)] \tag{3.75}$$

3.3.3 Numerical example: One-dimensional consolidation with top and bottom sand layers

For the case that both top and bottom of the clay layer are bounded by sand layers (Figure 3.12), the drainage of the excess pore pressure will occur on both sides. It is assumed that the initial excess pore pressure is u_0, the thickness of the clay layer is $2d$, and the z axis looks downward (Figure 3.12). The boundary conditions are: $u(0, t) = 0$ at $z = 0$ and $u(2d, t) = 0$ at $z = 2d$, which indicates that the pore pressure is always zero at the boundaries with sand layers. As this problem is time dependent, we have an initial condition in addition to boundary conditions: at $t = 0$, $u(z, 0) = u_0$, where u_0 is the initial excess pore pressure. A summary of the boundary and initial conditions is sketched in Figure 3.13.

It can be readily shown that the constant $-\lambda^2$ applies for this problem because of the boundary conditions. Therefore, the solution is of the form of Eq (3.74). Applying the first boundary condition, i.e., $u(0, t) = 0$ at $z = 0$, to Eq (3.74) results in $B = 0$. The second boundary condition, i.e., $u(2d, t) = 0$ at $z = 2d$, results in the following equation:

$$0 = \sum_{n=0}^{\infty} e^{-\lambda_n^2 t} A \sin(K_n 2d) \tag{3.76}$$

To hold this equation true, either the coefficient A or the term $\sin(K_n 2d)$ should be zero. It is clear that A cannot be zero because it will lead to a zero value for Eq (3.74). Therefore, $\sin(K_n 2d) = 0$, implying $K_n 2d = n\pi$; this can be rearranged to $\frac{\lambda_n}{\sqrt{C_v}} 2d = n\pi$, and

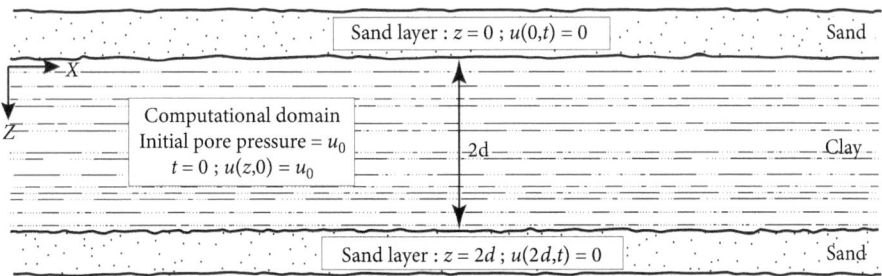

Figure 3.13 The computational domain and boundary conditions for the one-dimensional consolidation shown in Figure 3.12.

consequently we have $\lambda_n = \frac{\sqrt{C_v}}{2d} n\pi$. Therefore, the solution becomes

$$u(z, t) = \sum_{n=0}^{\infty} A_n e^{-\lambda_n^2 t} \sin\left(\frac{n\pi}{2d} z\right) \qquad (3.77)$$

Now we apply the initial condition, i.e., $t = 0$, $u(z, 0) = u_0$:

$$u_0 = \sum_{n=0}^{\infty} A_n \sin\left(\frac{n\pi}{2d} z\right) \qquad (3.78)$$

The coefficient A_n can be determined using the Fourier series formula, which yields $A_n = \frac{2u_0}{n\pi} \left[1 - (-1)^n\right]$. Finally, the solution for $u(z, t)$ takes the following form:

$$u(z, t) = \sum_{n=0}^{\infty} \frac{2u_0}{n\pi} \left[1 - (-1)^n\right] \cdot e^{-\left(\frac{\sqrt{C_v}}{2d} n\pi\right)^2 t} \sin\left(\frac{n\pi}{2d} z\right) \qquad (3.79)$$

This solution is plotted in Figure 3.14 using Code 3.3. For plotting purposes, the initial excess pore pressure (u_0), the thickness of the clay layer ($2d$), and the coefficient of consolidation (C_v) are assumed to be $u_0 = 10\,kN/m^2$, $2d = 10\,m$, and $C_v = 8 \times 10^{-3} m^2/s$.

The plot of the outcome in Figure 3.14 shows that the pore pressure is $10\,kN/m^2$ at the beginning all the way from the top to the bottom of the clay layer, but it becomes zero at

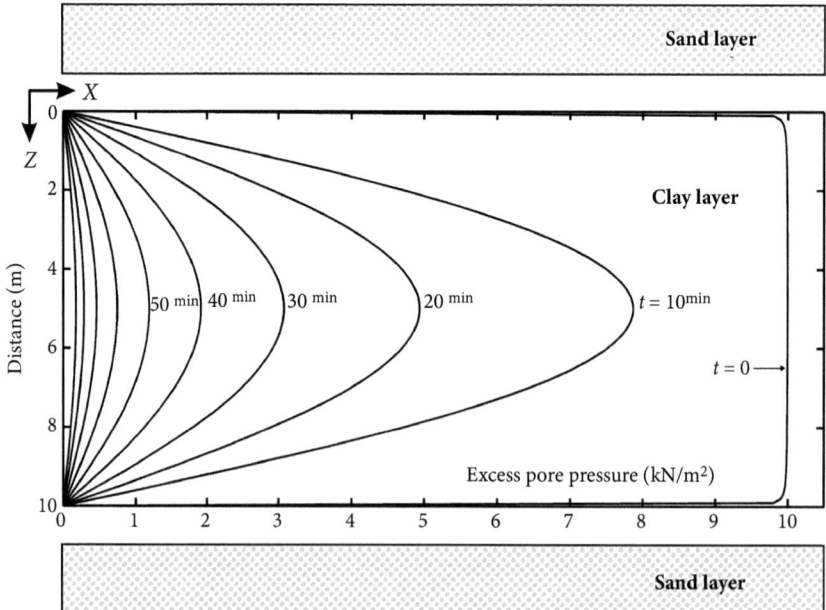

Figure 3.14 Plot of the solution of the double drainage one-dimensional consolidation problem as shown in Figures 3.12 and 3.13. The contours of pore pressure are in 10 min intervals. This figure is created using Code 3.3.

the top and bottom boundaries immediately because of the presence of sand layers at the boundaries. This problem is known as two-way or double drainage consolidation. According to Figure 3.14, most of the excess pore pressure is dissipated after approximately 90 min.

Code 3.3 *The MATLAB script used for plotting the solution of the one-dimensional consolidation problem shown in Figures 3.12 and 3.13 where both top and bottom boundary layers are sand. This code makes an animation of the pore pressure changes by time.*

```matlab
% Double-drainage consolidation
clear; clc; close all;
set(0,'DefaultAxesFontsize',20);
%
time=3250; % in seconds; total simulation time
d=5; % half length of the clay layer
u0=10; % initial pore pressure (KN/m^2)
N=1000; % number of terms in Fourier Series
dx=0.1; % in meter- space increment
dt=5; % in seconds-- time increment
cv=8.*10.^(-3); % in m^2/s coefficient of consolidation
k=1; m=1; % Counters
%
for t=0:dt:time;
    for x=0:dx:2*d;
        s=0.0;
        for n=1:N;
        z=2.*u0.*(1-(-1).^n)./(n.*pi);
        s=s+z.*sin(n.*pi.*x./(2.*d)).*...
            exp(-cv.*(n.*pi).^2.*t./(2.*d).^2);
        end
        h(k,m)=s;   k=k+1;
    end
    % Animations
    plot(h(:,m),0:dx:2*d,'LineWidth',3);
    ylim([0 2.*d]);
    xlim([0 u0.*1.25]);
    text(u0*0.9,d,'time =','FontSize',18)
    ylabel('Distance (m)');
    xlabel('Pore pressure (kN/m^2)');
    legend('1-D consolidation with double drainage');
    set(gca,'linewidth',1.5);
    %
    str = sprintf('%3.1f', t);
    text(u0.*1.0,d,str,'FontSize',18);
    text(u0.*1.1,d,'s','FontSize',18);
    %
    refreshdata; drawnow; clf;
    %
    m=m+1; k=1;
end
% End of the code %
```

3.3.4 Numerical example: One-dimensional consolidation with sand layer at one side

In this example, the clay layer is resting on an impermeable bed and is bounded by a sand layer at the top (Figure 3.15). The thickness of the clay layer is 5 m and the uniform external pressure at ground level is 100 kN/m². The coefficient of consolidation (C_v) is assumed to be $C_v = 3 \times 10^{-7} m^2/s$. The impermeable bed prevents the drainage of water content from the bottom of the clay layer, making drainage possible only from the top. This problem is known as one-way drainage consolidation. The boundary and initial conditions of the problem are summarized in Figure 3.16.

Considering Eq (3.70), it can be shown that the two cases of λ^2 and 0 lead to zero answers for Eq (3.70), and thus the only acceptable solution is $-\lambda^2$. Therefore, the solution for the one-dimensional consolidation equation takes the form of Eq (3.74). By applying

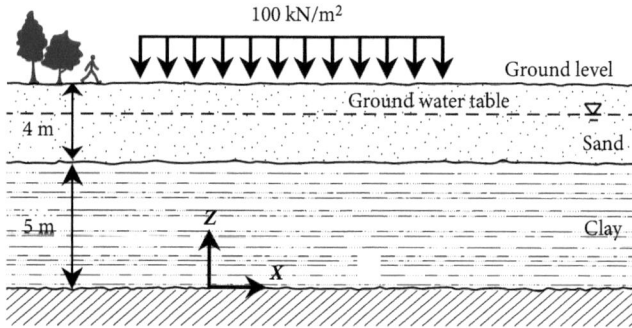

Figure 3.15 A multilayered foundation with a 5 m clay layer bounded by an impermeable bed at the bottom and a sand layer at the top under external loading of $\Delta P = 100$ kN/m².

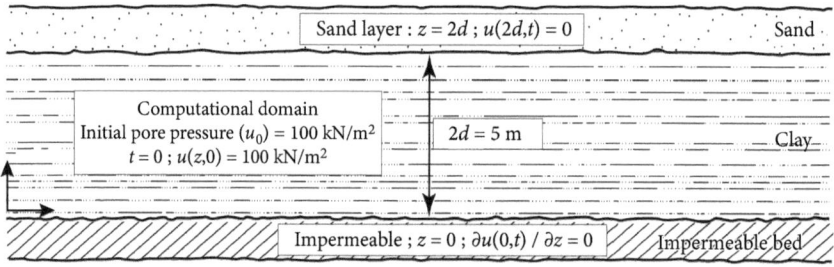

Figure 3.16 The computational domain and boundary conditions for the one-dimensional consolidation problem shown in Figure 3.15.

the bottom boundary condition $\left(z = 0, \ \frac{\partial u(0, t)}{\partial z} = 0\right)$ to Eq (3.74), we get

$$\frac{\partial u(z, t)}{\partial z} = \sum_{n=0}^{\infty} e^{-\lambda_n^2 t} [AK_n \cos(K_n z) - BK_n \sin(K_n z)] \tag{3.80}$$

By setting z and $\frac{\partial u(0, t)}{\partial z}$ as zero for the bottom boundary condition, Eq (3.80) requires that $A = 0$. Consequently, the solution takes the following form:

$$u(z, t) = \sum_{n=0}^{\infty} e^{-\lambda_n^2 t} B \cos(K_n z) \tag{3.81}$$

The top boundary condition is now applied to Eq (3.81):

$$0 = \sum_{n=0}^{\infty} B e^{-\lambda_n^2 t} \cos(2d \cdot K_n) \tag{3.82}$$

Since the coefficient B cannot be zero, the term $\cos(2d \cdot K_n)$ should be zero which implies that $2d \cdot K_n = \left(n - \frac{1}{2}\right) \pi$ and $K_n = \frac{\left(n - \frac{1}{2}\right)\pi}{2d}$. From earlier, we know that $K_n^2 = \frac{\lambda_n^2}{C_v}$, which gives $\lambda_n^2 = C_v \left[\frac{\left(n - \frac{1}{2}\right)\pi}{2d}\right]^2$. This turns the solution into the following form:

$$u(z, t) = \sum_{n=1}^{\infty} B e^{-\lambda_n^2 t} \cos\left[\frac{\left(n - \frac{1}{2}\right)\pi}{2d} z\right] \tag{3.83}$$

Only the initial condition $(t = 0, \ u(z, 0) = u_0)$ remains to be applied to Eq (3.83) at this stage, which results in

$$u_0 = \sum_{n=1}^{\infty} B \cos\left[\frac{\left(n - \frac{1}{2}\right)\pi}{2d} z\right] \tag{3.84}$$

By employing the Fourier series equations, the coefficient B can be determined as $B = \frac{2u_0}{\left(n - \frac{1}{2}\right)\pi} \sin\left[\left(n - \frac{1}{2}\right)\pi\right]$. Subsequently, incorporating this coefficient into the solution for $u(z, t)$ yields

$$u(z, t) = \sum_{n=1}^{\infty} \frac{2u_0}{\left(n - \frac{1}{2}\right)\pi} \sin\left[\left(n - \frac{1}{2}\right)\pi\right] e^{-C_v\left[\frac{\left(n - \frac{1}{2}\right)\pi}{2d}\right]^2 t} \cos\left[\frac{\left(n - \frac{1}{2}\right)\pi}{2d} z\right] \tag{3.85}$$

Equation (3.85) is the final solution, which can be plotted using Code 3.4 as shown in Figure 3.17. It can be seen that at the top boundary layer, which is adjacent to a sand layer, the excess pore pressure immediately becomes zero after the start of consolidation due to rapid drainage achieved by the sand formation. At the bottom boundary, i.e., neighbouring an impermeable layer, the excess pore pressure dissipates very slowly and takes more than 1600 days to disappear.

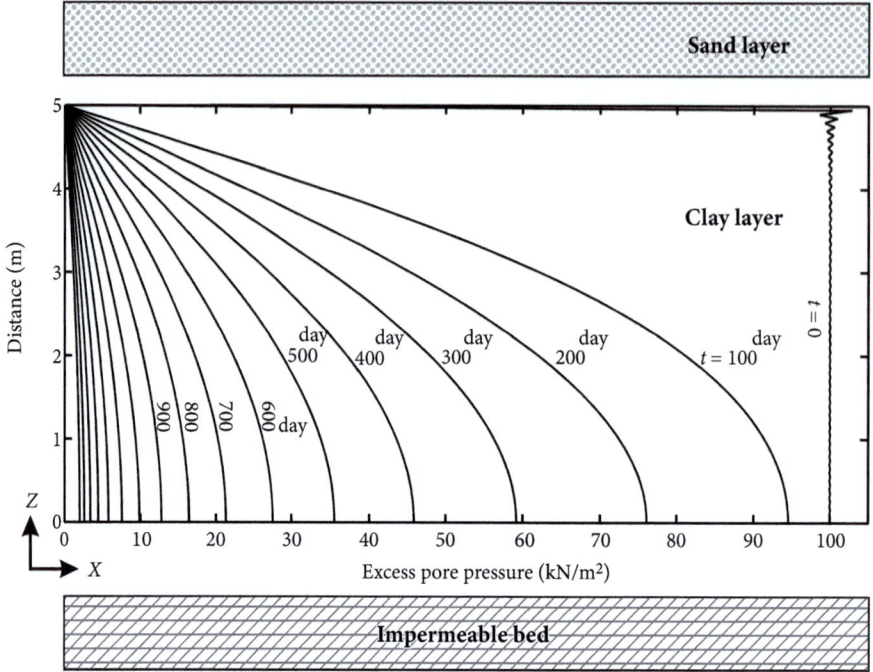

Figure 3.17 Plot of the solution of the one-way drainage one-dimensional consolidation problem as shown in Figures 3.15 and 3.16. The contours of pore pressure are in 100 day intervals. This figure is generated using Code 3.4.

Code 3.4 *The MATLAB script used for plotting the solution of the one-way drainage one-dimensional consolidation problem shown in Figures 3.15 and 3.16. This code makes an animation of the pore pressure changes by time. The result is shown in Figure 3.17.*

```
% 1D soil with one-way drainage consolidation
clear;
close all;
set(0,'DefaultAxesFontsize',20);
%
time=3600.*24.*1700; % seconds; simulation time
d=2.5; % half length of the clay layer
u0=100; % kN/m2; initial pore pressure
N=700; % number of terms in Fourier Series
dx=0.05; % in meter- space increment
dt=3600.*24; % in seconds; time increment
cv=3.*10.^(-7); % m^2/s consolidation coefficient
k=1; m=1; % Counters
%
for t=0:dt:time;
    for x=0:dx:2*d;
        %
        s=0.0;
        for n=1:N;
        temp=2.*u0.*sin((n-0.5).*pi)./((n-0.5).*pi);
        s=s+temp.*cos((n-0.5).*pi.*x./(2.*d)).*...
```

```
              exp(-cv.*((n-0.5).*pi).^2.*t./(2.*d).^2);
          end
        h(k,m)=0+s;   k=k+1;
      end
      % Animation
      plot(h(:,m),0:dx:2*d,'LineWidth',3);
      ylim([0 2.*d]);
      xlim([0 u0.*1.25]);
      text(u0*0.8,d,'time=','FontSize',18);
      set(gca,'linewidth',2.5);
      %
      str = sprintf('%3.1f', t);
      text(u0.*0.9,d,str,'FontSize',18);
      text(u0.*1.05,d,'s','FontSize',18);
      %
      refreshdata;   drawnow;
      clf;
    %
    m=m+1;   k=1;
end
% End of the code %
```

3.4 Groundwater flow in soil

Groundwater is part of the water cycle, resulting from the infiltration of precipitation into the subsurface. Usually, the term groundwater is used for water beneath the water table and is saturated (Freeze and Cherry 1979). Groundwater is stored in fractures and spaces in soil and rock beneath the land surface. The geologic layers that store water and supply them to water wells (e.g., Figure 3.18) are called aquifers.

The flow velocity of groundwater depends on the hydraulic gradient of the water and the hydraulic properties of the aquifer such as the transmissivity (T) and the storage coefficient (S). The storage coefficient is also known as storativity. Transmissivity is the product of the aquifer permeability (K) and its thickness (b), and reveals how easily water can move through the aquifer; it is usually reported in m^2/s. The storage coefficient is a dimensionless parameter which is defined as the volume of water that an aquifer releases (or absorbs) from storage per unit surface area of aquifer per unit decline (or rise) in the hydraulic head normal to that surface. Specific storage (S_s) is defined as $S_s = \frac{S}{b}$, where b is the thickness of the aquifer. Specific storage (S_s) is the volume of water that a unit volume of an aquifer releases (or absorbs) under a unit decrease (or rise) of hydraulic head (Freeze and Cherry 1979). The recharge of rainfall into groundwater $(R,$ Figure 3.19) will increase the groundwater level, while excessive pumping leads to a decrease of the groundwater level.

Figure 3.18 Photo showing the groundwater discharge downstream of a dam project through a water well. Workers place filter material around the well from height due to the presence of artesian flow conditions in the area. More details are given in Heidarzadeh et al. (2015).

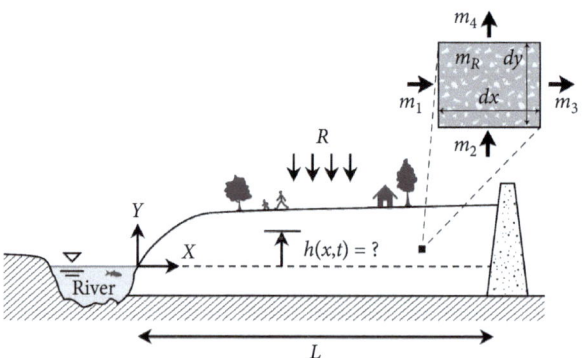

Figure 3.19 Cross section of an aquifer bounded by a river on the left and by an impervious dam on the right. The layer is under precipitation (or recharge, shown by R in this sketch) with a constant rate of R which results in an increase in the water table in the aquifer. In this figure, m is the mass inflow/outflow. The parameter m_R is the mass inflow from the recharge rate of R.

3.4.1 Governing equation for two-dimensional groundwater flow

Consider the soil cell shown in Figure 3.19. In a confined aquifer with certain transmissivity (T) and storage coefficient (S), the aquifer behaves as a compressible medium. The flow is assumed to be unsteady. Therefore, the amount of mass inflow and outflow to and from the cell is not equal. The difference between the mass inflow ($m_{in} = m_1 + m_2 + m_R$) and the outflow ($m_{out} = m_3 + m_4$) is calculated as the change of fluid mass storage over time (Δm). The equation for the conservation of mass takes the following form:

$$m_{in} - m_{out} = \Delta m \tag{3.86}$$

Based on Figure 3.19, the mass inflows and outflows over a time interval dt are $m_1 = \rho_w v_x dy\, dt$, $m_2 = \rho_w v_y dx\, dt$, $m_3 = \left[\rho_w v_x + \frac{\partial(\rho_w v_x)}{\partial x} dx \right] dy\, dt$, $m_4 = \left[\rho_w v_y + \frac{\partial(\rho_w v_y)}{\partial y} dy \right] dx\, dt$, and $m_R = \frac{\rho_w R}{b} dx\, dy\, dt$. Note that m_R is a volumetric change: the recharge exists in both directions dx and dy. Based on the definition of specific storage (S_s) and considering the change in hydraulic head (dh) over a time interval (dt), the absorbed or released volume of water can be expressed as $\Delta V = S_s\, dh\, dx\, dy$. Consequently, the water mass increase in the cell is $\Delta m = \rho_w S_s\, dh\, dx\, dy$. By substituting these expressions into Eq (3.86), the equation for conservation of mass becomes (Freeze and Cherry 1979)

$$-\frac{\partial(\rho_w v_x)}{dx} - \frac{\partial(\rho_w v_y)}{dy} + \frac{\rho_w R}{b} = \rho_w S_s \frac{dh}{dt} \tag{3.87}$$

This equation can be expanded as

$$-\rho_w \frac{\partial v_x}{dx} - v_x \frac{\partial \rho_w}{dx} - \rho_w \frac{\partial v_y}{dy} - v_y \frac{\partial \rho_w}{dy} + \frac{\rho_w R}{b} = \rho_w S_s \frac{dh}{dt} \tag{3.88}$$

Since the terms $\frac{\partial \rho_w}{dx}$ and $\frac{\partial \rho_w}{dy}$ are much smaller than the terms $\frac{\partial v_x}{dx}$ and $\frac{\partial v_y}{dy}$, we ignore the former two terms in Eq (3.88) and therefore get

$$-\frac{\partial v_x}{dx} - \frac{\partial v_y}{dy} + \frac{R}{b} = S_s \frac{dh}{dt} \tag{3.89}$$

Applying Darcy's law ($v_x = -K_x \frac{\partial h}{\partial x}$, $v_y = -K_y \frac{\partial h}{\partial y}$) and assuming a homogenous and isotropic medium with permeability coefficient K, this equation takes the following form:

$$K\frac{\partial^2 h}{\partial x^2} + K\frac{\partial^2 h}{\partial y^2} + \frac{R}{b} = S_s \frac{\partial h}{dt} \tag{3.90}$$

Multiplying both sides by b (the thickness of the aquifer),

$$bK\frac{\partial^2 h}{\partial x^2} + bK\frac{\partial^2 h}{\partial y^2} + R = bS_s \frac{\partial h}{dt} \tag{3.91}$$

Knowing that the transmissivity (T) and the storage coefficient (S) are $T = Kb$ and $S = S_s b$, Eq (3.91) can be rewritten as

$$T\frac{\partial^2 h}{\partial x^2} + T\frac{\partial^2 h}{\partial y^2} + R = S\frac{\partial h}{\partial t} \tag{3.92}$$

We divide both sides by T (the transmissivity) to get

$$\frac{\partial^2 h}{\partial x^2} + \frac{\partial^2 h}{\partial y^2} + \frac{R}{T} = \frac{S}{T}\frac{\partial h}{\partial t} \tag{3.93}$$

Equation (3.93) is the governing PDE for groundwater flow in aquifers.

3.4.2 Solution of one-dimensional groundwater flow in soil

The one-dimensional form of Eq (3.93) is considered here and is written as

$$T\frac{\partial^2 h}{\partial x^2} + R = S\frac{\partial h}{\partial t} \tag{3.94}$$

In the form shown in Eq (3.94), two boundary conditions and one initial condition are required to solve the equation. Equation (3.94) is a nonhomogeneous equation, unlike other examples seen earlier in this chapter. Thus, its solution is slightly different. First, we solve the homogeneous equation in the space domain (i.e., x) by ignoring the term R, and apply the boundary conditions.

Using separation of variables, we assume that the solution for the homogeneous equation (i.e., $\frac{\partial^2 h}{\partial x^2} = S\frac{\partial h}{\partial t}$) takes the form $h(x, t) = X(x) \cdot Y(t)$. By taking the derivatives of $h(x, t)$ with respect to x and t, and substituting them in the homogeneous equation, we get $TX''Y = SXY'$.

Dividing both sides by XY, we achieve

$$T\frac{X''}{X} = S\frac{Y'}{Y} \tag{3.95}$$

Since the left-hand side of Eq (3.95) is a function of only x, while the right-hand side depends only on t, this equation is constrained to be either zero, a positive value (e.g., λ^2), or a negative value (e.g., $-\lambda^2$).

It can be easily shown that the two cases of zero and a positive value (e.g. λ^2) do not lead to correct solutions. Therefore, assuming the solution is $-\lambda^2$, Eq (3.95) results in two ordinary equations in the x and t domains. We consider the equation in the x domain here because we assumed that Eq (3.94) is homogeneous in the space domain. The resulting space-domain equation from Eq (3.95) is

$$X'' + \frac{\lambda^2}{T} X = 0 \qquad (3.96)$$

The solution to Eq (3.96) has the following general form:

$$X(x) = A\cos(qx) + B\sin(qx) \qquad (3.97)$$

The coefficients A, B, and q are determined using the boundary conditions.

After obtaining the solution for $X(x)$ from the homogeneous equation, the subsequent step involves inserting this solution into Eq (3.94). The complete nonhomogeneous equation is then solved in the time domain (t), taking into account the initial conditions of the problem.

3.4.3 Numerical example: One-dimensional groundwater flow

Consider the aquifer shown in Figure 3.19, which is blocked on its right-hand boundary by an impermeable earth dam and is connected to a river on the left-hand side. Assume that a constant recharge is present into the aquifer on its entire length with a rate of $R = 0.001$ m/day. Due to this recharge rate, the water table in the aquifer rises. It is assumed that the water table in the aquifer is the same as that of the river before the start of the recharge. Other parameters are: length $L = 500$ m, transmissivity $T = 2000$ m^2/day, and storage coefficient $S = 0.07$. For this example, the status of the water table at different times is desired, i.e., $h(x, t) = ?$.

The governing equation for one-dimensional groundwater flow through an aquifer is given in Eq (3.94) and requires two boundary conditions and one initial condition to solve it. The boundary conditions are zero seepage through the right-hand boundary $(x = L)$, which implies $\frac{\partial h}{\partial x} = 0$, and at the left-hand boundary $(x = 0)$ the water level is always same as that of the river, $h = 0$. Note that the origin of the Cartesian coordinates in Figure 3.19 is placed on the river water surface. The only initial condition of this problem is that the water table in the aquifer at the beginning $(t = 0)$ is the same as that of the river level over its entire length: $t = 0$, $h(x, 0) = 0$. All the boundary and initial conditions are demonstrated in Figure 3.20.

As discussed earlier, we solve the homogeneous form of Eq (3.94) in the space domain by removing the R term and applying the boundary conditions. This results in Eq (3.97). By applying the left-hand boundary condition $(x = 0, h = 0)$ to Eq (3.97), we achieve $A = 0$; consequently, we have, $X(x) = B\sin(qx)$. This leads the equation for $h(x, t)$ to take the following form:

$$h(x, t) = X(x) \cdot Y(t) = Y(t)B\sin(qx) \qquad (3.98)$$

The derivative of Eq (3.98) with respect to x is $\frac{\partial h(x, t)}{\partial x} = Y(t)Bq\cos(qx)$. Now, the right-hand boundary condition $(x = L, \frac{\partial h}{\partial x} = 0)$ is applied, which leads to $0 = Y(t)Bq\cos(qL)$.

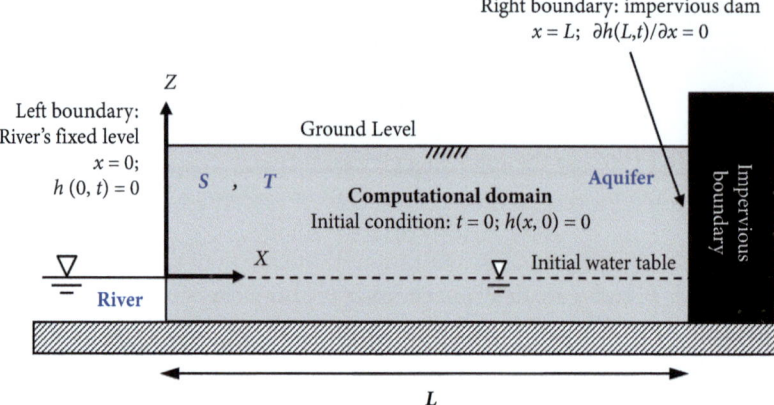

Figure 3.20 The boundary and initial conditions of a groundwater flow problem in an aquifer bounded by a river on the left-hand side and by an impervious dam on the right-hand side (see Figure 3.19).

This equation requires that $q = \dfrac{\left(n - 1/2\right)\pi}{L}$ for $n = 1, 2, 3, \ldots$ Subsequently, the solution takes the following form:

$$h(x,t) = \sum_{n=1}^{\infty} BY(t) \sin\left[\left(n - 1/2\right)\frac{\pi}{L}x\right] \tag{3.99}$$

Equation (3.99) is then substituted into the nonhomogeneous Eq (3.94) resulting in the following:

$$-T\sum_{n=1}^{\infty} BY(t)\left[\left(n - 1/2\right)\frac{\pi}{L}\right]^2 \sin\left[\left(n - 1/2\right)\frac{\pi}{L}x\right] + R = S\sum_{n=1}^{\infty} BY'(t)$$
$$\sin\left[\left(n - 1/2\right)\frac{\pi}{L}x\right] \tag{3.100}$$

In this equation, all the terms are of sinusoidal type, except for R. Here, we expand the term R using Fourier series formulas involving the sine terms of the form $\sin\left[\left(n - 1/2\right)\frac{\pi}{L}x\right]$, as follows:

$$R = \sum_{n=1}^{\infty} b_n \sin\left[\left(n - 1/2\right)\frac{\pi}{L}x\right] \tag{3.101}$$

The coefficients b_n are determined by employing Fourier series formulas:

$$b_n = \frac{2}{L}\int_0^L R \sin\left[\left(n - 1/2\right)\frac{\pi}{L}x\right] dx \tag{3.102}$$

$$b_n = \frac{2R}{\left(n - 1/2\right)\pi} \tag{3.103}$$

Therefore, R takes the following shape:

$$R = \sum_{n=1}^{\infty} \frac{2R}{(n - \frac{1}{2})\pi} \sin\left[(n - \frac{1}{2})\frac{\pi}{L}x\right] \tag{3.104}$$

We substitute the equation for R into Eq (3.100) to achieve the following:

$$-T\sum_{n=1}^{\infty} B Y(t)\left[(n - \frac{1}{2})\frac{\pi}{L}\right]^2 \sin\left[(n - \frac{1}{2})\frac{\pi}{L}x\right]$$

$$+\sum_{n=1}^{\infty} \frac{2R}{(n - \frac{1}{2})\pi} \sin\left[(n - \frac{1}{2})\frac{\pi}{L}x\right] = S\sum_{n=1}^{\infty} B Y'(t) \sin\left[(n - \frac{1}{2})\frac{\pi}{L}x\right] \tag{3.105}$$

The $\sin\left[(n - \frac{1}{2})\frac{\pi}{L}x\right]$ terms cancel out from both sides of the equation. Subsequently, the equation is reduced to the following form:

$$S B Y'(t) + T\left[(n - \frac{1}{2})\frac{\pi}{L}\right]^2 B Y(t) = \frac{2R}{(n - \frac{1}{2})\pi} \tag{3.106}$$

Dividing both sides by $S \cdot B$,

$$Y'(t) + \frac{T}{S}\left[(n - \frac{1}{2})\frac{\pi}{L}\right]^2 Y(t) = \frac{2R}{S B\pi(n - \frac{1}{2})} \tag{3.107}$$

Equation (3.107) is a first-order ODE, and $Y(t)$ can be readily solved as follows:

$$Y(t) = e^{-\frac{T}{S}[(n-1/2)\frac{\pi}{L}]^2 t} \cdot \left\{ \frac{2R L^2}{B T[(n - \frac{1}{2})\pi]^3} e^{\frac{T}{S}[(n-1/2)\frac{\pi}{L}]^2 t} + C \right\} \tag{3.108}$$

This equation for $Y(t)$ is then substituted into Eq (3.99) to achieve the solution for $h(x, t)$ as follows:

$$h(x, t) = \sum_{n=1}^{\infty} B e^{-\frac{T}{S}[(n-1/2)\frac{\pi}{L}]^2 t} \cdot \left\{ \frac{2R L^2}{B T[(n - 1/2)\pi]^3} e^{\frac{T}{S}[(n-1/2)\frac{\pi}{L}]^2 t} + C \right\}$$

$$\sin\left[(n - 1/2)\frac{\pi}{L}x\right] \tag{3.109}$$

The initial condition ($t = 0$, $h(x, 0) = 0$) is applied at this stage, which results in $B\left\{\frac{2R L^2}{B T[(n-1/2)\pi]^3} + C\right\} = 0$. This equation gives the value of C as follows:

$$C = \frac{-2R L^2}{B T[(n - \frac{1}{2})\pi]^3} \tag{3.110}$$

Finally, the solution for one-dimensional groundwater flow in the aquifer shown in Figure 3.19, with the boundary conditions illustrated in Figure 3.20, takes the following form:

$$h(x,t) = \sum_{n=1}^{\infty} \frac{2RL^2}{T\left[(n-\frac{1}{2})\pi\right]^3}\left\{1 - e^{-\frac{T}{S}\left[(n-1/2)\frac{\pi}{L}\right]^2 t}\right\} \sin\left[(n-\frac{1}{2})\frac{\pi}{L}x\right] \qquad (3.111)$$

Equation (3.111) provides the status of the groundwater level at different times (t) and at different locations (x) within the aquifer. From the onset of recharge at the rate of R to the system, the groundwater table will rise. Code 3.5 can be used to plot the solution, which is shown in Figure 3.21. According to Figure 3.21, the water table in the aquifer rises approximately 6 cm after 9 days of continuous recharge at the rate of $R = 0.001$ m/day.

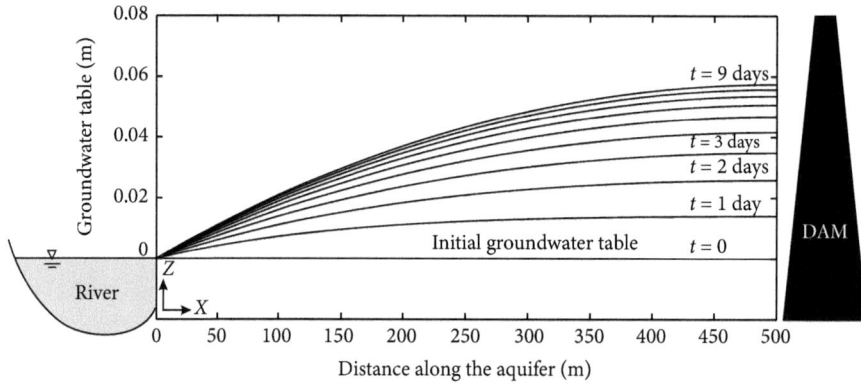

Figure 3.21 The solution for the groundwater flow problem in an aquifer of length $L = 500\ m$ bounded by a river on the left-hand side and by an impervious dam on the right-hand side which is recharged at constant rate of $R = 0.001\ m$/day. The curves show water table levels at different times following the onset of recharge. Other parameters are transmissivity $T = 2000\ m^2$/day, storage coefficient $S = 0.07$. The solution counter interval is one day. This figure is generated using Code 3.5.

Code 3.5 *The MATLAB script used for plotting the solution of the groundwater flow problem in an aquifer bounded by a river on the left-hand side and by an impervious dam on the right-hand side. Note that this code generates an animation for the variation of the water table at different times. The results are presented in Figure 3.21. This MATLAB code can be downloaded at: http://www.oup. com/AdvancedMathematicalModelling.*

```
% 1D groundwater flow in aquifer
clear; close all; clc;
set(0,'DefaultAxesFontsize',20);
%
R=0.001./(24.*3600); % Recharge rate (m/s)
L=500; % Lebgth of aquifer (m)
```

```
T=2000./(24.*3600); %% Transmissivity (m2/s)
S=0.07; %Storage coefficient
time=3600.*24.*50; %time of simulations
dt=3600.*6; %time interval
%
k=1; m=1;
for t=0:dt:time
for x=0:1:L
    s=0.0;
    for n=1:300
        temp1=2.*R.*L.^2./(T.*((n-0.5).*pi).^3);
        temp2=-1.*T.*(((n-0.5).*pi./L).^2)./S;
        s=s+temp1.*sin((n-0.5).*pi.*x./L).*...
            (1-exp(temp2.*t));
    end
    f(m,k)=s;
    k=k+1;
end
%
subplot(2,1,1);
plot(0:1:500,f(m,:),'b','LineWidth',3.0);
xlabel('Distance along aquifer (m)');
ylabel('Water table (m)');
xlim([0 500]);
ylim([0 0.1]);
set(gca,'linewidth',2.5);
%
    str = sprintf('%3.1f', t./3600);
    text(L.*0.5,0.08,str,'FontSize',22);
    text(L.*0.4,0.08,'Time = ','FontSize',22);
    text(L.*0.57,0.08,'hour','FontSize',22);
%
drawnow;
refreshdata; clf;
%
m=m+1; k=1;
end
% End of the code %%%
```

3.5 Problems for further study

Problem 3.1

For the concrete dam shown in Figure 3.22 with a reservoir water level of 10 m and downstream water level of 2 m, calculate the distribution of the hydrostatic water head in the foundation. The bottom, left, and right boundaries of the foundation are impermeable.

Problem 3.2

For the concrete dam shown in Figure 3.23, the bottom and left boundaries of the foundation are impermeable while the right boundary is a sand layer. Compute the distribution of the piezometric water head in the foundation.

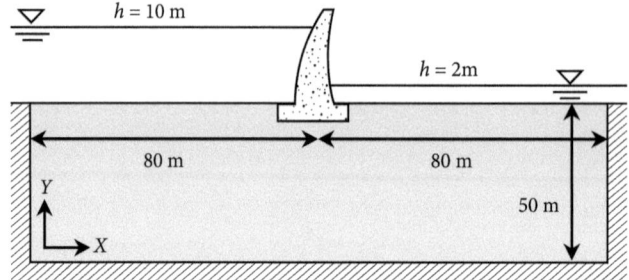

Figure 3.22 A concrete dam and its foundation with dimensions. The distribution of the piezometric water head in the foundation of the dam is desired. The bottom, left, and right boundaries of the foundation are impermeable.

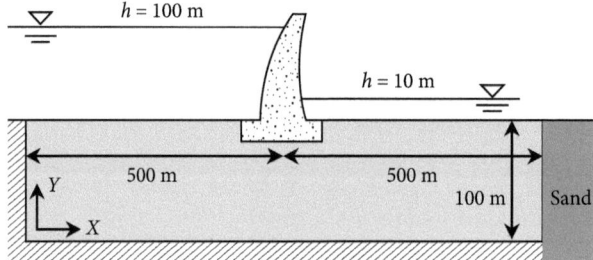

Figure 3.23 The distribution of the piezometric water head in the foundation of this concrete dam is desired. The bottom and left boundaries of the foundation are impermeable, while the right boundary is a sand layer.

Answer to Problem 3.2

$$h(x,y) = \sum_{n=1}^{\infty} \frac{1}{\cosh\left[\frac{\pi}{2}\left(n - \frac{1}{2}\right)\right]} \left\{ \frac{20(-1)^{(n+1)}}{\pi\left(n - \frac{1}{2}\right)} + \frac{180}{\pi\left(n - \frac{1}{2}\right)} \sin\left[\frac{\pi}{2}\left(n - \frac{1}{2}\right)\right]\right\}$$
$$\cos\left[\frac{\pi}{1000}\left(n - \frac{1}{2}\right)x\right] \cosh\left[\frac{\pi}{1000}\left(n - \frac{1}{2}\right)y\right] \tag{3.112}$$

A plot of the answer to Problem 3.2 is presented in Figure 3.24.

Problem 3.3

The foundation of a concrete dam is bounded by impermeable layers on its bottom and left boundaries, whereas the right boundary reaches a sand layer at a distance of infinity (Figure 3.25). Calculate the distribution of the piezometric water head in the foundation.

Answer to Problem 3.3

$$h(x,y) = \int_{\lambda=0}^{\lambda=\infty} \left\{ \frac{2H\sin(\lambda L)}{\pi\lambda\cosh(\lambda D)}\right\} \cos(\lambda x)\cosh(\lambda y)\, d\lambda \tag{3.113}$$

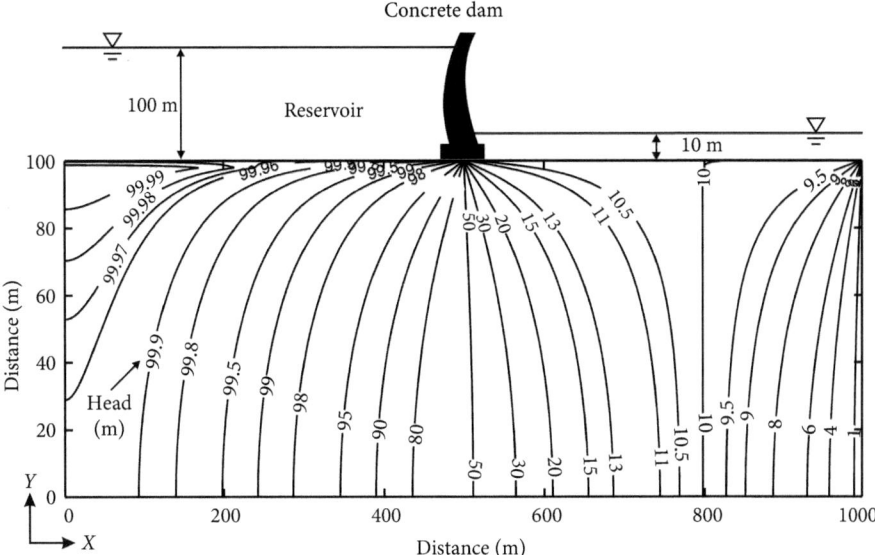

Figure 3.24 Plot of the answer to Problem 3.2 showing the contours of the piezometric water head in the foundation of this concrete dam. The bottom and left boundaries of the foundation are impermeable, while the right boundary is a sand layer.

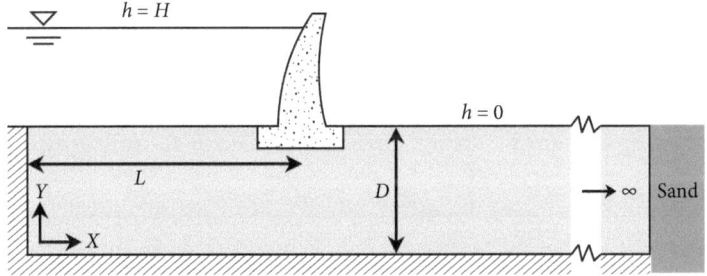

Figure 3.25 A concrete dam with its foundation adjacent to a sand layer at a distance of infinity. The bottom and left boundaries of the foundation are impermeable.

A plot of the answer to Problem 3.3 is given in Figure 3.26 for the case of $H = 10\,m$, $D = 5\,m$, and $L = 4\,m$.

Problem 3.4

For the soil consolation problem shown in Figure 3.27, the clay layer is saturated and is resting on an impermeable bed and is open to air from the top. Assume the coefficient of consolidation is $C_v = 3 \times 10^{-7}\,m^2/s$. Obtain the solution for the pore pressure changes in the clay layer over time and plot the pore pressure contours with three-month intervals.

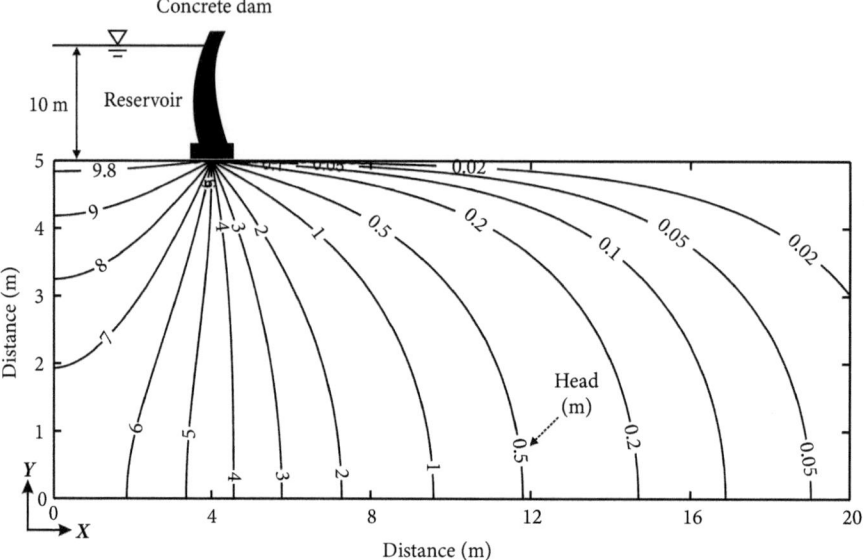

Figure 3.26 Contours of the piezometric water head in the foundation of the concrete dam of Problem 3.3. The bottom and left boundaries of the foundation are impermeable, while the right boundary is a sand layer at a distance of infinity.

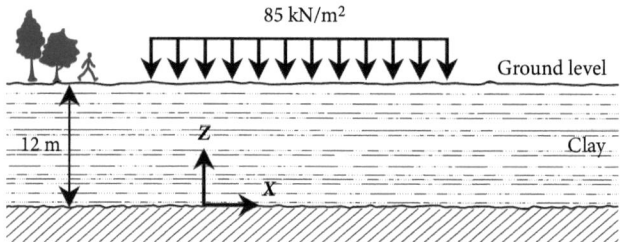

Figure 3.27 A 12 m clay layer bounded by an impermeable bed at the bottom and open to the air from the top, under external loading of $\Delta P = 85$ kN/m^2.

Problem 3.5

The aquifer shown in Figure 3.28 is bounded by rivers on both sides with a vertical separation of 10 m. A constant recharge is made into the aquifer over its entire length at a rate of $R = 0.001$ m/day. Assume that the initial water table in the aquifer is along the line connecting the surfaces of the two rivers (Figure 3.28). Calculate the status of the water table at different times and plot it at one-day intervals. The transmissivity and storage coefficients of the aquifer are $T = 2000$ m^2/day and $S = 0.07$, respectively.

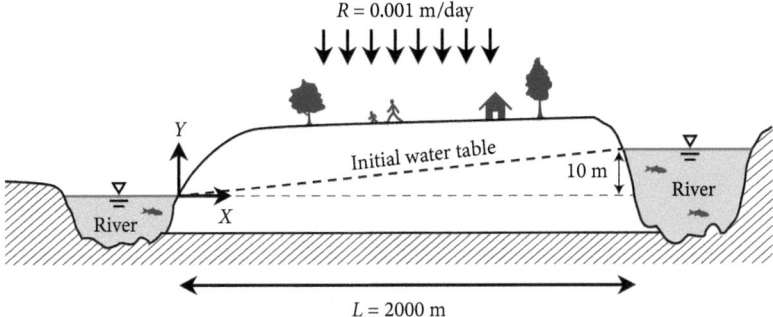

Figure 3.28 Cross section of a soil layer bounded by two rivers on both sides. The layer is under precipitation (or recharge) with constant rate of R.

References

Das, B.M. (2019). *Advanced Soil Mechanics*, 5th edn. CRC Press.

Freeze, R.A., and Cherry, J.A. (1979). *Groundwater*. Prentice-Hall Inc.

Heidarzadeh, M., Mirghasemi, A.A., and Niroomand, H. (2015). Construction of relief wells under artesian flow conditions at dam toes: Engineering experiences from Karkheh dam, Iran. *International Journal of Civil Engineering*, 13(1), 73–80.

Mahab Consulting Engineers (1998). *Karkheh project: Technical report of dam body and foundation*. Tehran, Iran.

Smith, I. (2014). *Elements of Soil Mechanics*. John Wiley & Sons.

Chapter 4
Coastal Engineering

4.1 Introduction

Coastal engineering, a subdiscipline of civil engineering, focuses on marine and coastal construction, management, marine energy harvesting, and coastal defence. The coastal environment plays a crucial role in contemporary times, significantly impacting national and international trade as well as the transportation of goods through ports and harbours. Unlike the relatively stable land environment, the coast is an ever-changing environment. Continuous wave action and variations in wave characteristics such as height, period, and direction make the coast a dynamic environment that undergoes changes. Consequently, managing and defending the coast necessitates well-calculated and proven approaches. This means, in general, that coastal design and management are more challenging than for the land environment.

Among topics addressed by coastal engineering are marine construction, coastal defence, sediment transport, tidal analysis, and port management. Coastal wave run-up and the associated design of defence systems are probably among the most important topics in coastal engineering. Examples of coastal defences in Japan and the UK are shown in Figures 4.1 and 4.2. The Sendai coast in Japan (Figure 4.1) was severely damaged by the March 2011 tsunami, which encouraged the Japanese authorities to design and construct a multilayer coastal defence along this coast after the 2011 disaster. Three layers of such a multilayer defence system are shown in Figure 4.1, including concrete armour (tetrapod units), a sea dyke, and coastal vegetation. The other defence layers that are not shown in Figure 4.1 are coastal landfill and a coastal highway.

A prominent example of coastal defence in the UK is the large pebble dyke at Chesil Beach, Portland. Although this structure appears to be much higher than the sea level (around 10–12 m above the mean sea level), it was overtopped by the extreme waves of the February 2014 UK storms, resulting in significant damage to this massive coastal defence.

Waves introduce significant challenges to the coastal environment, setting it apart from land-based construction. Their influence is pivotal in distinguishing the dynamics and construction requirements specific to coastal areas. Therefore, the gateway to understanding coastal phenomena and practising coastal engineering is to understand waves and their characteristics. In this chapter, we start with the derivation of the wave equation and continue with solving a few coastal problems such as tsunami, wave run-up, and wave spectral analysis.

A Practical Approach to Advanced Mathematical Modelling in Civil Engineering. Mohammad Heidarzadeh et al., Oxford University Press. © Mohammad Heidarzadeh, Theodosios K. Papathanasiou, Yurui Fan, Hamid Bahai (2025).
DOI: 10.1093/9780191888656.003.0004

Figure 4.1 Waves acting along the coast of Sendai in east of Japan, which was severely damaged during the March 2011 Japan mega-tsunami. This photo, taken in September 2019, shows the multilayer protection constructed here involving concrete armour (tetrapod units), a sea dyke, and coastal vegetation.

Figure 4.2 The 11 m-high pebble sea dyke at Chesil Beach (Portland, UK) to protect the coastal areas from extreme wave run-up during storms. This photo shows the beach in March 2017.

4.2 Wave equations

Consider a control volume of the sea at depth in two dimensions (X–Z plane) as shown in Figure 4.3a. The total water depth, $h(x, t)$, is the sum of the water depth, $d(x, t)$, and wave amplitude, $\eta(x, t)$ (Figure 4.3b). We apply Newton's second law on this control volume as follows:

$$F = m \cdot a \tag{4.1}$$

in which F is force, m is mass, and a is acceleration. Considering the control volume in Figure 4.3a, F is expressed using the following equation:

$$F = F(x, t) - F(x + \Delta x, t) \tag{4.2}$$

As velocity (u) is a function of both time (t) and space (x), acceleration is calculated in the following way:

$$a = \frac{du}{dt} = \frac{\left(\frac{\partial u}{\partial t} dt + \frac{\partial u}{\partial x} dx\right)}{dt} = \frac{\partial u}{\partial t} + \frac{\partial u}{\partial x}\frac{dx}{dt} \tag{4.3}$$

As $u = \frac{dx}{dt}$, Eq (4.3) is rewritten as follows:

$$a = \frac{\partial u}{\partial t} + u \frac{\partial u}{\partial x} \tag{4.4}$$

The mass of water (m) in a tiny control volume with a length Δx (Figure 4.3a), assuming a constant water depth, $h(x, t)$, is calculated using the formula $m = \rho h \Delta x$, where ρ represents the density of seawater. By substituting Eqs (4.4) and (4.2) into Eq (4.1), we derive:

$$F(x, t) - F(x + \Delta x, t) = \rho h \Delta x \left[\frac{\partial u}{\partial t} + u \frac{\partial u}{\partial x}\right] \tag{4.5}$$

(a) 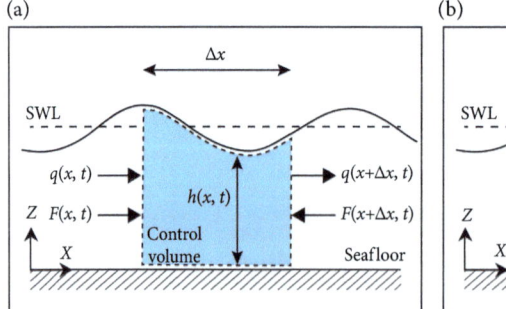   (b)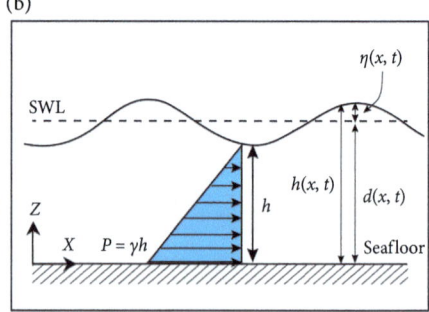

Figure 4.3 (a) Sketch showing a control volume of the sea water column with total water depth of $h(x, t)$, the forces acting, and the discharge volumes on both sides of the control volume. (b) The hydrostatic pressure assumption for the water column, which shows a linear pressure distribution at depth. Also shown is the total water depth, $h(x, t)$, the water depth, $d(x, t)$, and the wave amplitude $\eta(x, t)$.

Here, utilizing the notation $\Delta F = F(x + \Delta x, t) - F(x, t)$, Eq (4.5) can be rearranged to:

$$-\frac{\Delta F}{\Delta x} = \rho h \left[\frac{\partial u}{\partial t} + u \frac{\partial u}{\partial x} \right] \qquad (4.6)$$

By considering the limit $\Delta x \to 0$, Eq (4.6) transforms into the following:

$$-\frac{\partial F}{\partial x} = \rho h \left[\frac{\partial u}{\partial t} + u \frac{\partial u}{\partial x} \right] \qquad (4.7)$$

According to Eq (4.7), the calculation of the force exerted by water on the control volume is required. In the context of shallow water, characterized by a ratio of depth (d) to wavelength (L) of less than 0.05 (where L represents the wavelength), the prevailing condition is that of a hydrostatic pressure distribution, as depicted in Figure 4.3b. This condition leads to the following equation for force, as described in works such as Sorensen (2006):

$$F = \frac{1}{2} \rho g h^2 \qquad (4.8)$$

By taking the derivative of F relative to x, we derive:

$$\frac{\partial F}{\partial x} = \rho g h \frac{\partial h}{\partial x} \qquad (4.9)$$

Upon combining Eqs (4.7) and (4.9), the momentum equation for fluid flow is determined as follows:

$$-\rho g h \frac{\partial h}{\partial x} = \rho h \left[\frac{\partial u}{\partial t} + u \frac{\partial u}{\partial x} \right] \qquad (4.10)$$

This equation is further simplified into:

$$\frac{\partial u}{\partial t} + u \frac{\partial u}{\partial x} = -g \frac{\partial h}{\partial x} \qquad (4.11)$$

Equation (4.11) is the momentum equation for fluid flow.

After extracting the momentum equation, Eq (4.11), the continuity of flow and associated equations are discussed here. Applying the continuity concept to the water input to and output from the control volume (Figure 4.3a), we calculate the volume of stored water in the control volume (V_{st}) by subtracting the volume of the output water (V_o) from the input water (V_i):

$$V_{st} = V_i - V_o \qquad (4.12)$$

Referring to Figure 4.3a, it can be seen that $V_{st} = \Delta x \Delta h$, $V_i = q(x, t) \Delta t$, and $V_o = q(x + \Delta x, t) \Delta t$. These terms are substituted into Eq (4.12), which leads to the following equation:

$$\Delta x\, \Delta h = q(x, t)\, \Delta t - q(x + \Delta x, t)\, \Delta t \tag{4.13}$$

Since $\Delta q = q(x + \Delta x, t) - q(x, t)$, the above equation can be rearranged to:

$$-\frac{\Delta q}{\Delta x} = \frac{\Delta h}{\Delta t} \tag{4.14}$$

As we approach the limit $\Delta x \to 0$ and $\Delta t \to 0$, the outcome becomes:

$$-\frac{\partial q}{\partial x} = \frac{\partial h}{\partial t} \tag{4.15}$$

Given that discharge (q) is the product of water depth (h) and velocity (u), represented as $q = uh$, upon substituting this expression into Eq (4.15) it transforms into:

$$\frac{\partial h}{\partial t} + \frac{\partial (uh)}{\partial x} = 0 \tag{4.16}$$

This equation is called the continuity equation.

Thus far, two equations governing wave motion have been derived, Eqs (4.11) and (4.16), collectively known as the shallow-water equations (SWEs). These equations are shown in the system of differential equations below, encompassing the variables of velocity (u) and water depth (h):

$$\begin{cases} \dfrac{\partial h}{\partial t} + \dfrac{\partial (uh)}{\partial x} = 0 \\[2ex] \dfrac{\partial u}{\partial t} + u\dfrac{\partial u}{\partial x} = -g\dfrac{\partial h}{\partial x} \end{cases} \tag{4.17}$$

The above system of differential equations is a nonlinear system because it involves the nonlinear term $u\frac{\partial u}{\partial x}$. Remember that the system shown in Eq (4.17) was obtained assuming that the waves are shallow-water waves, which implies wavelengths (L) of approximately more than 20 times the water depth (d), i.e., $\frac{d}{L} < 0.05$. This is why the set of differential equations in Eq (4.17) are known as SWEs.

For the case of linear SWEs, it is possible to combine the SWEs in Eq (4.17) into a single equation for wave motion. By neglecting the nonlinear term (i.e., $u\frac{\partial u}{\partial x}$), Eq (4.17) reduces to the following form:

$$\begin{cases} \dfrac{\partial h}{\partial t} + \dfrac{\partial (uh)}{\partial x} = 0 \\[2ex] \dfrac{\partial u}{\partial t} = -g\dfrac{\partial h}{\partial x} \end{cases} \tag{4.18}$$

The equations in Eq (4.18) are called the linear shallow-water equations (LSWEs).

Looking at Figure 4.3b, the following relationship can be established between the total water depth, $h(x, t)$, the water depth, $d(x, t)$, and wave amplitude, $\eta(x, t)$:

$$h(x, t) = d(x, t) + \eta(x, t) \tag{4.19}$$

Equation (4.19) is introduced into Eq (4.18) to obtain the following system:

$$\begin{cases} \dfrac{\partial(d+\eta)}{\partial t} + \dfrac{\partial[u(d+\eta)]}{\partial x} = 0 \\ \dfrac{\partial u}{\partial t} = -g\dfrac{\partial(d+\eta)}{\partial x} \end{cases} \tag{4.20}$$

After taking the derivatives, the above system takes the following form:

$$\begin{cases} \dfrac{\partial \eta}{\partial t} + d\dfrac{\partial u}{\partial x} + \dfrac{\partial(u\eta)}{\partial x} = 0 \\ \dfrac{\partial u}{\partial t} = -g\dfrac{\partial \eta}{\partial x} \end{cases} \tag{4.21}$$

Since $\frac{\partial(u\eta)}{\partial x}$ is a nonlinear term, it is neglected here as this system of differential equation is for LSWEs. Therefore, we achieve:

$$\begin{cases} \dfrac{\partial \eta}{\partial t} + d\dfrac{\partial u}{\partial x} = 0 \\ \dfrac{\partial u}{\partial t} = -g\dfrac{\partial \eta}{\partial x} \end{cases} \tag{4.22}$$

In this system, we differentiate the top equation with respect to time (t) and the bottom equation relative to space (x) to achieve the following system:

$$\begin{cases} \dfrac{\partial^2 \eta}{\partial t^2} + d\dfrac{\partial^2 u}{\partial x \partial t} = 0 \\ \dfrac{\partial^2 u}{\partial t \partial x} = -g\dfrac{\partial^2 \eta}{\partial x^2} \end{cases} \tag{4.23}$$

This can be rearranged to:

$$\begin{cases} \dfrac{\partial^2 u}{\partial x \partial t} = -\dfrac{1}{d}\dfrac{\partial^2 \eta}{\partial t^2} \\ \dfrac{\partial^2 u}{\partial t \partial x} = -g\dfrac{\partial^2 \eta}{\partial x^2} \end{cases} \tag{4.24}$$

As the left-hand sides of both equations in the above system are the same, the system is thus reduced to a single equation as follows:

$$-g\dfrac{\partial^2 \eta}{\partial x^2} = -\dfrac{1}{d}\dfrac{\partial^2 \eta}{\partial t^2} \tag{4.25}$$

Table 4.1 Different types of boundary and initial conditions used for solving the wave propagation equation.

Type of condition	Description	Mathematical expression
Initial condition	Water particles are at rest at the beginning $(t = 0)$ and thus velocity is zero	$\dfrac{\partial \eta}{\partial t} = 0$
Initial condition	Water surface has a height of H_0 at the beginning $(t = 0)$	$\eta = H_0$
Boundary condition	There is no flow from the left boundary $(x = 0)$	$\dfrac{\partial \eta}{\partial x} = 0$
Boundary condition	There is no flow from the right boundary $(x = L)$	$\dfrac{\partial \eta}{\partial x} = 0$

Or,

$$\frac{\partial^2 \eta}{\partial t^2} = (g\,d)\,\frac{\partial^2 \eta}{\partial x^2} \tag{4.26}$$

Defining a new parameter, $C = \sqrt{g\,d}$, which is known as wave celerity (C), the previous equation transforms to:

$$\frac{\partial^2 \eta}{\partial t^2} = C^2 \frac{\partial^2 \eta}{\partial x^2} \tag{4.27}$$

This is the one-dimensional (1D) wave equation for shallow-water waves.

Equation (4.27) can be solved by considering boundary and initial conditions. As the wave equation involves both time (t) and space (x) variables, initial and boundary conditions are required to solve it. Table 4.1 gives a list of typical boundary and initial conditions used for modelling wave propagation.

4.3 Modelling tsunamis generated by earthquakes

Tsunami is considered among the most destructive coastal hazards, and has been responsible for many historical and modern natural disasters such as the December 2004 Indian Ocean tsunami, with a death toll of more than 220,000 (Heidarzadeh et al. 2020a), and the March 2011 Japan tsunami, which killed more than 20,000 people along the coasts of east Japan (Figure 4.4; Heidarzadeh and Satake 2013). Tsunamis are generated by various sources such as earthquakes, volcanoes, landslides (submarine and subaerial), and meteorites. Probably the most common tsunamis are those generated by submarine earthquakes where the rupture of a submarine fault generates instant seafloor uplift (Figure 4.5). The rapid uplift of the seafloor is nearly instantly transmitted to the sea surface, triggering a tsunami through the force of gravity.

11 March 2011
Japan tsunami

Figure 4.4 Photo showing the damage made by the 11 March 2011 Japan tsunami, where a large ship was moved inland and placed on the roof of a building. The author MH is shown here.

Tsunamis are considered long waves because their wavelengths (typically $L = 500$ km or more for large earthquakes in the middle of the ocean) are much greater than the average ocean water depth of $d = 5$ km, leading to a relative water depth of $\frac{d}{L} = \frac{5}{500} = 0.01 < 0.05$ (Heidarzadeh et al. 2014). Therefore, tsunami waves are considered shallow-water waves and thus the wave equation extracted in Eq (4.27) can be used to model them.

Here, a 1D ocean with a length of L is assumed, where a submarine earthquake generates an uplift of the seafloor whose height and length are H_0 and $L/20$, respectively (Figure 4.5). To solve the shallow-water wave equation, Eq (4.27), since there are two unknowns (x and t) and both terms involving x and t are second-order derivatives, two initial conditions and two boundary conditions are required.

The first initial condition concerns the zero-velocity condition at all water points at the beginning of the simulation, i.e., $\frac{\partial \eta(x,0)}{\partial t} = 0$. This basically means the ocean is at rest before the generation of the tsunami. The second initial condition is about the water surface situation right after the earthquake ($t = 0$). It is commonly accepted in coastal engineering that the initial sea surface displacement due to a submarine earthquake is the same as the seafloor crustal deformation, as indicated in Figure 4.5 (e.g., Heidarzadeh et al. 2017). This is because the process of the generation of coseismic seafloor uplift is very rapid (e.g., less than a minute). Therefore, by referring to Figure 4.5, the second initial condition is written as follows:

$$\eta(x,0) = \begin{cases} 0 & 0 < x < \dfrac{19}{40}L \\[2mm] H_0 & \dfrac{19}{40}L \leq x < \dfrac{21}{40}L \\[2mm] 0 & \dfrac{21}{40}L < x \leq L \end{cases} \tag{4.28}$$

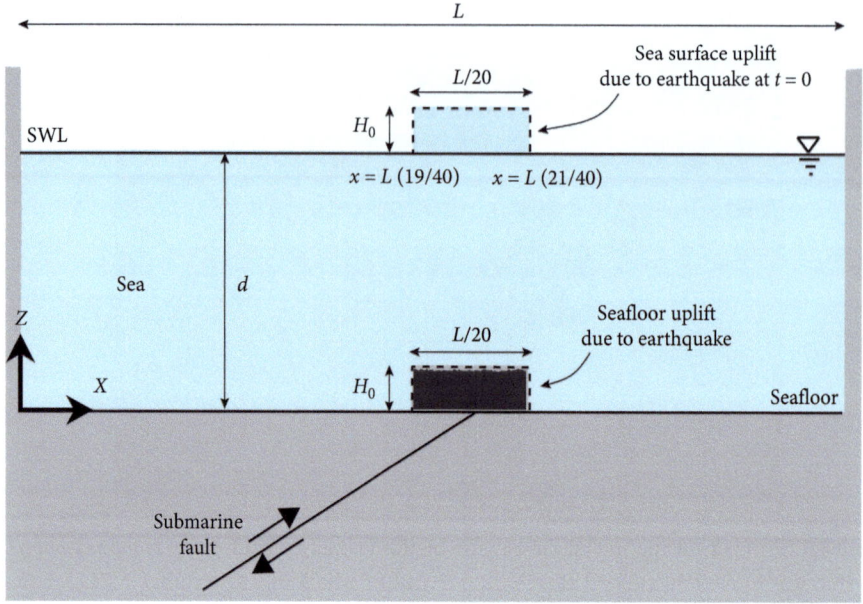

Figure 4.5 Sketch showing the generation of a tsunami due to a submarine fault in a one-dimensional ocean. SWL stands for still water level. The length of the coseismic uplift is assumed to be $L/20$, where L is the length of the water body (sea).

Regarding boundary conditions, it is assumed that there are no flows at the left $(x = 0)$ and right $(x = L)$ boundaries, which means $\frac{\partial \eta(0, t)}{\partial x} = 0$ and $\frac{\partial \eta(L, t)}{\partial x} = 0$. All the necessary initial and boundary conditions have now been established prior to solving the wave equation.

The separation of variables method is applied here to solve the wave equation, based on the assumption that the solution is the product of two independent functions $X(x)$ and $T(t)$, as follows:

$$\eta(x, t) = X(x) \cdot T(t) \tag{4.29}$$

in which $\eta(x, t)$ is the wave amplitude at any location (x) in the computational domain at time (t). $X(x)$ and $T(t)$ are functions which exclusively depend on x and t, respectively. By differentiating $\eta(x, t)$ respect to x and t, we obtain: $\frac{\partial \eta}{\partial x} = X' \cdot T$, $\frac{\partial \eta}{\partial t} = X \cdot T'$, $\frac{\partial^2 \eta}{\partial x^2} = X'' \cdot T$, and $\frac{\partial^2 \eta}{\partial t^2} = X \cdot T''$. These terms are substituted into Eq (4.27), leading to:

$$X'' \cdot T = \frac{1}{C^2} X \cdot T'' \tag{4.30}$$

Dividing both sides of Eq (4.30) by $X \cdot T$:

$$\frac{X''}{X} = \frac{1}{C^2} \frac{T''}{T} \tag{4.31}$$

In Eq (4.31), both sides must be a constant value because the left-hand side depends only on x and the right-hand side is a function of t only. Such a constant value could be zero, negative $\left(-\lambda^2\right)$, or positive $\left(\lambda^2\right)$, as indicated below:

$$\frac{X''}{X} = \frac{1}{C^2}\frac{T''}{T} = \begin{cases} -\lambda^2 \\ 0 \\ \lambda^2 \end{cases} \tag{4.32}$$

To explore potential solutions, Eq (4.32) must be solved for all three cases, involving constant values of 0, λ^2, and $-\lambda^2$. To start, we assume that the negative value $-\lambda^2$ is the correct answer. Consequently, Eq (4.32) leads to:

$$\frac{X''}{X} = \frac{1}{C^2}\frac{T''}{T} = -\lambda^2 \tag{4.33}$$

This equation produces two ODEs as follows:

$$X'' + \lambda^2 X = 0 \tag{4.34}$$

and

$$T'' + (C\lambda)^2 T = 0 \tag{4.35}$$

The general solutions for the aforementioned ODEs exhibit the following forms:

$$X(x) = d_1 \cos(\lambda x) + d_2 \sin(\lambda x) \tag{4.36}$$

and

$$T(t) = d_3 \cos(C\lambda t) + d_4 \sin(C\lambda t) \tag{4.37}$$

By substituting these functions for $X(x)$ and $T(t)$ into Eq (4.29), the solution for wave propagation, $\eta(x, t)$, takes the following form:

$$\eta(x, t) = \{d_1 \cos(\lambda x) + d_2 \sin(\lambda x)\} \times \{d_3 \cos(C\lambda t) + d_4 \sin(C\lambda t)\} \tag{4.38}$$

Considering that the constant value λ is arbitrary, this allows for various numerical values, such as λ_1, λ_2, λ_3, and so on to be potential solutions. Note that if P_1, P_2, P_3, and so forth are solutions for a differential equation, their linear combinations also serve as solutions. Hence, one potential solution for the differential equation may be the sum of all these solutions. Applying these fundamental principles, an answer inspired by the wave propagation equation from Eq (4.38) would be as follows:

$$\eta(x, t) = \sum_{n=0}^{n=\infty} \{d_{1n} \cos(\lambda_n x) + d_{2n} \sin(\lambda_n x)\} \times \{d_{3n} \cos(C\lambda_n t)$$
$$+ d_{4n} \sin(C\lambda_n t)\} \tag{4.39}$$

The coefficients d_{1n}, d_{2n}, d_{3n}, d_{4n}, and λ_n can be determined by applying the boundary conditions of the problem.

To apply the first initial condition, $t = 0$, $\frac{\partial \eta}{\partial t} = 0$, the derivative of $\eta(x, t)$ needs to be taken relative to t, leading to:

$$\frac{\partial \eta(x, t)}{\partial t} = \sum_{n=0}^{n=\infty} \{d_{1n} \cos(\lambda_n x) + d_{2n} \sin(\lambda_n x)\}$$
$$\times \{-C\lambda_n d_{3n} \sin(C\lambda_n t) + C\lambda_n d_{4n} \cos(C\lambda_n t)\} \tag{4.40}$$

Applying the first initial condition $\left(t = 0, \frac{\partial \eta}{\partial t} = 0\right)$, results in $d_{4n} = 0$.

To apply the two boundary conditions, the derivative of $\eta(x, t)$ relative to x is required, which is provided below:

$$\frac{\partial \eta(x, t)}{\partial x} = \sum_{n=0}^{n=\infty} \{-\lambda_n d_{1n} \sin(\lambda_n x) + \lambda_n d_{2n} \cos(\lambda_n x)\}$$
$$\times \{d_{3n} \cos(C\lambda_n t) + d_{4n} \sin(C\lambda_n t)\} \tag{4.41}$$

By applying the first boundary condition $\left(x = 0, \frac{\partial \eta}{\partial x} = 0\right)$, Eq (4.41) yields $d_{2n} = 0$. Subsequently, the next boundary condition $\left(x = L, \frac{\partial \eta}{\partial x} = 0\right)$ is applied, resulting in $\lambda_n = \frac{n\pi}{L}$. So far, the solution takes the following form:

$$\eta(x, t) = \sum_{n=0}^{n=\infty} K_n \cos\left(\frac{n\pi}{L} x\right) \cos\left(\frac{n\pi C}{L} t\right) \tag{4.42}$$

It is now time to apply the last condition, which is the initial condition of sea surface status at the beginning of the tsunami ($t = 0$) provided by Eq (4.28). Here, looking at Eq (4.42), it is helpful to expand the function $H(x)$ in Eq (4.28) as a Fourier series using the term $\cos\left(\frac{n\pi}{L} x\right)$, as follows:

$$\eta(x, 0) = \begin{cases} 0 & 0 < x < \dfrac{19}{40} L \\[2mm] H_0 & \dfrac{19}{40} L \leq x < \dfrac{21}{40} L \\[2mm] 0 & \dfrac{21}{40} L < x \leq L \end{cases} = a_0 + \sum_{n=1}^{n=\infty} a_n \cos\left(\frac{n\pi}{L} x\right) \tag{4.43}$$

Using the Fourier series formulae, we have:

$$a_0 = \frac{1}{L} \int_0^L \eta(x, 0)\, dx = \frac{1}{L} \int_{\frac{19}{40} L}^{\frac{21}{40} L} H_0\, dx = \frac{H_0}{20} \tag{4.44}$$

and

$$a_n = \frac{2}{L} \int_0^L \eta(x, 0) \cos\left(\frac{n\pi}{L} x\right) dx = \frac{2}{L} \int_{\frac{19}{40} L}^{\frac{21}{40} L} H_0 \cos\left(\frac{n\pi}{L} x\right) dx \qquad (4.45)$$

After integration, a_n is determined as follows:

$$a_n = \frac{2 H_0}{n\pi} \left[\sin\left(\frac{21\, n\pi}{40}\right) - \sin\left(\frac{19\, n\pi}{40}\right) \right] \qquad (4.46)$$

Using Eqs (4.42) and (4.43) and setting $t = 0$, we have:

$$\sum_{n=0}^{n=\infty} K_n \cos\left(\frac{n\pi}{L} x\right) = a_0 + \sum_{n=1}^{n=\infty} a_n \cos\left(\frac{n\pi}{L} x\right) \qquad (4.47)$$

This can be rewritten as:

$$K_0 + \sum_{n=1}^{n=\infty} K_n \cos\left(\frac{n\pi}{L} x\right) = a_0 + \sum_{n=1}^{n=\infty} a_n \cos\left(\frac{n\pi}{L} x\right) \qquad (4.48)$$

This equation leads to: $K_0 = a_0$ and $K_n = a_n$. Finally, by substituting these values into Eq (4.42), the solution takes the following form:

$$\eta(x, t) = \frac{H_0}{20} + \sum_{n=1}^{n=\infty} \frac{2 H_0}{n\pi} \left[\sin\left(\frac{21\, n\pi}{40}\right) - \sin\left(\frac{19\, n\pi}{40}\right) \right] \cos\left(\frac{n\pi}{L} x\right)$$
$$\cos\left(\frac{n\pi C}{L} t\right) \qquad (4.49)$$

The result of wave propagation modelling, Eq (4.49), is plotted in Figure 4.6 for the case of a 1D water basin with a length of $L = 100$ m, initial coseismic uplift of $H_0 = 3$ m due to an earthquake, and constant water depth of $d = 10$ m. The MATLAB script in Code 4.1 produces an animation of wave propagation over time. This and the other MATLAB scripts presented in this chapter can be downloaded from the following website: http://www.oup.com/AdvancedMathematicalModelling.

Note that the final solution presented in Eq (4.49) was derived by assuming that the value $-\lambda^2$ is the correct answer for Eq (4.32). It can be easily proven that the other two options in Eq (4.32) (i.e., 0 and λ^2) do not lead to any acceptable answers for Eq (4.32).

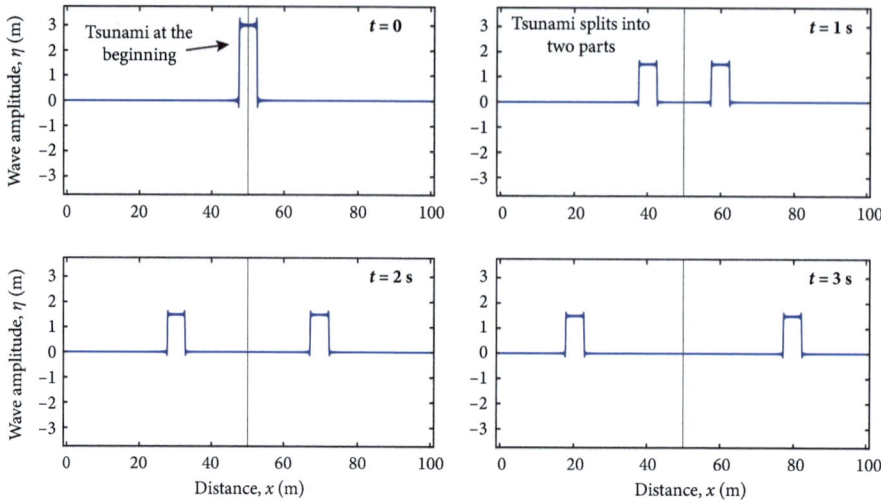

Figure 4.6 Snapshots of wave propagation at different times for the case of a 1D water basin (sea) with a length of $L = 100$ m, initial coseismic uplift of $H_0 = 3$ m due to an earthquake, and a constant water depth of $d = 10$ m. The length of the coseismic uplift is assumed to be $L/20$, where L is the length of the water body (sea). This figure was generated using Code 4.1.

Code 4.1 *The MATLAB script used for plotting the solution of wave propagation for the case of a 1D water basin (sea) with a length of $L = 100$ m, initial coseismic uplift of $H_0 = 3$ m, and constant water depth of $d = 10$ m. The length of the coseismic uplift is assumed to be $L/20$, where L is the length of the water body (sea). This MATLAB code (and other MATLAB scripts in this chapter) can be downloaded from the following website: http://www.oup.com/AdvancedMathematicalModelling. This code produces Figure 4.6.*

```
% Matlab code for making animations of wave
% propagation due to a tsunami in a 1D sea
clear; close all;
set(0, 'Defaultaxesfontsize', 15);
%
L=100; % in meter- length of the sea
time=100; % in seconds- total simulation time
H0=3; % in meters- initia earthquake uplift
g=9.81; % in m/s^2- gravitational acceleration
d=10; % in meter- sea depth
C=(g.*d).^0.5; % long wave celerity
N=400; % number of fourier terms
dt=0.1; % in seconds- time interval
dx= 0.01; % in meter- space interval
%
k=1; m=1;
for t=0:dt:time
    for x=0:dx:L
        s=0.0;
        for nx=1:N
            zar=2.*H0.*(sin(21.*nx.*pi./40)-...
```

```
            sin(19.*nx.*pi./40))./(nx.*pi);
        s=s+zar.*cos(nx.*pi.*x./L).*cos(nx.*pi.*t.*C./L);
     end
     etta(k,m)=H0./20+s;
     k=k+1;
     end
   plot(0:dx:L,etta(:,m),'b', 'linewidth',1.5);
   xlim([-1 L+1]);
   ylim([-1.25.*H0 1.25.*H0]);
   line([50 50],[-1.25.*H0 1.25.*H0],'color',[0 0 0]);
   set(gca, 'linewidth', 1.5);
   %
   refreshdata; drawnow
   k=1; m=m+1;
 end
 % End of the code %
```

4.4 Modelling tsunamis generated by landslides

Following earthquakes, landslides, whether submarine or subaerial, are the second most common cause of tsunamis and have been responsible for numerous disasters in recent decades. One such recent catastrophe occurred due to a landslide-generated tsunami following the collapse of the Anak Krakatau volcano in Indonesia in December 2018, resulting in a death toll of over 450 (Heidarzadeh et al. 2020b). A subaerial landslide triggers a tsunami by creating a disturbance at the sea surface through the sudden collapse of significant rock and soil materials into the ocean, as depicted in Figure 4.7a.

A simplified version of tsunami generation by subaerial landslide is shown in Figure 4.8 in a semi-infinite sea. For simplicity, the initial sea surface displacement resulting from this subaerial landslide is assumed to resemble a rectangular water hump (Figure 4.8). It

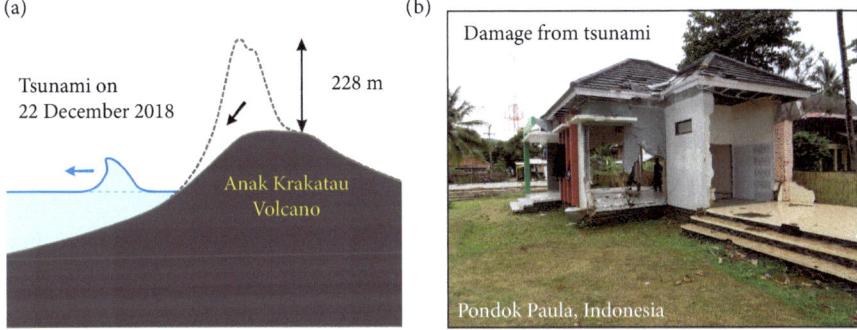

Figure 4.7 (a) Generation mechanism of a tsunami by a subaerial landslide with a sudden collapse of a large amount of rock and soil into the sea. (b) An example of damage made by the December 2018 Anak Krakatau tsunami to coastal communities in Indonesia. This figure is modified from Heidarzadeh et al. (2020b).

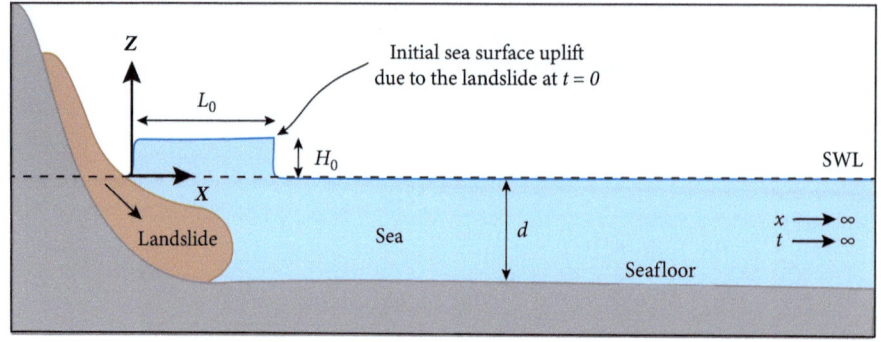

Figure 4.8 Boundary and initial conditions for a tsunami generated by a subaerial landslide.

is important to note that in reality, the initial sea surface disturbance caused by a land-slide might actually resemble a Gaussian surface (Heidarzadeh et al. 2020c, 2023). The governing equation for modelling tsunami propagation is given in Eq (4.27). However, the boundary and initial conditions are different this time. The new conditions are: at the beginning $(t = 0)$, all water particles are at rest and thus water particle velocity is zero for the entire computational domain: $\frac{\partial \eta(x, 0)}{\partial t} = 0$. At the left boundary $(x = 0)$, there is no flow: $\frac{\partial \eta(0, t)}{\partial x} = 0$. At the beginning $(t = 0)$, the water surface experiences an uplift of H_0 with the following equation:

$$\eta(x, 0) = \begin{cases} H_0 & 0 < x < L_0 \\ 0 & x \geq L_0 \end{cases} \tag{4.50}$$

Using the method of separation of variables and assuming that the solution of the wave equation is of the form $\eta(x, t) = X(x) \cdot T(t)$, similar to the previous problem (Section 4.2), we know that the answer must be $-\lambda^2$ (referring to Eq (4.32)). Therefore, the wave equation is reduced to the following two ordinary differential equations:

$$X(x) = d_1 \cos(\lambda x) + d_2 \sin(\lambda x) \tag{4.51}$$

and

$$T(t) = d_3 \cos(C\lambda t) + d_4 \sin(C\lambda t) \tag{4.52}$$

Here, we apply the initial and boundary conditions one by one. The first initial condition, $t = 0$, $\frac{\partial \eta(x, 0)}{\partial t} = 0$, leads to $d_4 = 0$. The boundary condition, $x = 0$, $\frac{\partial \eta(0, t)}{\partial x} = 0$, results in $d_2 = 0$. Therefore, up to this point, the solution is as follows:

$$\eta(x, t) = \int_{\lambda=0}^{\lambda=\infty} d_5 \cos(\lambda x) \cos(C\lambda t) \, d\lambda \tag{4.53}$$

Compared to Section 4.2, a change in the mathematical symbol used for the solution is evident here. The sum symbol (Σ) in Section 4.2 is replaced with an integral ($\int$) due to the infinite boundary on the right in this particular problem (Figure 4.8). Additionally, a new coefficient, $d_5 = d_3 d_4$, is introduced. Equation (4.53) involves only one unknown, which is d_5. This value can be obtained by applying the other initial condition, i.e., Eq (4.50).

Setting $t = 0$ in Eq (4.53) gives: $\eta(x, 0) = \int_{\lambda=0}^{\lambda=\infty} d_5 \cos(\lambda x) \, d\lambda$. On the other hand, $\eta(x, 0)$ is also an initial condition of the problem, as seen in Eq (4.50). Therefore, it can be written as:

$$\begin{cases} H_0 & 0 < x < L_0 \\ 0 & x \geq L_0 \end{cases} = \int_{\lambda=0}^{\lambda=\infty} d_5 \cos(\lambda x) \, d\lambda \tag{4.54}$$

We can expand the left-hand side as a Fourier integral using the term $\cos(\lambda x)$ where it takes the following form:

$$\begin{cases} H_0 & 0 < x < L_0 \\ 0 & x \geq L_0 \end{cases} = \int_{\lambda=0}^{\lambda=\infty} B(\lambda) \cos(\lambda x) \, d\lambda \tag{4.55}$$

Using the formulae for Fourier integral expansions, the coefficient $B(\lambda)$ is determined as follows:

$$B(\lambda) = \frac{2}{\pi} \int_{\lambda=0}^{\lambda=L_0} H_0 \cos(\lambda x) \, dx = \frac{2H_0}{\pi\lambda} \sin(\lambda L_0) \tag{4.56}$$

Consequently, Eq (4.54) can be rewritten as follows:

$$\int_{\lambda=0}^{\lambda=\infty} d_5 \cos(\lambda x) \, d\lambda = \int_{\lambda=0}^{\lambda=\infty} \frac{2H_0}{\pi\lambda} \sin(\lambda L_0) \cos(\lambda x) \, d\lambda \tag{4.57}$$

Therefore, $d_5 = \frac{2H_0}{\pi\lambda} \sin(\lambda L_0)$, and the solution of the wave propagation equation due to a landslide tsunami becomes:

$$\eta(x, t) = \int_{\lambda=0}^{\lambda=\infty} \frac{2H_0}{\pi\lambda} \sin(\lambda L_0) \cos(\lambda x) \cos(C\lambda t) \, d\lambda \tag{4.58}$$

Remember that this solution is valid for long waves or shallow-water waves (i.e., $\frac{d}{L} < 0.05$, where L is the wavelength and d the water depth). Snapshots of wave propagation at different times are shown in Figure 4.9. The MATLAB script in Code 4.2 generates an animation of wave propagation, and is available at: http://www.oup.com/AdvancedMathematicalModelling.

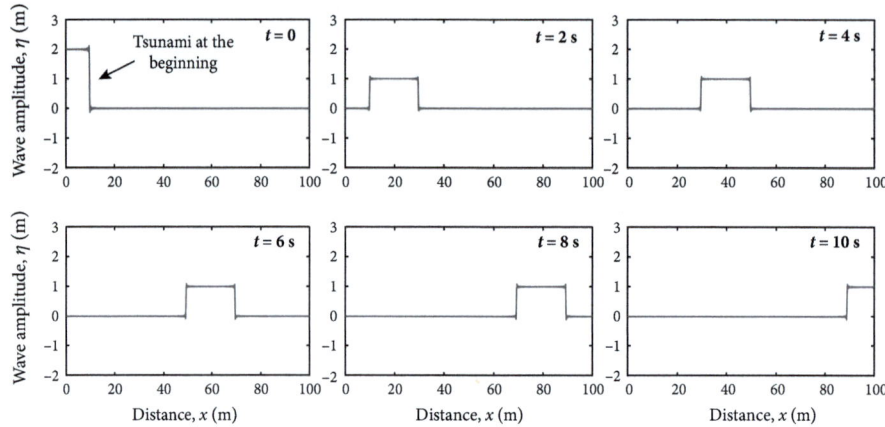

Figure 4.9 Snapshots of wave propagation, due to a subaerial landslide, at different times for the case of a 1D semi-infinite water basin (sea), an initial water surface uplift of $H_0 = 2\,m$, initial sea surface uplift length of $L_0 = 10\,m$, and with constant water depth $d = 10\,m$. This figure is generated using Code 4.2.

Code 4.2 *The MATLAB script used for plotting the solution of wave propagation for the case of a 1D semi-infinite water basin (sea), with an initial water surface uplift of $H_0 = 2\,m$, initial sea surface uplift length of $L_0 = 10\,m$ due to a subaerial landslide, and a constant water depth of $d = 10\,m$. This code generates Figure 4.9.*

```matlab
% Matlab code for making animations of wave
% propagation due to a tsunami in a semi-infinite 1D sea
clear; close all;
set(0,'DefaultAxesFontsize',20);
%
time=20; % total animation time - in seconds
L=200; % length of the sea - in meters
L0=10; % initial length of sea surface uplift - in meters
d = 10; % water depth - in meters
H0= 2; % initial height of sea surface uplift - in meters
C=sqrt(d.*9.81); % in m/s - wave celerity
dt=0.1; % time step - in seconds
dx=0.1; % spatial step - in meters
%
k=1; m=1;
for t=0:dt:time
    for x=0:dx:L
        lam=0.0000000001:0.01:10;
        cof=2.*H0.*sin(L0.*lam)./(pi.*lam);
        u=cof.*cos(lam.*x).*cos(C.*lam.*t);
        etta(m,k)=trapz(lam,u);
        k=k+1;
    end
%
clf;
subplot(2,1,1);
plot(0:dx:L,etta(m,:),'b','LineWidth',1.5);
xlabel('x (m)'); ylabel('etta (m)');
```

```
xlim([0 L]); ylim([-2 3]);
set(gca,'linewidth',1.5);
%
str = sprintf('%3.1f',t);
text(100,2.5,str,'FontSize',20);
text(110,2.5,'s','FontSize',20);
%
drawnow; refreshdata;
%
k=1; m=m+1;
end
% End of the code %
```

4.5 Wave run-up

Wave run-up is defined as the vertical distance between still water level (SWL) and the highest point along the beach or coastal structures that waves can reach (Heidarzadeh et al. 2020a). Run-up can lead to beach erosion, inundation, overtopping of coastal structures, and damage to coastal communities due to flooding and debris impacts (Figure 4.10). For instance, a wave run-up height of 3.3 m was recorded along the coast of Mumbai due to the 2004 Indian Ocean tsunami (Figure 4.10b) (Heidarzadeh et al. 2020a). Hurricane Maria in 2017 produced a run-up height of 1.8 m along the coast of Dominica (Figure 4.10c; Heidarzadeh et al. 2018).

Here, we apply the system of differential equations presented in Eq (4.22) for shallow-water waves to solve wave run-up along a sloping beach (Figure 4.11). The tsunami, which is generated by a submarine earthquake, propagates from the deep ocean towards the coast and runs up on the beach. In Figure 4.11, the water depth is constant in the first half of the computational domain, followed by variable water depth along the sloping beach. Therefore, it is challenging to solve this problem using the method of separation of variables. As a result, a numerical approach based on the finite difference method (FDM) is applied for solving the system of equations shown in Eq (4.22).

The FDM is based on discretizing the computational domain to finite differences (Δx) (Figure 4.12) and applying Taylor's theorem, which states that, for an n-times differentiable function $f(x)$, the Taylor series can be written as:

$$f(x + \Delta x) = f(x) + \frac{1}{1!}\frac{\partial f}{\partial x}(\Delta x)^1 + \frac{1}{2!}\frac{\partial^2 f}{\partial x^2}(\Delta x)^2 + \cdots + \frac{1}{n!}\frac{\partial^n f}{\partial x^n}(\Delta x)^n + \cdots \quad (4.59)$$

where $n!$ is the factorial of n. The first derivative of f can be approximated as:

$$f(x + \Delta x) \cong f(x) + \frac{1}{1!}\frac{\partial f}{\partial x}(\Delta x)^1 \quad (4.60)$$

(a)

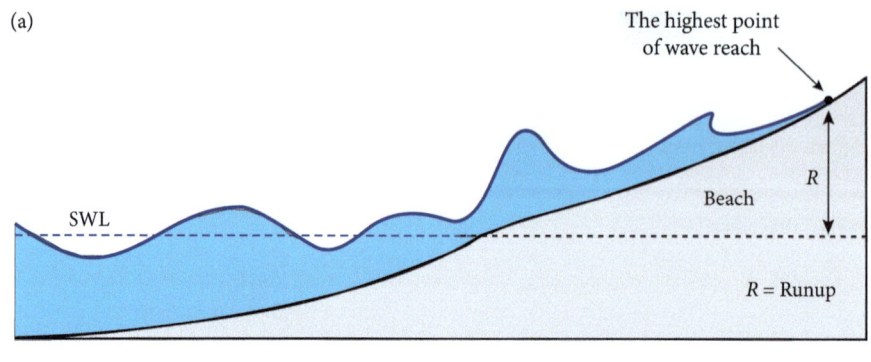

(b) Mumbai, India (2004 Indian Ocean tsunami)   (c) Dominica (2017 Hurricane Maria)

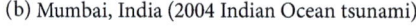

Figure 4.10 Definition of wave run-up (a) and two examples of wave run-ups along the coasts in Mumbai (India) during the December 2004 Indian Ocean tsunami (Heidarzadeh et al. 2020a) (b) and in Dominica during the September 2017 category 5 Hurricane Maria (Heidarzadeh et al. 2018) (c).

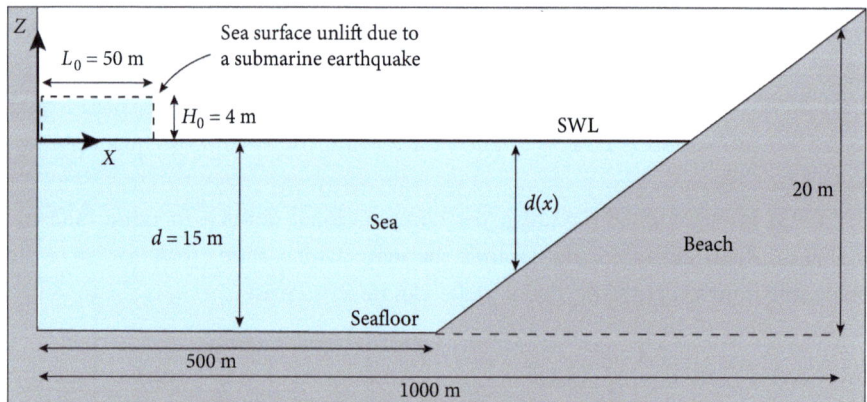

Figure 4.11 The boundary and initial conditions for modelling tsunami run-up on a sloping beach. Note that the figure is not to scale.

which can be rearranged as follows:

$$\frac{\partial f}{\partial x} = \frac{f(x + \Delta x) - f(x)}{\Delta x}$$

(4.61)

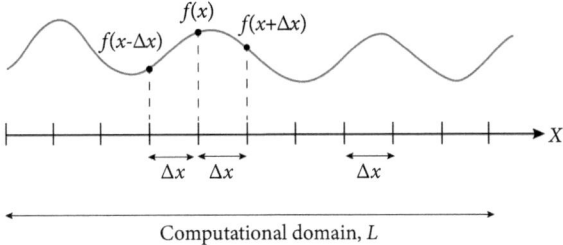

Figure 4.12 Discretizing a computational domain of length L to finite differences (Δx), which is the basis of the finite difference method.

Equation (4.61) is named the forward difference (see Figure 4.12).

The Taylor series in Eq (4.59) can be developed for $f(x - \Delta x)$ as follows:

$$f(x - \Delta x) = f(x) + \frac{1}{1!} \frac{\partial f}{\partial x}(-\Delta x)^1 + \frac{1}{2!} \frac{\partial^2 f}{\partial x^2}(-\Delta x)^2 + \ldots + \frac{1}{n!} \frac{\partial^n f}{\partial x^n}(-\Delta x)^n + \ldots$$

(4.62)

From this series, the first derivative of f can be approximated as follows:

$$\frac{\partial f}{\partial x} = \frac{f(x) - f(x - \Delta x)}{\Delta x}$$

(4.63)

which is known as backward difference (see Figure 4.12).

By combining the forward and backward differences, Eqs (4.61) and (4.63), the first derivative of f can alternatively be approximated using the following scheme:

$$\frac{\partial f}{\partial x} = \frac{f(x + \Delta x) - f(x - \Delta x)}{2\,\Delta x}$$

(4.64)

known as central difference (Figure 4.12).

Consequently, we can employ these approximations to discretize the governing equations of wave propagation, Eq (4.17). As the water depth (d) is not constant this time (see Figure 4.11), by ignoring the nonlinear terms the governing equations become:

$$\begin{cases} \dfrac{\partial (d + \eta)}{\partial t} + \dfrac{\partial (u\,h)}{\partial x} = 0 \\[2mm] \dfrac{\partial u}{\partial t} = -g\,\dfrac{\partial (d + \eta)}{\partial x} \end{cases}$$

(4.65)

The preceding equation can be simplified into:

$$\begin{cases} \dfrac{\partial \eta}{\partial t} + \dfrac{\partial (u\,h)}{\partial x} = 0 \\[2mm] \dfrac{\partial u}{\partial t} = -g\,\dfrac{\partial \eta}{\partial x} \end{cases}$$

(4.66)

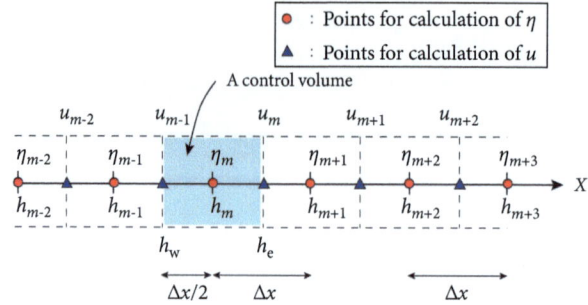

Figure 4.13 The staggered grid system used for velocity (u) and wave amplitudes (η) as they are calculated at different grid points spaced half a grid spacing, i.e., a distance of $\Delta x/2$, from each other.

To discretize Eq (4.66), it is common to use a staggered grid system (Press 2007; Kämpf 2009) where velocity (u) and wave amplitudes (η) are calculated at different grid points which are spaced half a grid spacing from each other; i.e., a distance of $\Delta x/2$ from each other (Figure 4.13).

For discretization of the differential equations, we use terms such as u_m^k and η_m^k where the subscript represents space (x) and the superscript k is for time (t). Applying a forward difference scheme, the bottom equation in the system of equations shown in Eq (4.66) can be discretized as follows:

$$\frac{u_m^{k+1} - u_m^k}{\Delta t} = -g\left(\frac{\eta_{m+1}^k - \eta_m^k}{\Delta x}\right) \tag{4.67}$$

Here, Δt is the time step, and Δx the grid spacing. From Eq (4.67), the velocity at the next time step (time $k + 1$) is given as:

$$u_m^{k+1} = u_m^k - \Delta t g\left(\frac{\eta_{m+1}^k - \eta_m^k}{\Delta x}\right) \tag{4.68}$$

For discretizing the other equation in the system of Eq (4.66), consider the control volume shown in Figure 4.13. Using a forward difference scheme, it can be written as:

$$\frac{\eta_m^{k+1} - \eta_m^k}{\Delta t} + \frac{(u_m^{k+1} h_e) - (u_{m-1}^{k+1} h_w)}{\Delta x} = 0 \tag{4.69}$$

Note the locations of the two parameters h_e and h_w in Figure 4.13. We use the upwind differencing scheme to calculate h_e and h_w (Press 2007; Kämpf 2009). According to the upwind scheme, the values of h_e and h_w depend on the flow directions as follows:

$$h_w = \begin{cases} h_{m-1}^k & u_{m-1}^{k+1} > 0 \\ h_m^k & u_{m-1}^{k+1} < 0 \end{cases} \tag{4.70}$$

and

$$
h_e =
\begin{cases}
h_m^k & u_m^{k+1} > 0 \\[2mm]
h_{m+1}^k & u_m^{k+1} < 0
\end{cases}
\tag{4.71}
$$

Using Eqs (4.70) and (4.71), we achieve:

$$
u_m^{k+1} \, h_e = 0.5 \left(u_m^{k+1} + \left| u_m^{k+1} \right| \right) h_m^k + 0.5 \left(u_m^{k+1} - \left| u_m^{k+1} \right| \right) h_{m+1}^k
\tag{4.72}
$$

and

$$
u_{m-1}^{k+1} \, h_w = 0.5 \left(u_{m-1}^{k+1} + \left| u_{m-1}^{k+1} \right| \right) h_{m-1}^k + 0.5 \left(u_{m-1}^{k+1} - \left| u_{m-1}^{k+1} \right| \right) h_m^k
\tag{4.73}
$$

Therefore, Eq (4.69) can be rewritten as follows:

$$
\frac{1}{\Delta t} \left(\eta_m^{k+1} - \eta_m^k \right) + \frac{1}{\Delta x} \left\{ \left[0.5 \left(u_m^{k+1} + \left| u_m^{k+1} \right| \right) h_m^k + 0.5 \left(u_m^{k+1} - \left| u_m^{k+1} \right| \right) h_{m+1}^k \right] \right.
$$
$$
\left. - \left[0.5 \left(u_{m-1}^{k+1} + \left| u_{m-1}^{k+1} \right| \right) h_{m-1}^k + 0.5 \left(u_{m-1}^{k+1} - \left| u_{m-1}^{k+1} \right| \right) h_m^k \right] \right\} = 0
\tag{4.74}
$$

From the previous equation, the updated wave amplitude $\left(\eta_m^{k+1} \right)$ is derived:

$$
\eta_m^{k+1} = \eta_m^k - \frac{\Delta t}{\Delta x} \left\{ \left[0.5 \left(u_m^{k+1} + \left| u_m^{k+1} \right| \right) h_m^k + 0.5 \left(u_m^{k+1} - \left| u_m^{k+1} \right| \right) h_{m+1}^k \right] \right.
$$
$$
\left. - \left[0.5 \left(u_{m-1}^{k+1} + \left| u_{m-1}^{k+1} \right| \right) h_{m-1}^k + 0.5 \left(u_{m-1}^{k+1} - \left| u_{m-1}^{k+1} \right| \right) h_m^k \right] \right\}
\tag{4.75}
$$

Up to this point, both the differential equations in Eq (4.66) have been discretized and can be applied in a numerical scheme to solve the wave propagation considering the initial and boundary conditions of the problem.

Numerical methods are sensitive to the choices of grid spacing (Δx) and time step (Δt), and there should be a certain relationship between Δx and Δt in order for the numerical scheme to converge to an acceptable solution. Such relationships are known as stability criteria. A popular stability criterion is that of Courant, Friedrichs, and Lewy (CFL; Courant et al. 1928):

$$
\Delta t \leq \frac{\Delta x}{\sqrt{g h_{max}}}
\tag{4.76}
$$

where h_{max} is the maximum total water depth in the computational domain, and g is gravitational acceleration.

The initial conditions are: the entire domain is at rest at the beginning, $t = 0$, $u_i^0 = 0$, and there is a positive water hump with height H_0 at a distance L_0 from the left side of the domain at the beginning (Figure 4.11). For boundary conditions, the left boundary is a vertical wall which fully returns the waves: $u_0^j = 0$.

To the right, there is a flooding boundary which is a moving boundary and allows the waves to flood the sloping beach. There are several methods for treating flooding boundaries in the literature. Here, we use the flooding boundary of Kämpf (2009), which is based on wet and dry cells. A wet cell is one having a total water depth $(h = d + \eta)$ of more than a certain small value $(h > h_{min})$, while a dry cell has $h < h_{min}$. Water depth values (d) are considered positive, while topography (also shown by parameter d) is marked with negative values. Wave amplitudes (η) include both sea level oscillations and land elevation. Knowing that the total water depth (h) is given by $h = d + \eta$, this will be zero for land areas because for a land area $d = -1 * \eta$ and thus $h = 0$ for land areas.

According to Kämpf's (2009) flooding algorithm, a dry cell will be flooded through calculating the speed at the boundary of wet and dry cells if the pressure-gradient force is directed towards the dry cell (see Code 4.3).

The following steps are taken to solve the wave propagation and run-up numerically:

- Discretize the domain and select the grid spacing (Δx).
- Calculate the time step (Δt) using the stability criteria, e.g., the CFL criterion shown in Eq (4.76).
- Give the initial and boundary conditions.
- Calculate the velocity (u) for the next time step for all grid points, Eq (4.68).
- Calculate wave amplitude (η) and total water depth (h) for the next time step for all grid points, Eq (4.75).
- Go to the next time step and repeat until the end of the simulation time.

The MATLAB code for a 1D wave run-up simulation is given in Code 4.3 for the run-up problem shown in Figure 4.11. The outcomes of simulations of wave run-up are presented in Figure 4.14 at different times.

Code 4.3 *The MATLAB script for numerical modelling of wave run-up on the sloping beach shown in Figure 4.11. The code is inspired by Kämpf (2009). This code generates Figure 4.14.*

```
% MATLAB code for modelling wave runup on a sloping beach
clear; close all; set(0,'DefaultAxesFontsize',15);
%%%%%%%%%%%%%%%%%%%%%%%%%%%%%%%%%%%%%%%%%%%%%%%
tic
lc=1000; % length of the channel
dx=10; % grid spacing
dt=0.25; % time increment-should be < dx/sqrt(g Hmax)
g=9.81; % acceleration due to gravity (9.81 m/s^2)
hmin=0.05; % minimum water height for a wet cell
t0=0.0; % start time of simulations
tend=5.*60; % final time of simulations in seconds
n=tend./dt; % total time steps
nx=103; % number of grid pints in x direction plus 2 for BCs
x0=1; % start of x axis for plotting the results
xend=103; % end of x axis for plotting the results
y0=-20; % start of y axis for plotting the results
yend=10; % end of y axis for plotting the results
```

```
%%%%%%%%%%%%%%%%%%%%%%%%%%%%%%%%%%%%%%%%%%
% bathymetry
for k=2:52
    x=(k-2).*dx;
    h0(k)=15;
end
for k=52:102
    x=(k-2).*dx;
    h0(k)=15-((2./50).*(x-500));
end
%%%%%%%%%%%%%%%%%%%%%%%%%%%%%%%%%%%%%%%%%%
% Boundary and Initial conditions
h0(1)=-20.0; % boundary = wall
h0(nx)=-20.0; % boundary = wall
eta(1:nx)=0.0; % at first the wave height field is zero
etan(1:nx)=0.0; % new water surface
for k=1:nx
    if(h0(k)<0.0), eta(k)=-1.*h0(k); end
    etan(k)=eta(k);
end
for k=1:nx
  h(k)=h0(k)+eta(k); % total water depth- (for dry land this is zero)
  wet(k)=1; % at first all grid are assummed to be wet
  if (h(k)<hmin),wet(k)=0, end; % cell is dry if total water level < hmin
  u(k)=0.0; % at first the velocity field is zero
  un(k)=0.0; % new velocity field is also zero
end
%%%%%%%%%%%%%%%%%%%%%%%%%%%%%%%%%%%%%%%%%%
% initial condition for tsunami
eta(2:7)=4.0;
%%%%%%%%%%%%%%%%%%%%%%%%%%%%%%%%%%%%%%%%%%
%   SIMULATION
%%%%%%%%%%%%%%%%%%%%%%%%%%%%%%%%%%%%%%%%%%
for it=1:n  % it is time incremnt- n is the number of time steps
% updating velocity in each time step (un)
    for k = 2:nx-1
        pgradx = -g*(eta(k+1)-eta(k))/dx;
        un(k) = 0.0;
        if(wet(k)==1)
            if(or(wet(k+1)==1,pgradx>0)), un(k) = u(k)+dt*pgradx,end
        else
            if(and(wet(k+1)==1,pgradx<0)), un(k) = u(k)+dt*pgradx,end
        end
    end
    % end of updating velocity field
    % updating wave hight field (etan)
    for k=2:nx-1
        hpe=0.5.*(un(k)+abs(un(k))).*h(k);
        hne=0.5.*(un(k)-abs(un(k))).*h(k+1);
        hue=hpe+hne;
        hpw=0.5.*(un(k-1)+abs(un(k-1))).*h(k-1);
        hnw=0.5.*(un(k-1)-abs(un(k-1))).*h(k);
        huw=hpw+hnw;
    etan(k)=eta(k)-dt.*(hue-huw)./dx;
    end
    % end of updating wave height field
    eta=etan;
    % updating variables for next iteration step
```

Code 4.3 *Continued*

```matlab
for k=1:nx
    h(k)=h0(k)+eta(k);
    wet(k)=1;
    if(h(k)<hmin), wet(k)=0;end
    u(k)=un(k);
end
% end of updating variables
% making animations
time=it.*dt;
subplot('position',[0.15 0.5 0.6 0.3]);
plot(eta(1:nx),'b','Linewidth',2.0); hold on;
plot(-h0(:),'k','LineWidth',3.0);
%
set(gca,'Xlim',[x0+1 xend-1]); set(gca,'Ylim',[y0 yend]);
set(gca,'Ytick',[y0:5:yend]); set(gca,'Xtick',[x0+1:10.0:xend-1]);
xlabels=[0 100 200 300 400 500 600 700 800 900 1000];
set(gca,'Xticklabel',xlabels); xlabel('Distance (m)');
ylabel('Wave amplitude (m)');
%
line([x0 xend],[0 0],'color',[0 0 0],'LineStyle','--');
str = sprintf('%3.1f', time);
text(30,-10,'Time =','FontSize',15); text(37,-10,str,'FontSize',15);
text(42.5,-10,'s','FontSize',15); set(gca,'linewidth',1.5);
%
refreshdata; drawnow; clf;
% end of animations
end
toc
% End of the code %
```

4.6 Wave spectra and Fourier analysis

For many engineering applications, it is helpful to convert the time series of ocean waves to the frequency domain for further analysis (Goda 2010; Figure 4.15). For example, the time series of the 21 May 2003 tsunami in the Mediterranean Sea recorded in Palma (Spain) and the corresponding spectrum are shown in Figure 4.15. Although the time series appears to be random and complicated (Figure 4.15b), the spectrum reveals that the tsunami is mainly made of a signal with a period of 21.7 min (or a frequency of 0.000768 Hz) (Figure 4.15c). This example may demonstrate that spectral analysis gives useful information about time series and makes them more accessible, understandable, and helpful.

Time series of real-world oceanic waves, $\eta(t)$, may appear to be complicated at first sight (Figure 4.15), but they can be decomposed into an infinite number of simple periodic sinusoidal or cosinusoidal waves with different frequencies f, or angular frequencies $\omega = 2\pi f$, by applying Fourier integrals (Kreyszig 2011):

$$\eta(t) = \int_{\omega=0}^{\infty} [A(\omega) \cos(\omega t) + B(\omega) \sin(\omega t)] \, d\omega \tag{4.77}$$

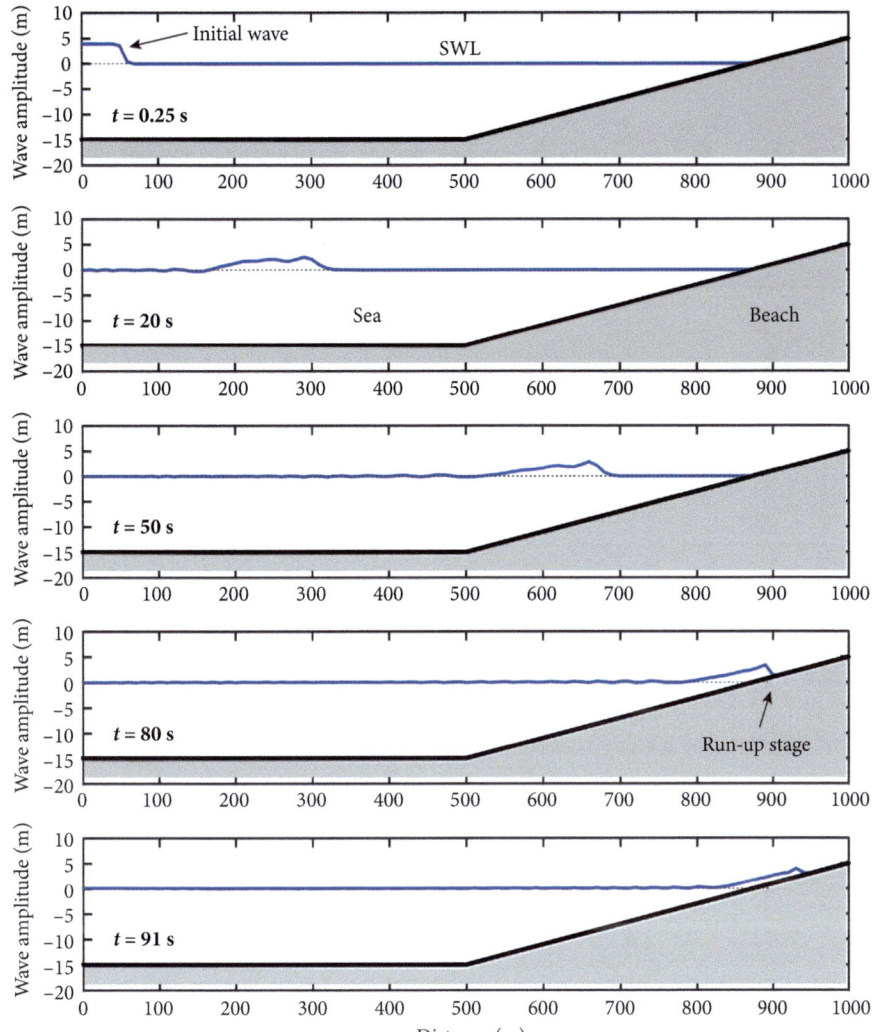

Figure 4.14 Snapshots of tsunami run-up simulation on a sloping beach at different times. The dimensions and other information for the problem are given in Figure 4.11. This figure is generated using Code 4.3.

where $A(\omega) = \frac{1}{\pi} \int_{\lambda=-\infty}^{\infty} \eta(\lambda) \cos(\omega\lambda) \, d\lambda$ and $B(\omega) = \frac{1}{\pi} \int_{\lambda=-\infty}^{\infty} \eta(\lambda) \sin(\omega\lambda) \, d\lambda$. By substituting these terms into Eq (4.77), we achieve the following:

$$\eta(t) = \frac{1}{\pi} \int_{\omega=0}^{\infty} \int_{\lambda=-\infty}^{\infty} \eta(\lambda) \left[\cos(\omega\lambda) \cos(\omega t) + \sin(\omega\lambda) \sin(\omega t)\right] d\lambda \, d\omega \qquad (4.78)$$

We understand that $\cos(\omega\lambda) \cos(\omega t) + \sin(\omega\lambda) \sin(\omega t) = \cos(\omega t - \omega\lambda)$. Therefore, we can rearrange the previous equation to the following form:

$$\eta(t) = \frac{1}{\pi} \int_{\omega=0}^{\infty} \left[\int_{\lambda=-\infty}^{\infty} \eta(\lambda) \cos(\omega t - \omega\lambda) \, d\lambda \right] d\omega \qquad (4.79)$$

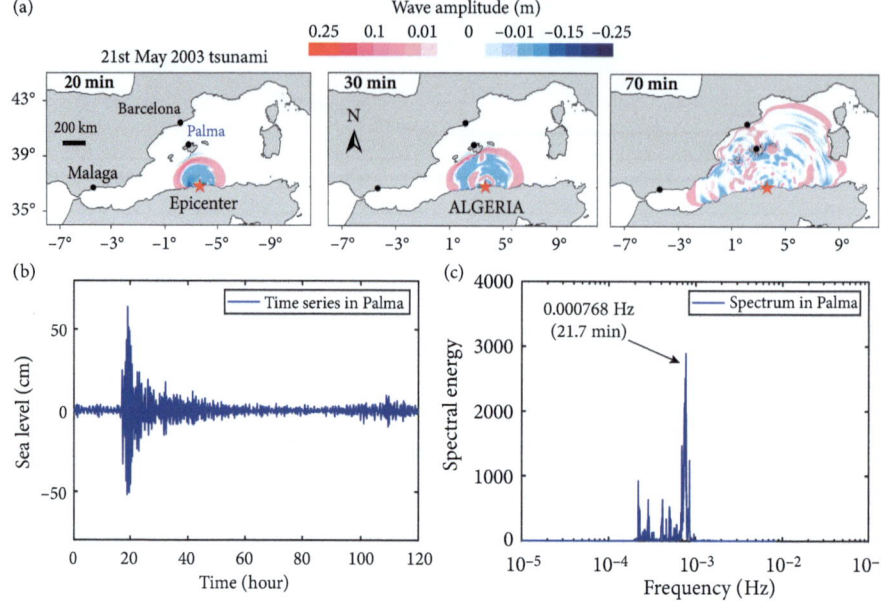

Figure 4.15 Snapshots of tsunami propagation at different times (a) and time series (b) of the 21 May 2003 tsunami in the Western Mediterranean Sea at Palma (Spain), due to an M6.8 earthquake offshore Algeria, along with the corresponding spectrum (c). This figure is modified from Heidarzadeh and Satake (2013).

By changing the domain of the first integral to $[-\infty, \infty]$, we achieve:

$$\eta(t) = \frac{1}{2\pi} \int_{\omega=-\infty}^{\infty} \left[\int_{\lambda=-\infty}^{\infty} \eta(\lambda) \cos(\omega t - \omega \lambda) \, d\lambda \right] d\omega \tag{4.80}$$

Because an integral of the form in Eq (4.80) but with a sinusoidal function instead of the cosinusoidal is zero, and taking into account that $e^{it} = \cos(t) + i \sin(t)$, where e is the base of the natural logarithm ($e \cong 2.7182818284$), and i is defined as $i = \sqrt{-1}$, Eq (4.80) can be rewritten as follows:

$$\eta(t) = \frac{1}{2\pi} \int_{\omega=-\infty}^{\infty} \left[\int_{\lambda=-\infty}^{\infty} \eta(\lambda) e^{i\omega(t-\lambda)} \, d\lambda \right] d\omega \tag{4.81}$$

By substituting angular frequency (ω) with oscillation frequency (f) using $\omega = 2\pi f$, we derive:

$$\eta(t) = \frac{1}{2\pi} \int_{f=-\infty}^{\infty} \left[\int_{\lambda=-\infty}^{\infty} \eta(\lambda) e^{i2\pi f(t-\lambda)} \, d\lambda \right] d(2\pi f) \tag{4.82}$$

This equation can be rearranged as follows:

$$\eta(t) = \int\limits_{f=-\infty}^{\infty} \left[\int\limits_{\lambda=-\infty}^{\infty} \eta(\lambda) e^{-i2\pi f\lambda} d\lambda \right] e^{i2\pi ft} df \tag{4.83}$$

Here, the Fourier transform of the function $\eta(t)$, which is shown as $F(f)$, is the term inside the brackets as follows (note that the parameter λ is replaced with t in the following):

$$F(f) = \int\limits_{t=-\infty}^{\infty} \eta(t) e^{-i2\pi ft} dt \tag{4.84}$$

The Fourier transform of a time function, $\eta(t)$, facilitates its transfer to the frequency domain, denoted as $F(f)$, where it decomposes the function into an infinite number of simple periodic waves represented in the form of $\cos(ft)$ and $\sin(ft)$.

As an example of applying a Fourier transform, we assume that the time series of the sea level during a storm is made of two sinusoidal waves with periods of 7 s (for short-period wind waves) and 13 s (for swell components). Therefore, the time series of the incident wave can be written as:

$$\eta(t) = \sin\left(\frac{2}{7}\pi t\right) + \sin\left(\frac{2}{13}\pi t\right) \tag{4.85}$$

To calculate the Fourier transform of this incident wave, Eq (4.84) is applied as follows:

$$F(f) = \int\limits_{t=-\infty}^{\infty} \left[\sin\left(\frac{2}{7}\pi t\right) + \sin\left(\frac{2}{13}\pi t\right) \right] e^{-i2\pi ft} dt \tag{4.86}$$

MATLAB can be used to solve the above integral using the code shown in Code 4.4 with the help of the integration function $trapz(x,y)$. The outcome is presented in Figure 4.16. Note that in Code 4.4, the amplitude of the Fourier transform is represented by $|F(f)|^2$, which is obtained by multiplying $F(f)$ by its conjugate value. According to Figure 4.16, although it is difficult to extract useful information from the time series, the Fourier transform clearly shows that two signals at frequencies of 0.077 Hz (a period of 13 s) and 0.143 Hz (a period of 7 s) are the dominant components of these sea level oscillations.

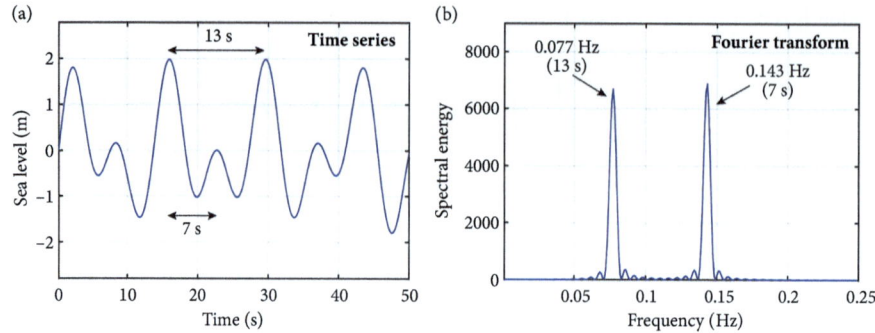

Figure 4.16 (a) The incident time series of oceanic waves made of two sinusoidal waves with periods of 7 s (for short-period wind waves) and 13 s (for swell components), and (b) the corresponding Fourier transform. This figure is created using MATLAB Code 4.4.

Code 4.4 *The MATLAB script to calculate the Fourier transform of the incident wave shown in Eq (4.84) and to plot it. This code generates Figure 4.16.*

```
%%% MATLAB code for calculating the Fourier Transform
%%% Incident oceanic wave is: eta(t)=sin(2*pi/7 t) + sin(2*pi/13 t)
clc; close all; clear;
set(0,'DefaultAxesFontsize', 16);
% Calculating Fourier Transform
k=1;
for f=0.0001:0.001:0.25
    lam=-80:0.001:80;
    eta=sin((2.*pi./7).*lam) + sin((2.*pi./13).*lam);
    y1=eta.*exp(-1i.*2.*pi.*f.*lam);
    z(k)=trapz(lam,y1);  % for integration
    %
    k=k+1;
end
%
% plotting the incident wave
t=0:0.01:50;
eta=sin((2.*pi./7).*t) + sin((2.*pi./13).*t);
%
figure(1);
subplot('position',[0.1 0.3 0.3 0.43]);
plot(t,eta,'b','linewidth',1.5);
set(gca,'xlim',[0 50],'ylim',[-2.8 2.8]);
set(gca,'linewidth',1.5); grid on;
%
% plotting the Fourier Transform
f=0.0001:0.001:0.25;
subplot('position',[0.48 0.3 0.3 0.43]);
plot(f,z.*conj(z),'b','linewidth',1.5);
set(gca,'xlim',[0.001 0.25],'ylim',[0 9000]);
set(gca,'linewidth',1.5); grid on;
%
%%% End of the code %%%
```

Time (min)	Sea level (cm)	Time (min)	Sea level (cm)	Time (min)	Sea Level (cm)	Time (min)	Sea level (cm)	Time (min)	Sea level (cm)
1	1.4	21	−3.8	41	−11.9	61	17.1	81	−5.3
2	1.7	22	24.6	42	−11.4	62	20.8	82	−1.6
3	2.0	23	−203.0	43	−94.9	63	33.4	83	3.2
4	2.3	24	−172.6	44	−74.4	64	7.0	84	−10.0
5	2.6	25	176.8	45	50.1	65	−33.3	85	0.7
6	1.9	26	−10.7	46	24.6	66	−12.7	86	13.5
7	3.2	27	4.7	47	−111.9	67	63.9	87	0.3
8	2.6	28	−8.9	48	−84.3	68	43.6	88	−21.9
9	2.9	29	−30.5	49	52.2	69	−25.7	89	1.8
10	3.2	30	−159.0	50	−1.2	70	−9.1	90	13.6
11	3.6	31	145.4	51	−59.7	71	70.6	91	−20.5
12	2.9	32	132.8	52	−50.1	72	41.3	92	−34.7
13	3.3	33	−39.7	53	25.4	73	−25.0	93	−4.9
14	3.6	34	−121.3	54	6.0	74	−8.3	94	8.9
15	3.0	35	−6.8	55	−46.4	75	60.4	95	−9.3
16	3.4	36	0.7	56	−17.8	76	23.1	96	−18.4
17	2.7	37	−120.9	57	38.7	77	−20.2	97	7.4
18	3.1	38	−60.4	58	14.3	78	−9.5	98	9.3
19	3.5	39	−36.9	59	0.92	79	17.2	99	3.1
20	10.9	40	−91.4	60	−2.5	80	−4.0	100	2.0

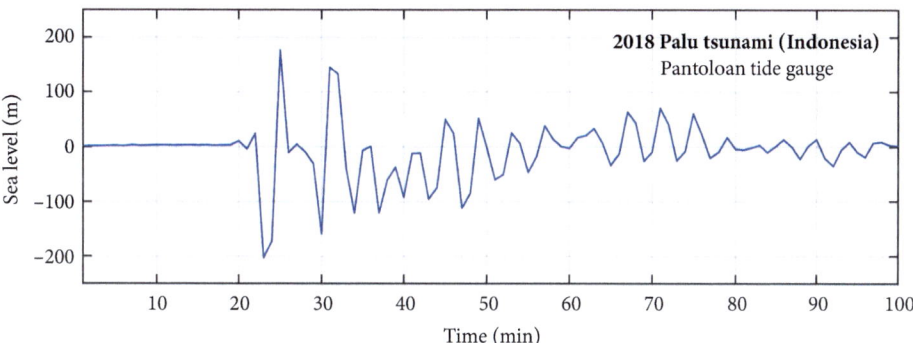

Figure 4.17 The actual time series data of the September 2018 Palu (Sulawesi) tsunami in Indonesia at the tide gauge station Pantoloan, and the wave plot (modified from Heidarzadeh et al. 2019). The table data file, named 'Chapter-4-input-Palu.txt,' can be downloaded at: http://www.oup.com/AdvancedMathematicalModelling.

Real records of oceanic waves are made of discrete time series at certain time intervals such as every 15 min (for tides), every 1 min (for storms and tsunamis), or every 0.1 s (for wind-generated waves). These types of time series data usually come in two columns where the first column gives time in seconds and the second column is for reporting the wave amplitudes in centimetres or metres. An example of such data and the corresponding wave plot are shown in Figure 4.17 for the record of the September 2018 Palu (Sulawesi) tsunami in Indonesia at the tide gauge station named Pantoloan (Heidarzadeh et al. 2019).

There are some algorithms for conducting Fourier transforms for discrete time series of ocean waves. Among them is the fast Fourier transform (FFT) method embedded in the MATLAB package, which is done using the command *FFT* in MATLAB. Code 4.5 performs an FFT on the records of the 2018 Palu (Indonesia) tsunami from the Pantoloan tide gauge station (Figure 4.18). To perform this FTT analysis, the data presented in the table at the top of Figure 4.17 needs to be

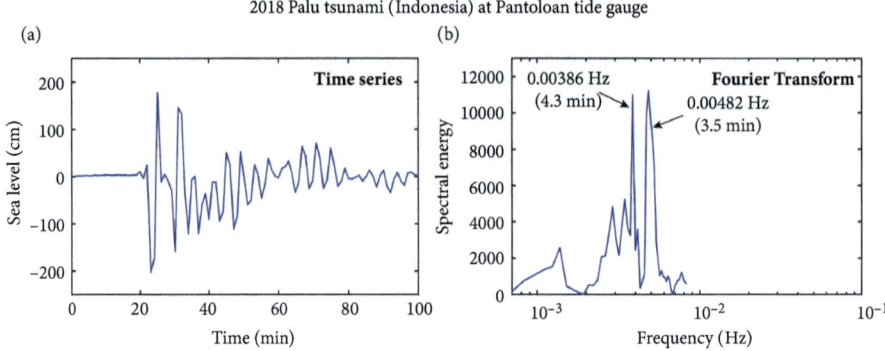

Figure 4.18 The Fourier transform for the time series of the September 2018 Palu (Sulawesi) tsunami in Indonesia at the tide gauge station Pantoloan (modified from Heidarzadeh et al. 2019). This figure is created using the FFT code presented in Code 4.5.

put in a text file called 'Chapter-4-input-Palu.txt' and should be used together with the MATLAB code (Code 4.5) in the same folder. This input data and the MAT-LAB code can be downloaded from the following website: http://www.oup.com/AdvancedMathematicalModelling.

The Fourier transform for the 2018 Palu tsunami reveals that most of the tsunami energy was concentrated in the period band 3.5–4.3 min (Figure 4.18). Obviously, the wave spectrum (the Fourier transform) involves several signals, but they are weak and do not carry much energy. Although the tsunami time series (Figure 4.18a) looks complicated and includes several signals, the Fourier transform is able to identify the most powerful signals, which belong to the period band 3.5–4.3 min (Figure 4.18b).

Code 4.5 *The MATLAB script used to conduct a fast Fourier transform on the records of the 2018 Palu (Indonesia) tsunami at the Pantoloan tide gauge station. To conduct this FFT analysis, the table of data displayed in Figure 4.17 (in a text file named 'Chapter-4-input-Palu.txt') should be placed in the same folder as this MATLAB code. This input data and the MATLAB code can be downloaded from the following website: http://www.oup.com/AdvancedMathematicalModelling. This code generates Figure 4.18.*

```
%%% MATLAB code to calculate Fourier Transform
%%% using FFT (fast Fourier Transform) algorithm
% Input wave is the 2018 Palu tsunami, ndonesia
clc; clear; close all;
set(0,'DefaultAxesFontsize',16); % fontsize
y=load('Chapter-4-input-Palu.txt'); % read input file
% input time srries is in two columns; time and amplitudes
dt=60; % time interval of input file in seconds
nx=length(y(:,2)); % number of input values
z=fft(y(:,2)); % applyng FFT algotithm
%
```

```
%%%%%%%%%%%%%%%%%%%%%%%%%%%%%%%
% plot the time series
figure (1)
subplot('position',[0.1 0.3 0.3 0.4]);
plot(y(:,1),y(:,2),'b','LineWidth',1.5);
xlabel('Time (min)');        % label on the X axis
ylabel('Sea level (cm)');  % label on the Y axis
set(gca,'ylim',[-250 250]);
set(gca,'xlim',[0 100]);
set(gca,'linewidth',1.5);
%%%%%%%%%%%%%%%%%%%%%%%%%%%%%%%
% plot the Fourier Transform
subplot('position',[0.48 0.3 0.3 0.4]);
semilogx((1:nx/2)./(dt.*nx),z(1:nx/2).*conj(z(1:nx/2))/nx,...
    'b','LineWidth',1.5);
xlabel('Frequency (Hz)');
ylabel('Spectral energy');
set(gca,'xlim',[10^(-4)*7 10^(-1)]);
set(gca,'XTick',[10^(-3) 10^(-2) 10^(-1)]);
set(gca,'ylim',[0 13000]);
set(gca,'linewidth',1.5);
%%% End of the code %%%
```

4.7 Problems for further study

Problem 4.1

Tsunamis generated by underwater volcanoes are often characterized by an initial Gaussian-shaped water hump, which can be approximated as a triangular sea surface uplift (Figure 4.19) (e.g., Heidarzadeh et al. 2020c). Solve the 1D propagation of the waves generated by the volcano eruption shown in Figure 4.19 in an enclosed sea with a length of 1000 m, assuming shallow-water wave conditions. Assume a constant water depth of d for the sea. Other relevant information is provided in Figure 4.19.

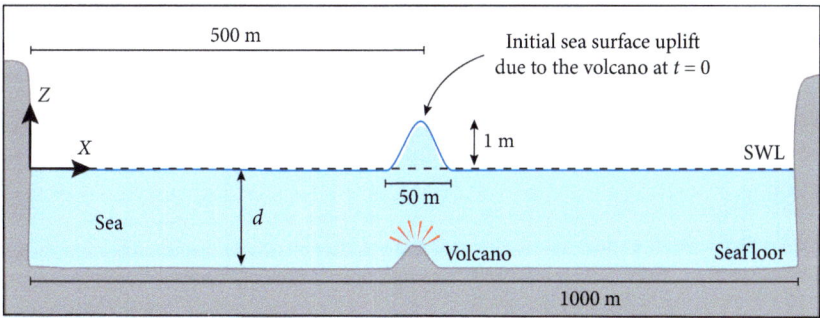

Figure 4.19 Boundary and initial conditions for a tsunami generated by a submarine volcano in an enclosed sea of length of 1000 m and an initial triangular sea surface uplift height of 1 m with a base length of 50 m.

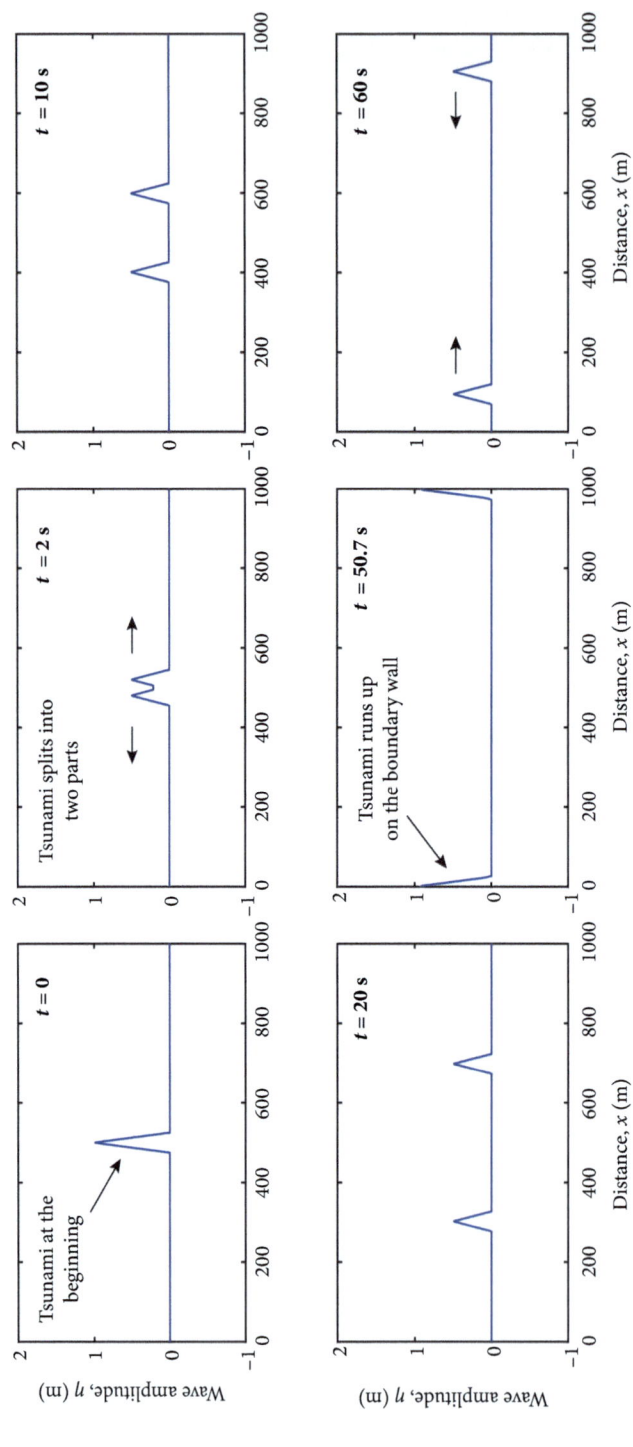

Figure 4.20 Snapshots of tsunami wave propagation due to a submarine volcano in an enclosed sea of length of 1000 m and an initial triangular sea surface uplift height of 1 m with a base length of 50 m. A constant water depth of $d = 10\,m$ is considered for the sea. This solution is made using Code 4.6.

Answer to Problem 4.1

Considering Figure 4.19, the initial sea surface uplift is determined by the following equation:

$$\eta(x,0) = \begin{cases} 0 & 0 \le x < 475 \\ \dfrac{x}{25} - 19 & 475 \le x < 500 \\ \dfrac{x}{25} + 21 & 500 \le x < 525 \\ 0 & 525 \le x \le 1000 \end{cases} \tag{4.87}$$

The wave propagation solution is presented below:

$$\eta(x,t) = \frac{25}{1000} + \sum_{n=1}^{n=\infty} \frac{80}{(n\pi)^2} \left[2\cos\left(\frac{n\pi}{2}\right) - \cos\left(\frac{19\,n\pi}{40}\right) \right.$$
$$\left. - \cos\left(\frac{21\,n\pi}{40}\right) \right] \cos\left(\frac{n\pi}{1000}x\right) \cos\left(\frac{n\pi C}{1000}t\right) \tag{4.88}$$

The wave propagation solution is shown in Figure 4.20 for a water depth of $d = 10\,m$. The MATLAB code for plotting the solution is given in Code 4.6.

Code 4.6 *The MATLAB script used for plotting the wave propagation solution for the case of a 1D enclosed water basin (sea), initial triangular water surface uplift of 1 m with base length of 50 m, due to a submarine volcano in an enclosed sea with constant water depth $d = 10\,m$ (see Figure 4.19). This code generates Figure 4.20.*

```
% Matlab code for making animations of wave
% propagation due to a volcanic tsunami in an enclosed 1D sea
% with length of 1000 m. Initial sea surface is like a triangle with
% height of 1 m and base of 50 m, located in the midddle of sdea.
clear; close all; clc;
set(0,'DefaultAxesFontsize',15);
%
d = 10; % water depth - in meters
C=sqrt(9.81.*d); % water celerity - in m/s
%
k=1; m=1;
for t=0:0.5:200
%t=100;
for x=0:0.5:1000
    s=0.0;
    for n=1:500
        cof=(80./(n.*pi).^2).*(2.*cos(n.*pi./2)-cos(19.*n.*pi./40)-...
            cos(21.*n.*pi./40));
        s=s+(cof).*cos(n.*pi.*x./1000).*cos(n.*pi.*t.*C./1000);
    end
    f(m,k)=s+25./1000;
    k=k+1;
end
```

Code 4.6 *Continued*

```
%
clf;
subplot(2,1,1);
plot(0:0.5:1000,f(m,:),'b','LineWidth',1.5);
xlabel('Distance (m)'); ylabel('Wave amplitude (m)');
xlim([0 1000]); ylim([-0.5 1.5]);
set(gca,'XTick',[0:100:1000]);
set(gca,'linewidth',1.5);
%
text(120,1.0,'t =','FontSize',18);
str = sprintf('%3.1f', t);
text(150,1.0,str,'FontSize',18);
text(200,1.0,'s','FontSize',18);
%
drawnow; refreshdata;
k=1; m=m+1;
end
% End of the code %
```

Problem 4.2

For the two-slope beach shown in Figure 4.21, calculate the wave run-up by applying 1D shallow-water equations and assuming an initial wave generated by seafloor uplift from a submarine earthquake as shown in Figure 4.21.

Problem 4.3

The time series of the 16 June 2021 tsunami offshore Seram Island (Indonesia) from the Tehoru tide gauge station is shown in Figure 4.22, along with the data. This data file is named 'Chapter-4-data-Tehoru-2021-tsu.txt' and is available at the following website: http://www.oup.com/AdvancedMathematicalModelling. Apply the FFT algorithm (Code 4.5) to generate the tsunami spectrum for this tsunami.

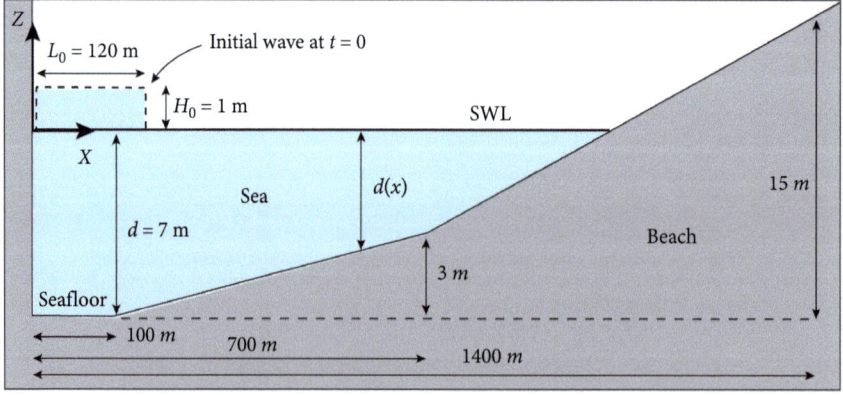

Figure 4.21 A two-slope beach under wave run-up attack due to an initial wave generated by seafloor uplift from a submarine earthquake.

Time (min)	Sea level (cm)	Time (min)	Sea level (cm)	Time (min)	Sea level (cm)	Time (min)	Sea level (cm)	Time (min)	Sea level (cm)
1	−1.7	21	31.0	41	−46.3	61	−6.2	81	30.7
2	−2.1	22	−42.6	42	5.6	62	24.1	82	−33.5
3	−1.3	23	−18.3	43	50.6	63	−18.4	83	12.2
4	−1.6	24	45.0	44	−25.5	64	−16.0	84	36.0
5	−1.8	25	−36.7	45	6.5	65	28.5	85	−49.3
6	−1.0	26	−16.6	46	35.4	66	−14.9	86	−1.5
7	−0.1	27	44.6	47	−28.7	67	−19.3	87	36.2
8	−0.1	28	−41.4	48	2.2	68	33.3	88	−35.0
9	−0.2	29	−40.4	49	32.1	69	−0.1	89	−5.3
10	0.9	30	45.4	50	−16.9	70	−25.5	90	32.4
11	1.0	31	0.2	51	−23.9	71	21.2	91	−24.9
12	2.1	32	−45.1	52	5.1	72	21.9	92	0.8
13	1.3	33	13.6	53	6.1	73	−28.4	93	24.5
14	1.6	34	24.1	54	−0.8	74	6.3	94	−23.8
15	1.9	35	−30.5	55	1.3	75	20.1	95	0.9
16	4.2	36	3.9	56	−0.6	76	−30.2	96	25.5
17	4.5	37	22.2	57	−8.4	77	11.6	97	−23.8
18	3.9	38	−29.6	58	4.8	78	22.4	98	−7.1
19	9.3	39	9.5	59	10.1	79	−35.9	99	21.6
20	20.7	40	31.6	60	−22.6	80	3.9	100	−12.7

June 2021 tsunami offshore Seram Island (Indonesia)

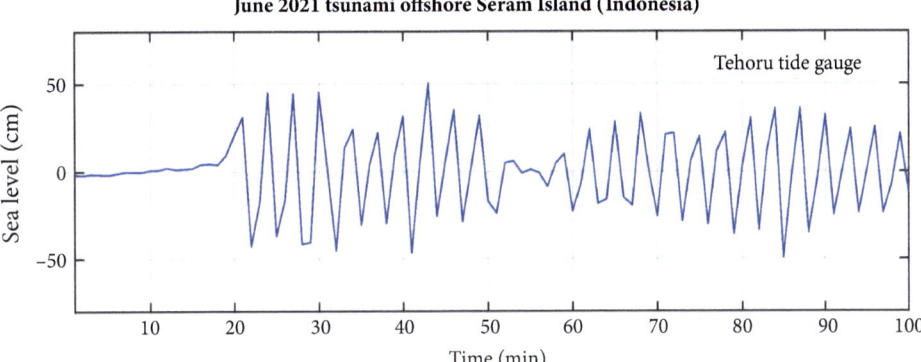

Figure 4.22 Time series of the 16 June 2021 tsunami offshore Seram Island (Indonesia) at the Tehoru tide gauge station. Data is provided from Heidarzadeh et al. (2022). This data file is named 'Chapter-4-data-Tehoru-2021-tsu.txt' and is available at the following website: http://www.oup.com/AdvancedMathematicalModelling.

References

Courant, R., Friedrichs, K., and Lewy, H. (1928). Über die partiellen Differenzengleichungen der mathematischen Physik. *Mathematische Annalen*, 100(1), 32–74.

Goda, Y. (2010). *Random Seas and Design of Maritime Structures* (*Advanced Series on Ocean Engineering*, Vol. 33). World Scientific Publishing Company.

Heidarzadeh, M., Gusman, A.R., and Mulia, I.E. (2023). The landslide source of the eastern Mediterranean tsunami on 6 February 2023 following the Mw 7.8 Kahramanmaraş (Türkiye) inland earthquake. *Geoscience Letters*, 10, 50. https://doi.org/10.1186/s40562-023-00304-8

Heidarzadeh, M., Gusman, A.R., Patria, A., and Widyantoro, B.T. (2022). Potential landslide origin of the Seram Island tsunami in Eastern Indonesia on 16 June 2021 following an Mw 5.9 earthquake. *Bulletin of the Seismological Society of America*, 112(5), 2487–2498. https://doi.org/10.1785/0120210274

Heidarzadeh, M., Rabinovich, A.B., Kusumoto, S., and Rajendran, C.P. (2020a). Field surveys and numerical modeling of the 26 December 2004 Indian Ocean tsunami in the area of Mumbai, west coast of India. *Geophysical Journal International*, 222(3), 1952–1964. https://doi.org/10.1093/gji/ggaa277

Heidarzadeh, M., Putra, P.S., Nugroho, H.S., and Rashid, D.B.Z. (2020b). Field survey of tsunami heights and runups following the 22 December 2018 Anak Krakatau volcano tsunami, Indonesia. *Pure and Applied Geophysics*, 177, 4577–4595. https://doi.org/10.1007/s00024-020-02587-w

Heidarzadeh, M., Ishibe, T., Sandanbata, O., Muhari, A., and Wijanarto, A.B. (2020c). Numerical modeling of the subaerial landslide source of the 22 December 2018 Anak Krakatoa volcanic tsunami, Indonesia. *Ocean Engineering*, 195, 106733. https://doi.org/10.1016/j.oceaneng.2019.106733

Heidarzadeh, M., Muhari, A., and Wijanarto, A.B. (2019). Insights on the source of the 28 September 2018 Sulawesi tsunami, Indonesia based on spectral analyses and numerical simulations. *Pure and Applied Geophysics*, 176, 25–43. https://doi.org/10.1007/s00024-018-2065-9

Heidarzadeh, M., Teeuw, R., Day, S., and Solana, C. (2018). Storm wave runups and sea level variations for the September 2017 Hurricane Maria along the coast of Dominica, eastern Caribbean Sea: Evidence from field surveys and sea level data analysis. *Coastal Engineering Journal*, 60(3), 371–384. https://doi.org/10.1080/21664250.2018.1546269

Heidarzadeh, M., Harada, T., Satake, K., Ishibe, T., and Takagawa, T. (2017). Tsunamis from strike-slip earthquakes in the Wharton Basin, northeast Indian Ocean: March 2016 Mw 7.8 event and its relationship with the April 2012 Mw 8.6 event. *Geophysical Journal International*, 47(3), 1601–1612. https://doi.org/10.1093/gji/ggx395

Heidarzadeh, M., Krastel, S., and Yalciner, A.C. (2014). The state-of-the-art numerical tools for modeling landslide tsunamis: A short review. In: *Submarine Mass Movements and Their Consequences*, S. Krastel et al. (eds), Chapter 43, 483–495, Springer International Publishing. https://doi.org/10.1007/978-3-319-00972-8_43

Heidarzadeh, M., and Satake, K. (2013). Waveform and spectral analyses of the 2011 Japan tsunami records on tide gauge and DART stations across the Pacific Ocean. *Pure and Applied Geophysics*, 170(6), 1275–1293. https://doi.org/10.1007/s00024-012-0558-5

Kämpf, J. (2009). *Ocean Modelling for Beginners: Using Open-Source Software*. Springer Science+Business Media.

Kreyszig, E. (2011). *Advanced Engineering Mathematics*. John Wiley & Sons.

Press, W.H. (2007). *Numerical Recipes: The Art of Scientific Computing*, 3rd edn. Cambridge University Press.

Sorensen, R.M. (2006). *Basic Coastal Engineering*. Springer Science+Business Media.

Chapter 5
Water and Environmental Engineering

5.1 Introduction

Water and environmental engineering is one of the key subdisciplines of civil engineering, and mainly addresses the quantity and quality of water systems. There are a number of urgent issues to be addressed in water and environmental engineering such as streamflow predictions, hydrological risk analyses, and pollution transport in river channels. Figure 5.1 presents a typical river system, in which the design of levees and the bridge is mainly subject to the flood magnitude with certain return periods.

There have been a number of approaches for streamflow predictions, which are mainly classified into physics-based and data-driven approaches. Physics-based approaches are developed relying on inherent physical laws, such as the principle of mass balance in hydraulic and hydrological processes. Moreover, due to the rapid development of computer techniques, data-driven approaches have been proposed in the last two decades. The data-driven approaches, especially machine learning (ML) techniques, are proposed to map the relationships between inputs (e.g., temperature, precipitation, evapotranspiration, etc.) and outputs (e.g., river flows) without considering the physical processes between these variables. The ML approaches include regression-based methods, artificial neural networks, support vector machine, and deep learning techniques.

Hydrological risk analysis is one of the most important issues to be considered in water resource management and hydrological designs. The risk analysis for hydrological extreme events is to identify the magnitude of an extreme event (e.g., floods) with a specific return period (e.g., 100 years). Statistical analysis is adopted to describe the probabilistic features

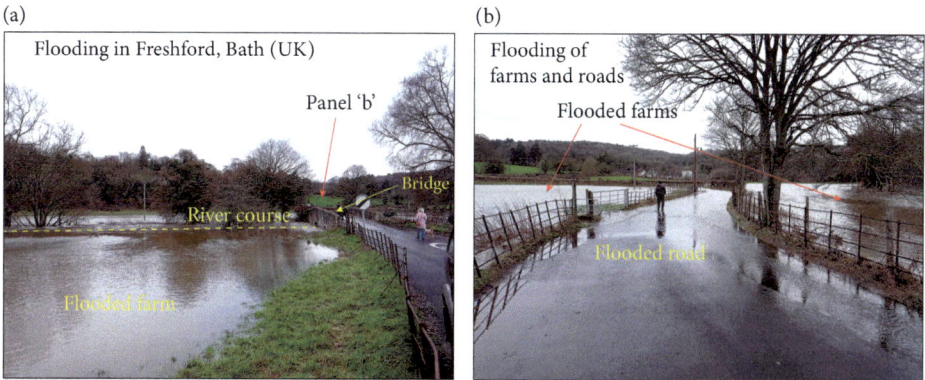

Figure 5.1 A typical river system with a bridge in Freshford, Bath (UK). These two photos show the area after a flood in January 2023 where the adjacent farms and roads are flooded.

A Practical Approach to Advanced Mathematical Modelling in Civil Engineering. Mohammad Heidarzadeh et al.,
Oxford University Press. © Mohammad Heidarzadeh, Theodosios K. Papathanasiou, Yurui Fan, Hamid Bahai (2025).
DOI: 10.1093/9780191888656.003.0005

of historical extreme events and then to derive the event magnitude under a predefined return period. However, many hydrological extremes are correlated with each other, such as heavy rainfall and flood peaks, as well as flood peaks and flood volumes. Consequently, univariate hydrological risk analysis, which focuses on a single variable, often fails to offer a comprehensive description of hydrological extreme events. As a result, advanced approaches have emerged over the past decade to address risk analysis for hydrological extremes within a multivariate context.

The water quality in a river or lake concerns different stakeholders, from water managers to local residents. Good quality water is essential for the well-being of the human population, as well as for our environment. The deterioration of water quality can lead to problems such as unsafe water supplies and aquatic environment deterioration. There are numerous pollutants with various physical, chemical, and/or biological properties. Generally, these pollutants can be classified into two categories: nonconservative (nonpersistent) and conservative (persistent). Nonconservative pollutants have the capacity to undergo changes facilitated by specific biological or chemical processes upon contact with water. These alterations may lead to the formation of other substances through reactions or mutations (Benedini and Tsakiris 2013). In contrast, the pollutants that do not undergo similar alterations are named conservative or persistent (Benedini and Tsakiris 2013). Mathematical models have been formulated specifically to simulate the transport of pollutants within river or lake systems. These models integrate a combination of analytical and numerical methodologies to effectively solve the complex mathematical equations governing pollutant dispersion.

This chapter addresses the three aspects described above in water and environmental engineering. Basic machine learning techniques including regression methods and artificial neural networks are introduced for streamflow predictions. The basic concepts of hydrological risk analysis are subsequently illustrated within both the univariate and multivariate contexts. Pollutant transport as well as its modelling process and solution methods are also described.

5.2 Hydrological predictions through machine learning techniques

5.2.1 Machine learning techniques in water and environmental engineering

Machine learning (ML) techniques have been widely used in water and environmental engineering, in which they can capture the mapping relationships between input and output variables without considering the physical laws underneath these variables. One typical application of ML techniques concerns predictions of streamflow based on some meteorological inputs such as rainfall, evapotranspiration, and temperature. Many ML techniques have been applied for forecasting and simulating streamflows. The statistical regression technique is a more traditional ML method routinely applied in hydrological predictions, which mainly consists of multiple linear regression (Sadri and Burn 2012; Adamowski and Karapataki 2008; Sachindra et al. 2013) and nonlinear regression approaches (Adamowski et al. 2012).

Moreover, due to the advancement of computer science, novel ML approaches have been developed for hydrological predictions, such as generalized regression neural networks (Cigizoglu 2005), artificial neural networks (Turan and Yurdusev 2009; Wang et al. 2009), support vector machine (Sujay Raghavendra and Deka 2014), and least squares support vector machines (Okkan and Serbes 2012). In addition, there are also nonparametric statistical approaches for hydrological predictions. Among them, the stepwise cluster analysis (SCA) method has been widely used for many hydrological applications such as streamflow predictions (Fan et al. 2015) and climate downscaling (Wang et al. 2013). The main advantage of SCA is that the inherent relationship between the explanatory and response variables is reflected through cluster trees, which are derived through cutting or merging the sample sets of response variables into new sets based on given criteria. These cluster trees, instead of specific mathematical functions, can establish the relationship between explanatory variables and response variables.

5.2.2 Regression-based methods

Given a training dataset comprising N observations $\{x_n\}$, where $n = 1, 2, 3, ..., N$, together with corresponding target values $\{y_n\}$, the goal is to predict the value of y for a new value of x. In the simplest approach, this can be done by directly constructing an appropriate function $y(x)$ whose values for new inputs x constitutes the predictions for the corresponding values of y as follows:

$$y = a_0 + a_1 x_1 + a_2 x_2 + ... + a_d x_d \tag{5.1}$$

where y is the output to be forecast from Eq (5.1), $x_1, x_2, ..., x_d$ are the inputs to be employed, and $a_0, a_1, ..., a_d$ are the unknown parameters to be estimated based on real observations (either historical or contemporary). In the problem of hydrological prediction, y can denote the streamflow at a gauge station, while x_i ($i = 1, 2, ..., d$) are some hydrological and climatological variables such as past streamflow, rainfall, and evapotranspiration.

One of the key steps for the establishment of the regression model is to estimate the unknown parameters, namely a_i ($i = 0, 1, 2, ..., d$), in Eq (5.1). Different approaches have been developed to do this task, including both deterministic and probabilistic estimation methods. The least squares method is one of the standard approaches in regression analysis. It estimates the unknown parameters by minimizing the sum of the squares of the residuals (defined as the difference between predictions and observations). Consider the simplest linear regression function with only one input variable:

$$y = a_0 + a_1 x + \varepsilon \tag{5.2}$$

where ε in Eq (5.2) is the random error with $E(\varepsilon) = 0$ and $V(\varepsilon) = \sigma^2$. If there are n observations available for x and y, namely $(x_1, y_1), (x_2, y_2), ..., (x_n, y_n)$, these observations are adopted to estimate the parameters a_0 and a_1 in Eq (5.2). For any input x_i, the corresponding prediction from the regression model, $\hat{y}_i$, is

$$\hat{y}_i = a_0 + a_1 x_i \tag{5.3}$$

The residual between the real observation (i.e., y_i) and the model prediction (i.e., $\hat{y}_i$) can be obtained as follows:

$$\varepsilon_i = y_i - \hat{y}_i = y_i - (a_0 + a_1 x_i) \tag{5.4}$$

For n observations, the sum of square errors (denoted as S^2) can be calculated as

$$S^2 = \sum_{i=1}^{n} \varepsilon_i^2 = \sum_{i=1}^{n} \left[y_i - (a_0 + a_1 x_i) \right]^2 \tag{5.5}$$

The least squares method aims to find the values for a_0 and a_1 resulting in S^2 being minimized. Note that S^2 is a function of a_0 and a_1 and its minimum value should be located at its stationary points, which leads to its partial derivatives being 0:

$$\frac{\partial S^2}{\partial a_0} = 0, \qquad \frac{\partial S^2}{\partial a_1} = 0 \tag{5.6}$$

Equation (5.6) leads to a linear equation system:

$$\begin{cases} a_0 n + a_1 \sum_{i=1}^{n} x_i = \sum_{i=1}^{n} y_i \\ a_0 \sum_{i=1}^{n} x_i + a_1 \sum_{i=1}^{n} x_i^2 = \sum_{i=1}^{n} x_i y_i \end{cases} \tag{5.7}$$

By solving the system shown in Eq (5.7), we can obtain the parameter values of a_0 and a_1 as follows:

$$a_1 = \frac{n \sum_{i=1}^{n} x_i y_i - \left(\sum_{i=1}^{n} x_i \right) \left(\sum_{i=1}^{n} y_i \right)}{n \sum_{i=1}^{n} x_i^2 - \left(\sum_{i=1}^{n} x_i \right)^2} \tag{5.8a}$$

$$a_0 = \frac{\left(\sum_{i=1}^{n} x_i^2 \right) \left(\sum_{i=1}^{n} y_i \right) - \left(\sum_{i=1}^{n} x_i \right) \left(\sum_{i=1}^{n} x_i y_i \right)}{n \sum_{i=1}^{n} x_i^2 - \left(\sum_{i=1}^{n} x_i \right)^2} \tag{5.8b}$$

Equations (5.3)–(5.8) provide the estimation process for the parameters of a linear regression function with only one input variable. Such a process can also be extended to estimate the unknown parameters for linear regression functions with multiple input variables as presented in Eq (5.1).

5.2.3 Artificial neural networks

Artificial neural networks (ANNs) are developed based on the structure of human brains. They have been utilized for various complex tasks including pattern recognition, clustering, classification, and simulation, demonstrating their ability to map intricate functions (Araghinejad 2014). In the last three decades, ANNs have been widely used in water and environmental engineering. Typical applications of ANN can be found in contexts such as rainfall–runoff modelling, streamflow forecasting, and groundwater modelling.

5.2.3.1 Components of an ANN

Neuron
The neuron is the basic unit of a neural network (Figure 5.2); it takes inputs (x_1 and x_2) with some mathematical operation (f) and produces one output (y).

Consider a neuron with two inputs as illustrated in Figure 5.2. There are three operations happening in the neuron shown in Figure 5.2, as described by Zhou (2019):

- Each input is multiplied by its weight (i.e., w_1 for x_1 and w_2 for x_2), resulting in the following terms:

$$x_1 \rightarrow w_1 \times x_1 \quad \text{and} \quad x_2 \rightarrow w_2 \times x_2 \tag{5.9a}$$

- The weighted terms in Eq (5.9a) are added together with a bias b:

$$w_1 \times x_1 + w_2 \times x_2 + b \tag{5.9b}$$

- The sum presented in Eq (5.9b) is passed through an activation function (f in Figure 5.2) to generate the output of the neuron:

$$y = f(w_1 \times x_1 + w_2 \times x_2 + b) \tag{5.9c}$$

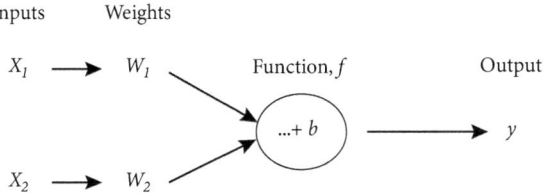

Figure 5.2 Sketch showing an artificial neuron and its main elements. This figure is inspired from Zhou (2019). Here, b represents the bias.

Table 5.1 Examples of activation functions.

Name	Function	Derivative
Linear	$f(x) = x$	$f'(x) = 1$
Symmetric saturating linear	$f(x) = \begin{cases} \delta & x \geq \theta \\ x & -\theta \leq x \leq \theta \\ -\delta & x \leq -\theta \end{cases}$	$f'(x) = \begin{cases} 0 & x \geq \theta \\ 1 & -\theta \leq x \leq \theta \\ 0 & x \leq -\theta \end{cases}$
Log sigmoid	$f(x) = \dfrac{1}{1 + e^{-\alpha x}}, \quad \alpha > 0$	$f'(x) = \alpha f(x)(1 - f(x))$
Tangent sigmoid	$f(x) = \dfrac{2}{1 + e^{-2x}} - 1$	$f'(x) = 1 - f(x)^2$
Radial basis	$f(x) = e^{-\frac{x^2}{\delta^2}}$	$f'(x) = -\dfrac{2x}{\delta^2} e^{-\frac{x^2}{\delta^2}}$

Activation function

As presented in Eq (5.9c), an activation function is adopted to generate the output of a neuron. The activation function defines how the weighted sum of the input is transformed into an output from a node or nodes in a layer of the network. A commonly used activation function is the sigmoid function,

$$f(x) = \frac{1}{1 + e^{-x}} \tag{5.10}$$

A sigmoid function has a characteristic S-shaped curve that can map the entire number line, i.e., $x \in (-\infty, \infty)$, into a small range such as between 0 and 1, or -1 and 1. Other activation functions are also available, as presented in Table 5.1.

Establishment of a neural network

A single neuron is insufficient for solving many practical problems, and thus a network of neurons is frequently used in parallel or series; this is called a neural network. A simple neural network can be formulated as shown in Figure 5.3. This neural network has two inputs, denoted x_1 and x_2, a hidden layer with two neurons, denoted h_1 and h_2, and an output layer with one neuron (i.e., y).

The weights and biases have also been defined in Figure 5.3. Then, the mapping between the inputs (i.e., x_1, x_2) and the output (i.e., y) can be expressed as:

$$I_1 = x_1 w_{11} + x_2 w_{21} + b_1 \tag{5.11a}$$

$$I_2 = x_1 w_{12} + x_2 w_{22} + b_2 \tag{5.11b}$$

$$h_1 = f_h (I_1) \tag{5.11c}$$

$$h_2 = f_h (I_2) \tag{5.11d}$$

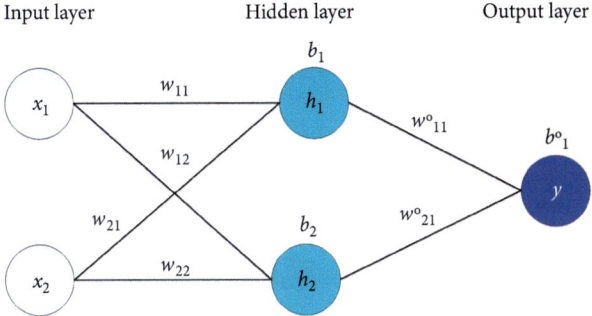

Figure 5.3 A typical neural network consisting of two neurons in the input layer (x_1 and x_2), one hidden layer with two neurons (h_1 and h_2), and one neuron in the output layer (y). Note that various weights (w_{ij}) and biases (b_i) are applied between layers and neurons. This figure is inspired by Zhou (2019).

$$I_1^o = h_1 w_{11}^o + h_2 w_{21}^o + b_1^o \tag{5.11e}$$

$$y = f_o\left(I_1^o\right) \tag{5.11f}$$

where $f_h\,(\cdot)$ and $f_o\,(\cdot)$ are the activation functions to be used in the hidden and output layers, respectively, w_{ij} are various weights between the layers and neurons, and b_i are biases. Equation (5.11) can also be expressed in vector format as follows:

$$I_{2\times1} = IW_{2\times2}^T \times X_{2\times1} + B_{2\times1} \tag{5.12a}$$

$$I_{1\times1}^o = \left[IW_{2\times1}^o\right]^T \times f_h(I)_{2\times1} + B_{1\times1}^o \tag{5.12b}$$

$$y = f_o(I^o)_{1\times1} \tag{5.12c}$$

where

$$IW = \begin{bmatrix} w_{11} & w_{12} \\ w_{21} & w_{22} \end{bmatrix}, \quad B = \begin{bmatrix} b_1 \\ b_2 \end{bmatrix}, \quad IW^o = \begin{bmatrix} w_{11}^o \\ w_{21}^o \end{bmatrix}, \quad B^o = b_1^o$$

The hidden layer in a neural network is any layer between the input (first) layer and the output (last) layer. A neural network can have multiple hidden layers. Moreover, a neural network can have any number of neurons in the hidden layers, but the basic idea stays the same: feed the input(s) forward through the neurons in the network to get the output(s) at the end (Zhou 2019).

5.2.3.2 Training algorithms for ANNs

The weights and biases are the parameters of a network that should be estimated (or determined) before using an ANN for predictions. As presented in Eq (5.11), there are many unknown parameters to be estimated. For a three-layer neural network with N_I nodes for the input layers, N_H nodes for the hidden layer, and N_o nodes for the output layer, there would be a total of $(N_I + 1) \times N_H + (N_H + 1) \times N_o$ parameters to be estimated. For the neural network described in Figure 5.3, there are a total of 9 unknown parameters (i.e., $(2+1) \times 2 + (2+1) \times 1$) to be determined before using the model for predictions. The weight and bias matrices of an ANN could be obtained by either supervised or unsupervised approaches. Supervised approaches employ a training process to estimate the weights and biases of a network based on available observations or datasets. Unsupervised learning approaches determine the weights and biases of a network without relying on labelled data.

The training process for a neural network is to determine the weights and biases to make the network fit well the mapping relationship in the datasets used for training. Loss functions are commonly used to quantify how well the network models the relationship embedded in the training datasets. Different loss functions are available, with the mean squared error (MSE) being one of the commonly used functions:

$$E = \frac{1}{n} \sum_{s=1}^{n} \left(y_{true,s} - y_{pred,s}\right)^2 \tag{5.13}$$

where E is the MSE, n is the number of observations, $y_{pred,s}$ represents the predictions obtained from the neural network, and $y_{true,s}$ represents the corresponding true observations. $e_s = \left(y_{true,s} - y_{pred,s}\right)^2$ is known as the squared error. Based on the loss function, defined in Eq (5.13), better predictions from the neural network will lead to lower values for the loss function. Consequently, the neural network training process is to update the weights and biases in the network to minimize the loss function values. For the loss function expressed by Eq (5.13), if $y_{pred,s}$ is obtained from the neural network with the structure presented in Figure 5.3, the loss function can be expressed as a multivariate function subject to the weights and biases of the network:

$$E = L\left(w_{11}, w_{12}, w_{21}, w_{22}, w_{11}^o, w_{21}^o, b_1, b_2, b_1^o\right) \tag{5.14}$$

For parameters of this equation, refer to Figure 5.3. How the loss function will change with a change of any parameter (e.g., w_{11}) can be determined by the partial derivative $\partial E/\partial w_{11}$. Since $E = \frac{1}{n} \sum_{s=1}^{n} \left(y_{true,s} - y_{pred,s}\right)^2$, the partial derivative $\partial E/\partial w_{11}$ can be decomposed based on the chain rule (for simplicity, we assume $n = 1$ here):

$$\frac{\partial E}{\partial w_{11}} = \frac{\partial E}{\partial y_{pred,s}} \times \frac{\partial y_{pred,s}}{\partial w_{11}} \tag{5.15a}$$

where

$$\frac{\partial E}{\partial y_{pred}} = -2\left(y_{true,s} - y_{pred,s}\right) \tag{5.15b}$$

$y_{pred,s}$ can be obtained through Eq (5.11). Therefore, using Eq (5.11), the partial derivative $\frac{\partial y_{pred,s}}{\partial w_{11}}$ can be further decomposed based on the chain rule again:

$$\frac{\partial y_{pred,s}}{\partial w_{11}} = \frac{\partial y_{pred,s}}{\partial h_1} \times \frac{\partial h_1}{\partial w_{11}} \tag{5.15c}$$

Based on Eqs (5.11e) and (5.11f), we can have

$$\frac{\partial y_{pred,s}}{\partial h_1} = w_{11}^o f_o' \left(I_1^o \right)$$
(5.15d)

Similarly, through Eqs (5.11a) and (5.11c), we can obtain $\frac{\partial h_1}{\partial w_{11}}$ as follows:

$$\frac{\partial h_1}{\partial w_{11}} = x_1 f_h' \left(I_1 \right)$$
(5.15e)

Consequently, the partial derivative of E with respect to w_{11} can be obtained as

$$\frac{\partial E}{\partial w_{11}} = \frac{\partial E}{\partial y_{pred,s}} \times \frac{\partial y_{pred,s}}{\partial h_1} \times \frac{\partial h_1}{\partial w_{11}} = \left[2 \left(y_{true,s} - y_{pred,s} \right) \right] \times \left[w_{11}^o f_o' \left(I_1^o \right) \right] \times \left[x_1 f_h' \left(I_1 \right) \right]$$
(5.15f)

We know that, if $\partial E / \partial w_{11} > 0$, the loss function E will increase with an increase in w_{11}; otherwise, the loss function E will decrease with an increase in w_{11} when $\partial E / \partial w_{11} < 0$.

Based on the partial derivative of the loss function with respect to the weights and biases of the network, a gradient descent method has been developed to train the neural network. This is expressed as

$$w_{ij}^{new(l)} = w_{ij}^{old(l)} + \eta \left(-\frac{\partial e_s}{\partial w_{ij}^{(l)}} \right)$$
(5.16)

where $w_{ij}^{(l)}$ is the weight or bias that links the ith neuron in the $(l - 1)$th layer to the jth neuron in the lth layer. $e_s = \left(y_{true,s} - y_{pred,s} \right)^2$ is the square error for the sth sample. η is the learning rate selected between 0 and 1.

Following Araghinejad (2014), the partial derivative with respect to a particular weight $w_{ij}^{(l)}$ can be obtained through the chain rule as follows:

$$\frac{\partial e_s}{\partial w_{ij}^{(l)}} = \frac{\partial e_s}{\partial I_{sj}^{(l)}} \frac{\partial I_{sj}^{(l)}}{\partial w_{ij}^{(l)}}$$
(5.17a)

where $I_{sj}^{(l)}$ are the inputs to the jth neuron in the lth layers from the sth sample, and

$$\frac{\partial I_{sj}^{(l)}}{\partial w_{ij}^{(l)}} = \frac{\partial}{\partial w_{ij}^{(l)}} \left(\sum_k w_{kj}^{(l)} y_{sk}^{(l-1)} \right) = y_{si}^{(l-1)}$$
(5.17b)

The term $\frac{\partial e_s}{\partial I_{sj}^{(l)}}$ is defined as $-\delta_{sj}$; therefore, we can have

$$-\frac{\partial e_s}{\partial w_{ij}^{(l)}} = \delta_{sj}^{(l)} y_{si}^{(l-1)}$$
(5.17c)

and thus Eq (5.16) can be reformulated as

$$w_{ij}^{new(l)} = w_{ij}^{old(l)} + \eta \delta_{sj}^{(l)} y_{si}^{(l-1)} \tag{5.17d}$$

Based on the chain rule, the term $\delta_{sj}^{(l)}$ can be obtained as

$$\delta_{sj}^{(l)} = -\frac{\partial e_s}{\partial I_{sj}^{(l)}} = -\frac{\partial e_s}{\partial y_{sj}^{(l)}} \frac{\partial y_{sj}^{(l)}}{\partial I_{sj}^{(l)}} = -\frac{\partial e_s}{\partial y_{sj}^{(l)}} f_j'\left(I_{sj}^{(l)}\right) \tag{5.17e}$$

where $y_{sj}^{(l)}$ is the output of the jth neuron in the lth layer, and f_j is the activation function used in the jth neuron. The chain rule is adopted again to get the derivative of $\frac{\partial e_s}{\partial y_{sj}^{(l)}}$ as follows:

$$\frac{\partial e_s}{\partial y_{sj}^{(l)}} = \sum_k \frac{\partial e_s}{\partial I_{sk}^{(l+1)}} \frac{\partial I_{sk}^{(l+1)}}{\partial y_{sj}^{(l)}} = -\sum_k \delta_{sj}^{(l+1)} w_{jk}^{(l+1)} \tag{5.17f}$$

Thus,

$$\delta_{sj}^{(l)} = \sum_k \delta_{sj}^{(l+1)} w_{jk}^{(l+1)} f_j'\left(I_{sj}^{(l)}\right) \tag{5.17g}$$

If l is the last layer,

$$\frac{\partial e_s}{\partial y_{sj}^{(l)}} = -2\left(y_{true,s} - y_{pred,s}\right) \tag{5.17h}$$

and therefore

$$\delta_{sj}^{(l)} = 2\left(y_{true,s} - y_{pred,s}\right) f_j'\left(I_{sj}^{(l)}\right) \tag{5.17i}$$

Based on Eqs (5.16) and (5.17), a backpropagation (BP) algorithm is developed in which network weights are moved along the negative of the gradient of the performance function through each iteration (which is usually called an epoch) in the steepest descent direction. The detailed procedures for the BP algorithm are as follows, based on Araghinejad (2014):

Step 1: The weights and biases are initialized as w_{ij}^{old}.
Step 2: The output signals are generated by applying the input vectors

$$y_{sj}^{(l)} = f_j\left[\sum w_{ij}^{(l)} y_{si}^{(l-1)}\right]$$

where $y_{si}^{(l-1)} = x_{si}$ for $l = 1$.

Step 3: $\delta_{sj}^{(l)}$ is firstly determined for the output layer expressed by Eq (5.17i).

Step 4: The partial derivative term $\delta_{sj}^{(l)}$ is then backpropagated through the output layer through Eq (5.17g).

Step 5: The weights (including biases) are updated based on the results of Steps 3 and 4:

$$w_{ij}^{new} = w_{ij}^{old} + \eta\delta_{sj}y_{si}^{(l-1)} \text{ for the last layer}$$

$$w_{ij}^{new(l)} = w_{ij}^{old(l)} + \eta\delta_{sj}^{(l)}y_{si}^{(l-1)} \text{ for the hidden layers}$$

5.2.4 Applications to streamflow prediction

5.2.4.1 Overview of the study area

The streamflow record at the Kingston river-flow gauging station in the catchment of the River Thames (UK) is adopted to illustrate the applicability of machine learning techniques in hydrological predictions. This gauging station is situated on the River Thames, about 33 feet (approximately 10 m) above sea level and 10 miles (approximately 16 km) southwest of Charing Cross (deemed the geographical centre of London). At this location, the river Thames is draining an upstream catchment area of 9948 km^2, as shown in Figure 5.4. The drainage area is mainly covered by arable/horticultural land (35.62%) and grassland (32.17%), followed by woodland (16.09%) and urban extent (14.04%). The precipitation and streamflow records are obtained from the National River Flow Archive (https://nrfa.ceh.ac.uk/data/station/spatial/39001), while the potential evapotranspiration for the studied catchment is collected from Tanguy et al. (2017).

5.2.4.2 Data correlation

The daily flow records in 2015 are employed to demonstrate the applicability of ML in hydrological predictions. Figure 5.5 presents the observations of rainfall (P), potential evapotranspiration (PET), and streamflow (Q) for the catchment in 2015. The streamflow at time t (i.e., Q_t) may be determined by a number of variables such as the past flow rate (i.e., Q_{t-1}), the current and past rainfall rates (i.e., P_t, P_{t-1}), and the current and past rates for potential evapotranspiration (i.e., PET_t, PET_{t-1}). Table 5.2 presents the correlation matrix illustrating the relationship between the inputs and output, based on correlation coefficients. The MATLAB script for producing Figure 5.5 and Table 5.2 is presented in Code 5.1. This MATLAB code, and other MATLAB scripts presented in this chapter, can be downloaded from the following website: http://www.oup.com/AdvancedMathematicalModelling.

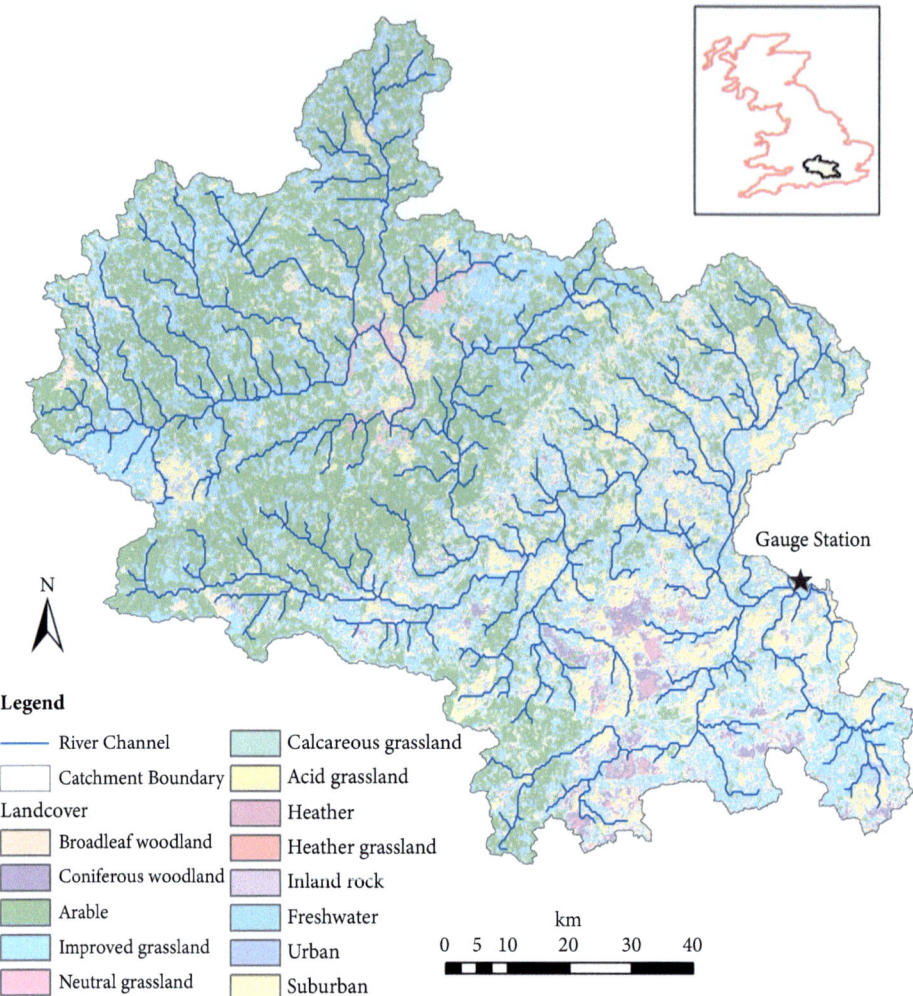

Figure 5.4 Catchment boundary of the River Thames. Reproduced from Gabriel and Fan (2022). Creative Commons Attribution (CC BY) license (https://creativecommons.org/licenses/by/4.0/).

Code 5.1 *The MATLAB script used for data visualization and correlation analysis. Note that to run the code, a data file named 'Chapter-5.2-Data.txt' should be located in the same folder as this code. The input file and this MATLAB code (and other MATLAB scripts from this chapter) can be downloaded from the following website: http://www.oup.com/AdvancedMathematicalModelling.*

```
clc; clear all; close all;
set(0,'defaultaxesfontsize', 16);
% to run the code, please firstly import the data file
% names as "Chapter-5.2-Data.txt"
rawdata = load('Chapter-5.2-Data.txt');
% colnames = Year,Month,Day,Precpitation,Streamflow,Pontential
% Evaporation
```

```matlab
RawP = rawdata(:,4);
RawQ = rawdata(:,5);
RawPet = rawdata(:,6);
n = length(RawP);
figure; subplot(3,1,1)
plot(RawP,'k-','linewidth',1.25); ylim([0 25]);
xlabel('Day'); ylabel('Rainfall (mm/d)')
title('(a) Rainfall'); set(gca,'linewidth', 1.5);
%-------------------------------------
subplot(3,1,2)
plot(RawPet,'b--','linewidth',1.25);
xlabel('Day'); ylabel('PET (mm/d)'); set(gca,'linewidth', 1.5);
title('(b) Potential Evapotranspiration'); ylim([0 3]);
%-------------------------------------
subplot(3,1,3);
plot(RawQ,'mo','markersize',5,'linewidth',1.25)
xlabel('Day'); ylabel('Streamflow (m^3/s)')
title('(c) Streamflow'); set(gca,'linewidth', 1.5);
ylim([0 300]);
%%==================Correlation Matrix==================
Pt = RawP(2:n);
Pt_1 = RawP(1:n-1);
Qt_1 = RawQ(1:n-1);
Pett = RawPet(2:n);
Pett_1=RawPet(1:n-1);
%Establish the input matrix
InputMatrix = [Qt_1,Pt,Pt_1,Pett,Pett_1];
% Output variable Qt
Qt = RawQ(2:n);
AllDataSets = [InputMatrix, Qt];
coefmatrix = corrcoef(AllDataSets);
display('Correlation Coefficients are:');display(coefmatrix);
```

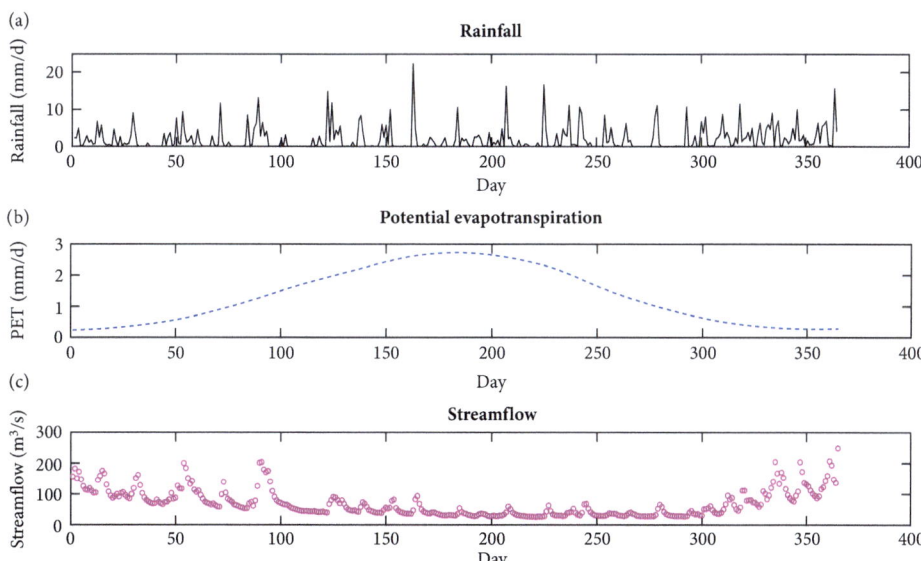

Figure 5.5 Observations of rainfall, potential evapotranspiration, and streamflow for the catchment of the River Thames (UK) in 2015. This figure is generated using Code 5.1.

Table 5.2 Correlation matrix between input and output variables, known as correlation coefficients, for the numerical example presented in Section 5.2.3.1. Note that this is a symmetric matrix. This table is generated using Code 5.1.

	Q_{t-1}	P_t	P_{t-1}	PET_t	PET_{t-1}	Q_t
Q_{t-1}	1.00	0.03	0.09	−0.60	−0.60	0.92
P_t		1.00	0.13	−0.06	−0.06	0.09
P_{t-1}			1.00	−0.06	−0.06	0.33
PET_t				1.00	1.00	−0.60
PET_{t-1}					1.00	−0.60
Q_t						1.00

5.2.4.3 Model evaluation

The performance or accuracy for any prediction model can be evaluated by different indices. The root mean square error (RMSE) and Nash–Sutcliffe efficiency (NSE) coefficient are two commonly used indices for evaluating the accuracy of hydrological predictions. RMSE is the standard deviation of prediction errors, exhibiting how far from the regression line data points are; it is a measure of how spread out these residuals are. In other words, it tells you how concentrated the data is around the line of best fit. The lower the RMSE, the better the model performance. The RMSE can be formulated as follows:

$$RMSE = \sqrt{\frac{1}{N} \sum_{i=1}^{N} (Q_{obs,i} - Q_{pred,i})^2} \qquad (5.18)$$

where $Q_{obs,i}$ is the observed value and $Q_{pred,i}$ is the predicted one.

The NSE is defined as one minus the sum of the absolute squared differences between the predicted ($Q_{pred,i}$) and observed ($Q_{obs,i}$) values normalized by the variance of the observed values during the period under investigation (Parajuli 2007). It is calculated as follows:

$$NSE = 1 - \frac{\sum_{i=1}^{N} (Q_{obs,i} - Q_{pred,i})^2}{\sum_{i=1}^{N} (Q_{obs,i} - \overline{Q})^2} \qquad (5.19)$$

where $\overline{Q}$ is the average value of the observations.

The NSE is used to assess the predictive power of hydrological models. An efficiency of one ($NSE = 1$) corresponds to a perfect match of forecasted (or predicted) data to the observed data. An efficiency of zero ($NSE = 0$) indicates a total mismatch between the model predictions and observations. The MATLAB functions for calculating RMSE and NSE are presented in Code 5.2 and Code 5.3, respectively. Note that Code 5.2 and

Code 5.3 are MATLAB functions. They therefore do not run independently, but need to be called by other MATLAB code, as shown later in this chapter.

Code 5.2 *The MATLAB function for RMSE. Note that this MATLAB code is a function. It therefore does not run independently, but needs to be called by other MATLAB code. Save this code as 'RMSE.m' on your computer. This MATLAB code (and other MATLAB code from this chapter) can be downloaded from the following website: http://www.oup.com/ AdvancedMathematicalModelling.*

```
%%--------------------RMSE ----------------------------
function [rmse] = RMSE( Qobs,Qpre)
%root mean square error
% Qobs: the real observations
% Qpre: the model predictions
n = length(Qobs);
Qmean = mean(Qobs);
E1 = 0;
for i = 1:n
  E1 = E1+(Qobs(i)-Qpre(i))^2;
end
rmse = sqrt((1/n)*E1);
end
%%% End.
```

Code 5.3 *The MATLAB function for NSE. Note that this MATLAB code is a function. It therefore does not run independently, but needs to be called by other MATLAB code. Save this code as 'NSE.m' on your computer.*

```
%%---------------------------NSE----------------------------
function [E ] = NSE(Qobs,Qpre )
%Nash-Sutcliffe efficiency coefficient (NSE)
n = length(Qobs);
Qmean = mean(Qobs);
E1 = 0;
E2 = 0;
for i = 1:n
  E1 = E1+(Qobs(i)-Qpre(i))^2;
  E2 = E2+(Qobs(i)-Qmean)^2;
end
E = 1-E1/E2;
end
```

5.2.4.4 Results from multivariate linear regression and ANN methods

Figures 5.6 and 5.7 present comparisons between predictions and observations in the training and testing periods for the multivariate linear regression (multilayer perceptron, or MLP) and ANN models. Here, the training allows the process to estimate unknown parameters (e.g., weights and biases in ANN) based on known datasets. After the training

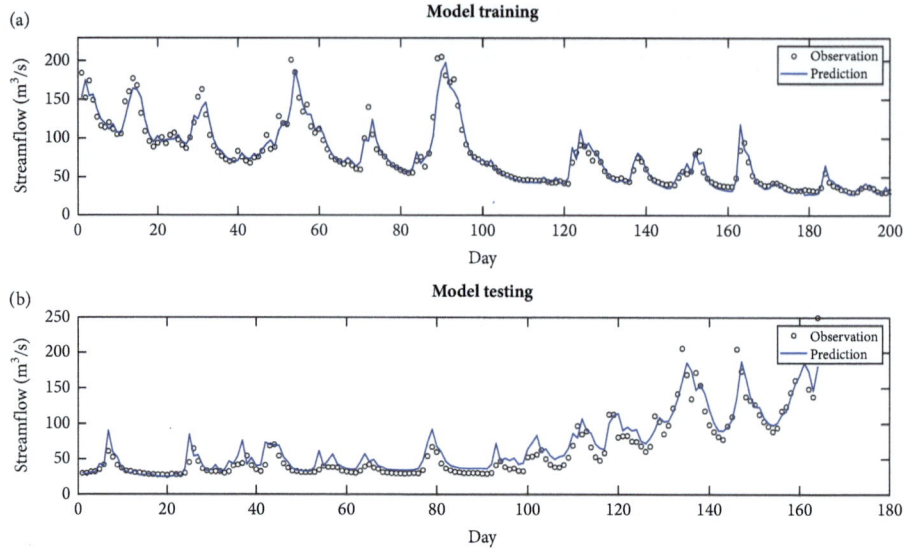

Figure 5.6 Performance of the linear regression model in the training and testing periods. This figure is produced using the MATLAB script in Code 5.4.

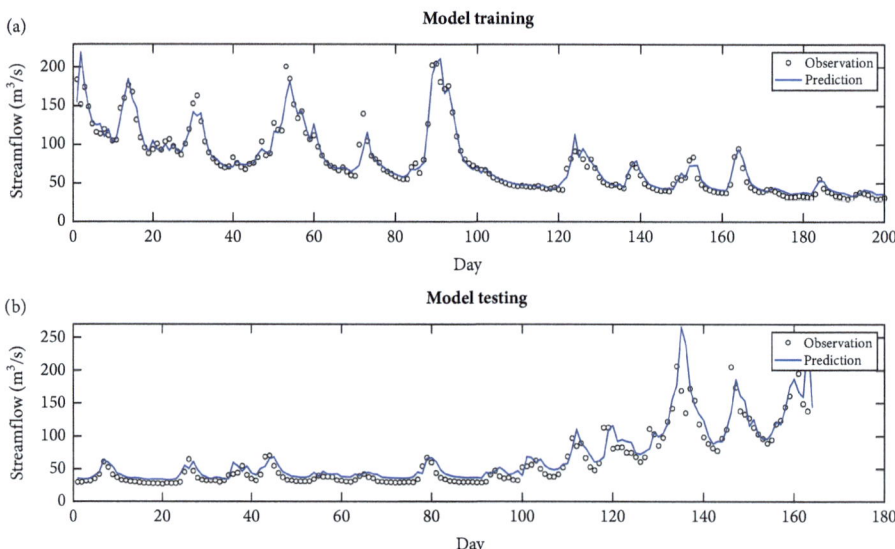

Figure 5.7 Performance of the ANN model in the training and testing periods. This figure is produced using the MATLAB script in Code 5.5.

phase, the testing phase assesses the model's performance on new and unseen data. The testing phase uses a separate dataset known as the testing dataset, which also consists of inputs and expected outputs but was not used during the training phase. For one-year observations of streamflow, rainfall, and PET, the first 200 sample observations are adopted to train the MLP and ANN models while the rest of the dataset is employed to validate the performance of the trained MLP and ANN models.

Table 5.3 The performance of the two ML techniques for hydrological predictions in terms of NSE and RMSE during training and testing processes.

		MLP	ANN
Training	NSE	0.9322	0.9595
	RMSE	10.6163	8.2108
Testing	NSE	0.8839	0.9226
	RMSE	15.396	12.5717

Table 5.3 shows the evaluations for the two models in both training and testing periods. The results indicate that both these ML methods can work well for hydrological predictions for the River Thames. Moreover, the MATLAB scripts for the MLP and ANN models are presented in Code 5.4 and Code 5.5. These two MATLAB scripts produce Figures 5.6 and 5.7, respectively.

Code 5.4 *MATLAB script for a linear regression model for hydrological predictions. To run this code, the data file 'Chapter-5.2-Data.txt' and the two functions 'RMSE.m' and 'NSE.m' should be located in the same folder. The input file and this code are available at the following website: http://www.oup.com/AdvancedMathematicalModelling. This code produces Figure 5.6.*

```
clc;clear all;close all;set(0,'defaultaxesfontsize',16);
% to run the code, please firstly import the data file
% named as "Chapter-5.2-Data.txt" and put it in the same folder
% also Functions "RMSE.m" and "NSE.m" should be in the same folder
rawdata = load('Chapter-5.2-Data.txt');
% colnames = Year, Month,Day,Precipitation,Streamflow,Potential Evaporation
RawP = rawdata(:,4);   RawQ = rawdata(:,5); RawPet = rawdata(:,6);
n = length(RawP);
% Define inputs and outputs Q(t) = f(Q(t-1),P(t), P(t-1), Pet(t), Pet(t-1))
Pt = RawP(2:n);   Pt_1 = RawP(1:n-1);
Qt_1 = RawQ(1:n-1);   Pett = RawPet(2:n);
Pett_1=RawPet(1:n-1);
%Establish the input matrix
InputMatrix = [Qt_1,Pt,Pt_1,Pett,Pett_1];
% Output variable Qt
Qt = RawQ(2:n);
%%% Model Training
% Define the training datasets
nTrain = 200
InputMatrix_Train = [ones(nTrain,1),InputMatrix(1:nTrain,:)];
Qt_Train = Qt(1:nTrain);
%---------------------Train multivariate linear regression model------
[b,bint,r,rint,stats] = regress(Qt_Train,InputMatrix_Train);
% b is the parameters in the regression model;
y2 = InputMatrix_Train*b; % get the prediction after training process
NSEtrain = NSE(Qt_Train, y2);   RMSEtrain = RMSE(Qt_Train, y2)
figure; subplot(2,1,1) %comparison between prediction and observation
plot(Qt_Train,'ko','linewidth',1.0,'markersize',5);
```

Code 5.4 *Continued*

```matlab
hold on; plot(y2,'b-','linewidth',1.5);
legend('Observation','Prediction');ylim([0 230]);
title('(a) Model Training');set(gca,'linewidth',1.5);
xlabel('Day');ylabel('Streamflow (m^3/s)');
%%% Testing Process % Validate the model by the Testing datasets
[nrow_test, ncol_test] = size(InputMatrix((nTrain+1):(n-1),:));
InputMatrix_Test = [ones(nrow_test,1),InputMatrix((nTrain+1):(n-1),:)];
Qt_Test = Qt((nTrain+1):(n-1));
ytest =  InputMatrix_Test*b; % get the prediction for testing process
NSEtest = NSE(Qt_Test, ytest); RMSEtest = RMSE(Qt_Test, ytest);
subplot(2,1,2);
plot(Qt_Test,'ko','linewidth',1.0,'markersize',5);
hold on;plot(ytest,'b-','linewidth',1.5);legend('Observation', 'Prediction');
title('(b) Model Testing');set(gca,'linewidth',1.5);
xlabel('Day');ylabel('Streamflow (m^3/s)');
```

Code 5.5 *MATLAB script for the ANN model for hydrological predictions. To run this code, the data file 'Chapter-5.2-Data.txt' and the two functions 'RMSE.m' and 'NSE.m' should be located in the same folder. This input file and the MATLAB code can be downloaded from the following website: http://www.oup.com/AdvancedMathematicalModelling. This code produces Figure 5.7.*

```matlab
clc;clear all;close all;set(0,'defaultaxesfontsize',16);
% to run the code, please import the data file named
% as "Chapter-5.2-Data.txt" and put it in the same folder
% also Functions "RMSE.m" and "NSE.m" should be in the same folder
rawdata = load('Chapter-5.2-Data.txt');
% colnames = Year,Month,Day,Precpitation,Streamflow,Potential Evaporation
RawP = rawdata(:,4);RawQ = rawdata(:,5);RawPet = rawdata(:,6);
n = length(RawP);
% Define inputs & outputs: Q(t)=f(Q(t-1),P(t), P(t-1), Pet(t), Pet(t-1))
Pt = RawP(2:n);  Pt_1 = RawP(1:n-1);
Qt_1 = RawQ(1:n-1);   Pett = RawPet(2:n);
Pett_1=RawPet(1:n-1);
%Establish the input matrix
InputMatrix = [Qt_1,Pt,Pt_1,Pett,Pett_1];
% Output variable Qt
Qt = RawQ(2:n);
%%% Model Training Define the training datasets
nTrain = 200;
InputMatrix_Train = InputMatrix(1:nTrain,:);
Qt_Train = Qt(1:nTrain);
  %---------------------Train ANN ------------------------------------
inputs = InputMatrix_Train';
targets = Qt_Train';
net = newff(inputs,targets,10); % 10 is nodes/neurons in the hidden layer;
net.trainParam.lr=0.05; % set the learning rate
net.trainParam.epochs=5000; % maximum training iteration
net.trainParam.goal=1e-5; % the learning goal or the loss to be achieved.
% This means the training process will stop when loss is less than 1e-5
[net,tr]=train(net,InputMatrix_Train',Qt_Train'); % training neural network
y2 = sim(net, InputMatrix_Train'); % get prediction after training process
```

```
NSEtrain = NSE(Qt_Train, y2);  RMSEtrain = RMSE(Qt_Train, y2)
figure; subplot(2,1,1)
%comparison between prediction and observation
plot(Qt_Train,'ko','markersize',5,'linewidth',1.0);hold on;
plot(y2,'b-','linewidth',1.5);ylim([0 230]);
legend('Observation','Prediction'); title('(a) Model Training');
set(gca,'linewidth',1.5); xlabel('Day');ylabel('Streamflow (m^3/s)');
%%% Testing Process, Validate the model by the Testing datasets
InputMatrix_Test = InputMatrix((nTrain+1):(n-1),:);
Qt_Test = Qt((nTrain+1):(n-1));
ytest = sim(net, InputMatrix_Test'); % get prediction for testing process
NSEtest = NSE(Qt_Test, ytest);  RMSEtest = RMSE(Qt_Test, ytest);
subplot(2,1,2); plot(Qt_Test,'ko','markersize',5,'linewidth',1.0);
hold on; plot(ytest,'b-','linewidth',1.5);
set(gca,'linewidth',1.5);ylim([0 270]);
legend('Observation','Prediction'); title('(b) Model Testing');
xlabel('Day');ylabel('Streamflow (m^3/s)');
```

5.3 Hydrological risk analysis

5.3.1 Extreme hydrological events

Excessive or inadequate water resources can lead to extensive loss of property and even human life. Scarcity in precipitation or streamflows could result in the occurrence of droughts, while excess rainfall and river discharges are associated with floods. Severe floods and droughts are more devastating in terms of deaths, suffering, and economic damage than other natural disasters such as earthquakes and volcanoes (Xu et al. 2001). Humans, in both developed and developing countries, are vulnerable to both floods and droughts despite significant progress in science and technology.

A drought is a natural hazard that stems from a prolonged deficiency of precipitation in comparison to the expected or normal amount of precipitation, which can result in insufficient water to meet the demands of human activities and the environment (Estrela and Vargas 2012; Chen et al. 2013). The term drought can be characterized as meteorological drought (precipitation well below average), hydrological drought (low river flows and water levels in rivers, lakes, and groundwater), agricultural drought (low soil moisture), or environmental drought (a combination of the above; IPCC 2007). The socioeconomic impacts of drought may arise from the interaction between natural conditions and human factors, such as changes in land use and land cover, as well as water demand and use. Excessive water withdrawal by humans can exacerbate the impact of drought (IPCC 2007).

Compared with droughts, floods are often recognized as an immediate hazard, being very apparent and often eliciting a striking consequence in a short period. Floods and their negative consequences are a natural phenomenon, independent of man's will, and are difficult to control. Floods involve river floods, flash floods, urban floods, and sewer floods, and can be caused by intense and/or long-lasting precipitation, snowmelt, dam break, or reduced conveyance due to ice jams or landslides (IPCC 2007). The severities of floods are influenced by many factors, such as the intensity, volume, and timing of precipitation, and the antecedent conditions of rivers and their drainage basins (e.g., the presence of snow and ice, soil character, wetness, urbanization, and the existence of dikes, dams, or reservoirs; IPCC 2007). Moreover, human activities can affect the severity of floods. For instance, human encroachment into plains and lack of flood response plans increases the damage of floods (IPCC 2007), while some hydraulic projects (e.g., dam construction) may mitigate the negative effects of floods. For instance, the Yangtze River floods in 1998 in China led to more than 14 million people being homeless. The Pakistan floods in 2010 directly affected about 20 million people and caused a death toll of nearly 2000.

Hydrologic risk analysis is an essential tool which can analyse and predict long-term flood risks and provide decision support for engineering design and flood management. Hydrologic risk is the probability of failure occurring on any hydraulic structure attributable to extremely low or high water fluxes. These may be grouped in two categories: (i) structural failure and (ii) performance failure (Gebregiorgis and Hossain 2012). At the drainage basin scale, consideration of flood risk plays a necessary role in the planning of water infrastructure projects, for example in the design of hydraulic structures (e.g., dam spillways, diversion canals, dikes, and river channels), urban drainage systems, cross drainage structures (e.g., culverts and bridges), reservoir management, and flood hazard mapping (Reddy and Ganguli 2012).

A flood is usually characterized by multidimensional characteristics, including flood peak, flood volume, and flood duration. Consequently, both univariate and multivariate flood risk analyses have been developed. Univariate flood frequency analyses mainly focus on the study of flood peaks for designing most hydraulic structures (Requena et al. 2013). However, a full screening of flood occurrence probability cannot be procured only through frequency analysis of flood peaks. For practical flood control, the volume and duration of a flood can provide additional information for decision makers, in which flood volume can be correlated with flood diversion, and flood duration can be used for flood supervision. Moreover, the full hydrograph is valuable in the case of dam design, as the inflow peak is converted into a different outflow peak during the routing process in the reservoir (Requena et al. 2013). In recent decades, a number of studies have reported multivariate hydrologic risk analyses for different flood variables such as flood peak, volume, and duration.

5.3.2 Univariate hydrological risk analysis

5.3.2.1 Flood series and frequency analysis

The hydrologist defines two data series of flood events: the annual maximum series (often called the AMAX series) and the peaks over threshold series (called the POT series).

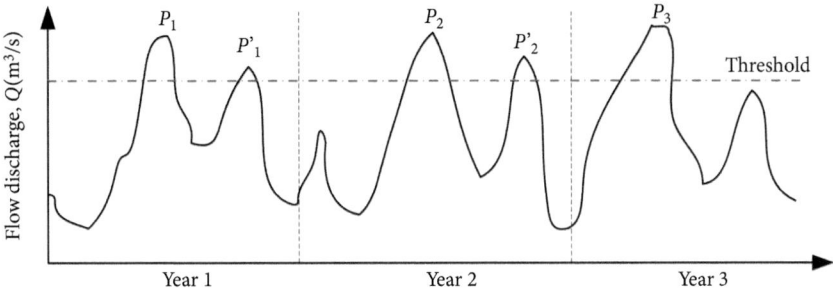

Figure 5.8 Definition of annual maximum discharges (P_1, P_2, P_3) and peaks over threshold $(P_1, P_1', P_2, P_2', P_3)$ over a three-year period.

As presented in Figure 5.8, the annual maximum series takes the single maximum peak discharge in each year of record so that the number of data values equals the record length in years. For statistical purposes, it is necessary to ensure that the selected annual peaks are independent of one another. This can be sometimes challenging, e.g., when an annual maximum flow early in January may be related to an annual maximum flow at the end of the previous December. For this reason, it is generally advisable to use the water year rather than the calendar year. The definition of the water year depends on the seasonal climatic and flow regimes. In humid temperate areas, such as the UK, it is often taken to be October to September because in October, soils should normally have rewetted following the summer dry period. In comparison, the POT series takes all the peaks over a selected threshold level of discharge. There might be more data values for analysis in this series than in the annual series, but there is more chance of successive peaks being related such that the assumption of statistical independence of the data might be compromised.

Risk analysis of extreme hydrological events is based on hydrological frequency analysis, which includes the estimation of the extreme variable (e.g., peak discharge) to be equalled or exceeded, on average, once in a specified period, i.e., T years. This is called the T-year design event and the peak, Q_T, is said to have a return period or recurrence interval of T years. This is the long-term average of the waiting time between successive exceedances of a flood magnitude, Q_T, and is effectively a shorthand way of referring to the probability that Q_T might be expected to be exceeded in any one year (see the next section for more details). It is important not to misunderstand the return period concept. The intervals in which Q_T is exceeded might vary considerably around the average value T. Therefore, a very long record could show 10-year events, Q_{10}, occurring at intervals much greater or much less than 10 years, and even in successive years.

Many hydrological extremes (e.g., the annual peak discharge) are considered to be a random variable. Probability and statistical methods are employed to analyse these random variables. To get meaningful estimates from flood frequency analysis, the following assumptions are implicit:

- The data describe random events.
- The data is homogeneous.
- The population parameters can be estimated from the sample data.
- The data is of good quality.

Moreover, the data should be (i) relevant, (ii) adequate, and (iii) accurate. The term 'relevant' means that the data must deal with the problem. For example, if the problem is the duration of flooding, then the data series should represent the duration of flows in excess of some critical value. The term 'adequate' primarily refers to the length of the data. The length of the data primarily depends on the variability of data, and hence there is no guideline for the length of data to be used for frequency analysis, though typically record lengths of more than 15–20 years are preferred. The term 'accurate' primarily refers to the homogeneity of data and the accuracy of the discharge figures. The data used for analysis should ideally have little or no effect from man-made changes. Changes in the stage–discharge relationship (i.e., correlation between the water level, or stage, in a river and the volume flowing past a reference point per unit of time or discharge) may render stage records nonhomogeneous and unsuitable for frequency analysis. It is therefore preferable to work with discharges; if stage frequencies are required, then the most recent rating curve is used.

5.3.2.2 Probabilistic distributions for flood variables

For hydrologic risk analysis, probabilistic models are desired to quantify the occurrence frequency of flood events and further derive the magnitude of a flood with a specific return period. There are several parametric or nonparametric models available to quantify the probabilistic features for a flood variable. Parametric distributions for flood frequency analysis mainly include extreme value type 1 (EV1), extreme value type 2 (EV2), generalized extreme value (GEV), two-parameter lognormal (LN2), Pearson type III (P3), and log-Pearson type III (LP3). Also, some nonparametric techniques, such as kernel density estimation, and entropy-based approaches have also been proposed for flood risk analysis (e.g., Sun et al. 2019). Qi et al. (2016) summarized the three main approaches for distribution selection: (i) official recommendation, (ii) experience-/knowledge-based selection and (iii) statistical-test-based selection. This study also concluded that many candidate distributions can be considered as applicable models, leading to uncertainty in probability distribution selection. In the following, some of the most popular probabilistic distributions for floods are introduced.

Lognormal distribution
Many hydrologic processes are positively skewed and are not normally distributed. However, for a strictly positive random variable $X > 0$, the logarithm-transformed equivalent $Y = \ln(X)$ can be described by a normal distribution. This is particularly true if the hydrologic variable results from some multiplicative process such as floods. If $y = \ln(x)$ follows a normal distribution, then x is said to follow a lognormal distribution. The probability density function (PDF) and cumulative distribution function (CDF) of the

lognormal distribution are given as:

$$\text{PDF}: f(x) = \frac{1}{x\delta_L\sqrt{2\pi}} \exp\left[-\frac{1}{2}\left(\frac{\ln x - \mu_L}{\delta_L}\right)\right] \tag{5.20a}$$

$$\text{CDF}: F(x) = \int_0^x \frac{1}{x} \exp\left[-\frac{1}{2}\left(\frac{\ln x - \mu_L}{\delta_L}\right)\right] dx \tag{5.20b}$$

where μ_L is the mean of logarithms of x, and δ_L is the standard deviation of $\ln(x)$.

Gumbel distribution

One of the distributions in flood frequency analysis is the double exponential distribution (also known as the Gumbel distribution or extreme value type 1 or Gumbel EV1 distribution). The PDF and CDF of the Gumbel maximum distribution are:

$$\text{PDF}: f(x) = \frac{1}{\beta} \exp\left[-\left(\frac{x-\mu}{\beta}\right) - \exp\left(-\frac{x-\mu}{\beta}\right)\right] \tag{5.21a}$$

$$\text{CDF}: F(x) = \exp\left\{-\exp\left[-\left(\frac{x-\mu}{\beta}\right)\right]\right\} \tag{5.21b}$$

where μ is the location parameter and β the shape parameter. If a reduced variate z is adopted which is expressed as $z = \frac{x-\mu}{\beta}$, the PDF and CDF can be reformulated as

$$\text{PDF}: f(z) = e^{-z-(e^{-z})} \tag{5.21c}$$

$$\text{CDF}: F(z) = e^{-e^{-z}} \tag{5.21d}$$

Generalized extreme value distribution

The GEV distribution is a family of continuous probability distributions that combines the Gumbel (EV1), Fréchet, and Weibull distributions. GEV makes use of three parameters: location, scale, and shape. The location parameter describes the shift of a distribution in a given direction on the horizontal axis. The scale parameter describes how spread out the distribution is and defines where the bulk of the distribution lies. The shape parameter strictly affects the shape of the distribution and governs the tail of each distribution. The CDF and PDF of the GEV distribution are formulated as:

$$\text{PDF}: f(x) = \left(\frac{1}{\sigma}\right) \exp\left\{-\left[1 + k\frac{(x-\mu)}{\sigma}\right]^{-\frac{1}{k}}\right\}\left[1 + k\frac{(x-\mu)}{\sigma}\right]^{-1-\frac{1}{k}} \tag{5.22a}$$

$$\text{CDF}: F(x) = \exp\left\{-\left[1 + k\frac{(x-\mu)}{\sigma}\right]^{\frac{1}{k}}\right\} \tag{5.22b}$$

where k is the shape parameter, μ is the location parameter, and σ is the scale parameter. For $k < 0$, the distribution has a finite upper bound at $\mu + \sigma/k$, while the distribution has a thicker right-hand tail for $k > 0$.

Binomial distribution

The binomial distribution is based on the binomial theorem which states that the probability of exactly r successes in n trials is

$$P_{r,n} = {}^nC_r (p^r)(q^{n-r}) = \frac{n!}{r!(n-r)!} p^r q^{n-r} \qquad (5.23)$$

where $P_{r,n}$ is the probability of a random hydrologic event of a given magnitude and exceedance occurring r times in n successive years, p is the probability of exceedance of a single event, q is the probability of nonexceedance, which is equal to $(1-p)$, r is the number of exceedances (or successes), and n is the total number of events.

The assumptions for the binomial distribution are based on tossing a coin or drawing a card from a pack, which operate under the following three conditions:

- The outcome of any trial can be two mutually exclusive events only, i.e., any trial can have either success or failure, true or false.
- Successive trials are independent.
- The probabilities are stable.

The binomial distribution is one kind of discrete probability distribution restricted generally to those random events for which the outcome can be described as a success or a failure. Discrete probability distributions are useful for those experiments where only two mutually exclusive events exist. Furthermore, successive trials are independent, and the probability of success remains constant from one trial to another.

The application of the binomial distribution can also be explained as follows:

(i) The probability of an event of exceedance occurring z times in n successive years will be

$$P_{z,n} = \frac{n!}{z!(n-z)!} p^z q^{n-z} \qquad (5.24a)$$

(ii) The probability of the event not occurring at all in n successive years will be

$$P_{0,n} = q^n = (1-p)^n \qquad (5.24b)$$

(iii) The probability of the event occurring at least once in n successive years will be

$$p_1 = P_{1,n} + P_{2,n} + \cdots + P_{n,n} = 1 - q^n = 1 - (1-p)^n \qquad (5.24c)$$

Risk and reliability concepts

Based on the probabilistic distributions, risk analysis studies can be conducted for a specific flood event. If this design event X_T has a return period of T years, the corresponding

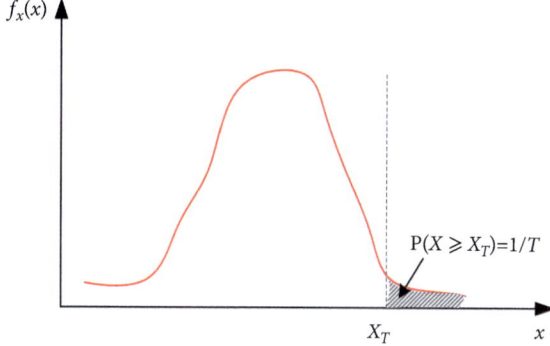

Figure 5.9 Explanation of flood return period.

annual probability of exceedance of p is the probability that the event will be equalled or exceeded in any given year, as shown in Figure 5.9:

$$p = P(X \geq X_T) = \frac{1}{T} \tag{5.25}$$

Consequently, the probability of nonexceedance in any one year is formulated as

$$q = P(x \leq X_T) = F(X_T) = 1 - p = 1 - \frac{1}{T} \tag{5.26}$$

Based on Eq (5.24b), the probability of nonexceedance in n years is

$$q' = \left(1 - \frac{1}{T}\right)^n \tag{5.27}$$

Therefore, the probability that X_T will occur at least once in n years, i.e., the risk of failure R, is defined based on Eq (5.24c):

$$R = 1 - \left(1 - \frac{1}{T}\right)^n \tag{5.28}$$

where n is the design life of the structure.

5.3.2.3 Parameter estimation and goodness of fit

Parameter estimation
Parameter estimation is important in flood frequency analysis. Any set of flood occurrences can be described using some statistical distribution, which may be mathematically represented through functional relationships that may contain a set of parameters, such as the shape, location, and scale parameters in the GEV distribution. Several approaches are available to estimate parameters in a specific distribution based on historical (or past) flood event records. Some commonly used approaches include graphical methods, the method of moments (MOM), and maximum likelihood estimation (MLE).

MLE is a technique used for estimating the parameters of a given distribution using some observed data. For example, if a population is known to follow a normal distribution but the mean and variance are unknown, MLE can be used to estimate them using a limited sample of the population by finding particular values of the mean and variance so that the observation is the most likely result to have occurred.

Let x_1, x_2, ..., x_n be observations from n independent and identically distributed random variables drawn from a probability distribution f_0, where f_0 is known to be from a family of distributions f that depend on some parameters θ. For example, f_0 could be known to be from the family of normal distributions f, which depend on parameters σ (standard deviation) and μ (mean). Therefore, $\theta = (\mu, \sigma)$, and x_1, x_2, ..., x_n would be observations from f_0. The goal of MLE is to maximize the likelihood function as follows:

$$L(\theta) = f(x_1, x_2, ..., x_n | \theta) = f(x_1 | \theta) \times f(x_2 | \theta) \times ... \times f(x_1 | \theta) \qquad (5.29a)$$

Often, the log-likelihood function is easier to work with:

$$l(\theta) = \log L(\theta) = \sum_{i=1}^{n} \log f(x_i | \theta) \qquad (5.29b)$$

The goal of maximum likelihood estimation is to find the values of θ that maximize the likelihood (or log-likelihood) function over the parameter space:

$$\hat{\theta} = \arg \max_{\theta \in \Theta} l(\theta) \qquad (5.29c)$$

Through the MLE process in Eq (5.29), the best approximation or model of the true distribution f_0 can be obtained as $f(x|\theta)$.

In MATLAB, the function `mle` can be used to estimate the parameters for a specified distribution through the MLE method. The command can be formulated as

$$\text{phat = mle (data, 'distribution' ,dist)} \qquad (5.30)$$

where `phat` holds the parameters of the `dist` (e.g., `phat` = (μ, β) for Gumbel), `data` is the sample data used for estimating the parameters, and `dist` is the name of the distribution to be estimated (e.g., lognormal or other distributions defined by the user).

Goodness of fit
The goodness of fit of a statistical model describes how well it fits a set of observations. of Goodness-of-fit measures typically summarize the discrepancy between observed values and the values expected under the model in question. Such measures can be used in statistical hypothesis testing, e.g., to test for normality of residuals, or to test whether two samples are drawn from identical distributions.

The Kolmogorov–Smirnov statistic (known as the K–S test) quantifies the distance between the empirical distribution function of the sample and the cumulative distribution function of the reference distribution, or between the empirical distribution functions of two samples. The null distribution of this statistic is calculated under the null hypothesis that the sample is drawn from the reference distribution (in the one-sample case) or that the samples are drawn from the same distribution (in the two-sample case).

Suppose that we have observations $x_1, x_2, ..., x_n$, which we think come from a distribution P. The K–S test is used to test the null hypothesis H_0, which indicates the samples come from P, against the alternative hypothesis H_1, which states the samples do not come from P.

The cumulative distribution function $F(x)$ of a random variable X is $F(x) = P[X \leq x]$. Given observations $x_1, x_2, ..., x_n$, the empirical distribution function $F_e(x)$ can be obtained through the Gringorten plotting position formula:

$$F_e(x) = \frac{i_x - 0.44}{n + 0.12} \tag{5.31}$$

where i_x is the position for the observation x in the ascending sequence of the observations. For the minimum observation, we have $i_x = 1$. For the maximum observation, it is $i_x = n$. The K–S test compares the empirical distribution function of the data, F_e, with the cumulative distribution function associated with the null hypothesis, F_c (calculated CDF from the distribution). The Kolmogorov–Smirnov statistic is

$$D_n = \max_x |F_c(x) - F_e(x)| \tag{5.32}$$

The K–S test can be easily implemented in MATLAB. The MATLAB command for performing the K–S test has the following formula:

$$[h,p] = \text{kstest}(x, Name, Value) \tag{5.33}$$

where x is the sample data used for the K–S test and Name is the distribution (either CDF or PDF) which is going to be tested. Here, Name is usually characterized as 'CDF'. Value here consists of the sample data and their corresponding CDF values. Generally, Value = [x, F(x)]; h = 0 to accept the obtained distribution, or 1 to reject it; p = p-value. Note that $p \geq 0.05$ implies that the difference between the sample data and the tested distribution is not statistically significant and the distribution is accepted as a plausible model for the data.

Alongside rigorous statistical tests like the K–S test, which evaluates whether a set of observations aligns with a specific distribution, there exist additional indices for comparing the effectiveness of various distributions. RMSE and the Akaike information criterion (AIC) are two commonly used approaches. The RMSE can be expressed as follows:

$$RMSE = \sqrt{E[F_c(x) - F_e(x)]} = \sqrt{\frac{1}{n-k}\sum_{i=1}^{n}[F_c(x_i) - F_e(x_i)]^2} \tag{5.34}$$

where n is the total sample size, k is the number of parameters in the distribution, $F_c(x)$ is the calculated CDF value from the distribution, and $F_e(x)$ is the corresponding empirical CDF obtained through Eq (5.31).

The AIC can also be adopted to identify the most appropriate probability distribution. It can consider (i) the lack of fit of the model, and (ii) the unreliability of the model due to the number of model parameters (Zhang and Singh 2006), and can be expressed as follows:

$$AIC = n \, log(MSE) + 2k \tag{5.35a}$$

where

$$MSE = \frac{1}{n-k} \sum_{i=1}^{n} [F_c(x_i) - F_e(x_i)]^2 \tag{5.35b}$$

A lower AIC value generally indicates better performance for a specific distribution.

5.3.2.4 Numerical example

The AMAX flood series at the Zhangjiashan station in the Jing River basin in China is adopted to illustrate flood risk analysis (Figure 5.10). The Jing River is one of the main tributaries of the Weihe River, which is the largest tributary of the Yellow River. It is located in the middle of the Loess Plateau (longitude 100.9 E–114.55 E, latitude 33.72 N–41.27 N), with a drainage area of 45,421 km^2. The flood records from 1960 to 2014 at the Zhangjiashan station (longitude 108.60 E, latitude 34.63 N) are presented in Table 5.4 and are used in this analysis.

Table 5.5 presents the performance of the Gumbel, GEV, and LN distributions in modelling the probabilistic features of the AMAX flood series. The results indicate that all three distributions are acceptable for further flood risk analysis. Nevertheless, the GEV distribution performs better while the Gumbel distribution is the worst among the three candidate distributions. Figure 5.11 compares the theoretical CDFs of the Gumbel, GEV, and LN with the empirical CDFs, which also demonstrate the best performance of the GEV distribution. Consequently, the GEV distribution is adopted for further hydro-logical designs. Table 5.6 presents the flood magnitudes under different return periods obtained from the GEV distribution. The MATLAB script in Code 5.6 can be used to produce Figure 5.11 and the results presented in Tables 5.5 and 5.6.

Code 5.6 *The MATLAB script for flood risk analysis using three different types of distribution (i.e., Gumbel, GEV, and LN). This code produces Figure 5.11 and Tables 5.5 and 5.6.*

```
clc;clear all;close all;set(0,'defaultaxesfontsize',17);
%Step 1 Load the flood data
flood_data = [1380, 792, 1391.3, 775, 540.1, 473, 1690.4, 262, 3730, 555.99, ...
    1044.2, 609, 1350, 929, 272, 2250, 548, 913, 1060, 2580, 831, 580, 547, ...
    914, 392, 515, 674, 438, 781, 398, 1427, 509.8, 631.3, 927,1560, 752, 1100, ...
    875, 3080, 976, 835, 629, 217.3, 573.13, 476.13, 1388.1, 865.13, 809.27,...
    317, 249, 300, 297, 828, 607, 301];
%% ========== Maximum Likelihood Estimation============
```

```
x = flood_data;
Pln = mle(x,'distribution','lognormal'); % lognormal, Pln include model parameters
Pev = mle(-x,'distribution','ev');% Gumbel (or EV), Pev includes two parameters
% In matlab, the Gumbel distribution (i.e. EV) is for minima. Therefore, the annual
% maxima is converted to minima firstly and then estimate the location and
% shape parameters. %Fmax(x0) = P(X<= x0) = P(-X>=-x0) = 1-Fmin (-x0)
Pgev = mle(x,'distribution','gev'); % GEV distribution, Pgev includes three parameters
%% --=================Goodness of fit=========
TestMatrix = zeros(3,4); %define the matrix to store the test results;
% rownames = {'LN', 'Gumbel', 'GEV'}; colnames = {'h','p','RMSE','AIC'}
rmseFUN = @(n,k, Fc, Fe)sqrt((1/(n-k))*sum(Fc-Fe).^2); %define RMSE function
aicFUN = @(n,k,MSE)n*log(MSE)+2*k; %define the AIC function
% ---------------empirical probability----------------------
SortX = sort(x);
for i = 1:length(x)
    Fe(i) = (i-0.44)/(length(x)+0.12);
end
% ---------goodness-of-fit for LN-------------------------------
LN_cdf = cdf('lognormal',x,Pln(1), Pln(2));
[h_ln,p_ln] = kstest(x,'CDF',[x',LN_cdf']); % matrix has 2 columns;
TestMatrix(1,1) = h_ln;  TestMatrix(1,2) = p_ln;
% calculated probability from LN Fc_LN
Fc_LN = cdf('lognormal',SortX,Pln(1), Pln(2)); % the SortX is adopted here;
TestMatrix(1,3) = rmseFUN(length(x),2,Fc_LN, Fe); % RMSE
MSE = TestMatrix(1,3)^2;
TestMatrix(1,4) = aicFUN(length(x),2, MSE); % AIC
% -----------------goodness-of-fit for Gumbel------------
EV_cdf = cdf('ev',-x,Pev(1), Pev(2)); % -x for minima
[h_ev,p_ev] = kstest(-x,'CDF',[-x',EV_cdf']); % matrix has 2 columns;
TestMatrix(2,1) = h_ev;   TestMatrix(2,2) = p_ev;
% calculated probability from Gumbel Fc_EV
Fc_EV_min = cdf('ev',-SortX,Pev(1), Pev(2)); % the -SortX is adopted here;
Fc_EV = sort(1-Fc_EV_min); % get CDF for maxima SortX and store in ascending order;
TestMatrix(2,3) = rmseFUN(length(x),2,Fc_EV, Fe); % RMSE
MSE = TestMatrix(2,3)^2;
TestMatrix(2,4) = aicFUN(length(x),2, MSE); % AIC
%-----------------goodness-of-fit for GEV------------
GEV_cdf = cdf('gev',x,Pgev(1), Pgev(2), Pgev(3)); % three parameters for GEV
[h_gev,p_gev] = kstest(x,'CDF',[x',GEV_cdf']); % matrix has 2 columns;
TestMatrix(3,1) = h_gev;   TestMatrix(3,2) = p_gev;
% calculated probability from LN Fc_LN
Fc_GEV = cdf('gev',SortX,Pgev(1), Pgev(2), Pgev(3)); % the SortX is adopted here;
TestMatrix(3,3) = rmseFUN(length(x),2,Fc_GEV, Fe); % RMSE
MSE = TestMatrix(3,3)^2;
TestMatrix(3,4) = aicFUN(length(x),2, MSE); % AIC
%%%%%%%%%%%%%%%%%%%%%%%%%%%%%%%%%%%%%%%%%%%%%%%
% PRINT the outcomes of performance test:
display ('Performances of different distributions on modelling AMAX flood series');
display(TestMatrix);
%% ==============Plot Comparison ==============
dx = 0.8*min(x):10:1.2*max(x); % discretize the the flood variable
for i = 1:length(dx)
    P_LN(i) = cdf('lognormal',dx(i),Pln(1), Pln(2));
    P_GEV(i) = cdf('gev',dx(i),Pgev(1), Pgev(2), Pgev(3));
    P_Gumbel_min(i) = cdf('ev',-dx(i),Pev(1), Pev(2)); % cdf for Gumbel minima;
    P_Gumbel(i) =1-P_Gumbel_min(i); % cdf for Gumbel maxima;
end
figure; plot(SortX,Fe,'ko','markersize',5,'linewidth',1.0)
```

Code 5.6 *Continued*

```
hold on; plot(dx,P_LN,'k','linewidth',1.0)
hold on; plot(dx,P_GEV,'m-.','linewidth',1.0);
hold on; plot(dx,P_Gumbel,'b--','linewidth',1.0);
xlabel('Peak flow (m^3/s)'); ylabel('Cumulative Distribution Function (CDF)')
legend('Empirical CDF','LN','GEV','Gumbel','location','southeast');
title('Comparison for different distributions');
set(gca,'linewidth',1.5);
%% ===================Flood design============================
T = [10, 50, 100, 200, 500];
P = 1 - 1./T;
Design_flood = icdf('gev', P, Pgev(1), Pgev(2), Pgev(3));
% PRINT the design floods using the results of GEV distribution
Design_flood_out(:,1)=T; Design_flood_out(:,2)= Design_flood;
display ('Design floods using GEV model are:'); display(Design_flood_out);
```

5.3.3 Multivariate hydrological risk analysis

5.3.3.1 Multidimensional features for flood events

A flood is usually characterized by multidimensional features, such as flood peak, volume, and duration. Univariate flood frequency analyses mainly address the occurrence of flood peaks, which may not develop a full screening of the flood event with a specific return period. For practical flood control, the volume and duration of a flood can provide additional information for decision makers, in which flood volume can be correlated with flood diversion, and flood duration can be used for flood supervision. Moreover, the full hydrograph is valuable in the case of dam design, as the inflow peak is converted into a different outflow peak during the routing process in the reservoir (Requena et al. 2013). Therefore, as a result of the multidimensional nature of flood events, multivariate hydrologic risk analysis is required for different flood variables such as flood peak, volume, and duration to achieve practical flood control.

Figure 5.12 describes a typical single-peak flood hydrograph, with the flood peak identified as the highest discharge of this flood event (i.e., Q_{Pi}). Flood duration (D) can be determined by identifying the times of the rise (point 's' in Figure 5.12) and fall (point 'e' in Figure 5.12) of the flood hydrograph. The start of the flood is marked by the sharp rise of the hydrograph and the end of the flood runoff is identified by the inflection point on the receding limb of the hydrograph. Between these two points, the total flood volume is estimated. Referring to Figure 5.12, if the rise time of the flood hydrograph is denoted by SD (day) and the fall by ED (day), the flood volume (V) of each flood event is determined using the following expression (Yue 2000, 2001):

$$V_i = \left(V_i^{total} - V_i^{baseflow}\right) = \sum_{j=SD_i}^{ED_i} Q_{ij} - \frac{1}{2}(Q_{is} + Q_{ie})(1 + D_i) \qquad (5.36a)$$

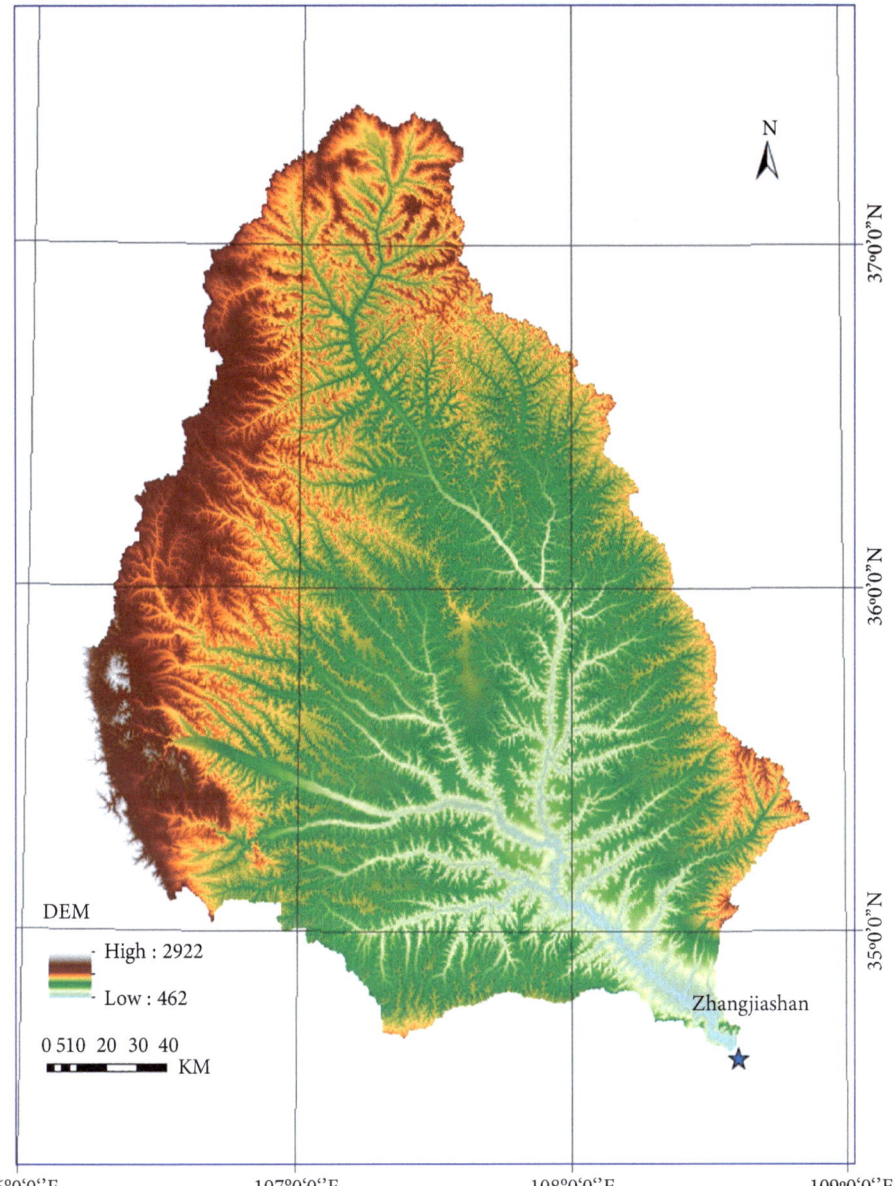

Figure 5.10 Location of the studied catchment, topographic map, and the location of the Zhangjiashan gauge station in China.

where Q_{ij} is the jth-day observed streamflow value for the ith year. Q_{is} and Q_{ie} are the observed daily streamflow value on start and end days of the flood hydrograph for the ith year, respectively. SD_i and ED_i are the start and end days of a flood event in the ith year, respectively. D_i is the flood duration in the ith year. The annual flood peak (Q_i) is obtained by the following equation:

$$Q_i = \max\left\{Q_{ij}, j = SD_i, SD_{i+1}, \ldots, ED_i\right\}, \qquad i = 1, 2, 3, \ldots\ldots n \qquad (5.36b)$$

Table 5.4 The AMAX flood series for the Zhangjiashan station.

Year	AMAX	Year	AMAX	Year	AMAX	Year	AMAX	Year	AMAX
1960	1380	1971	609	1982	547	1993	927	2004	476.13
1961	792.	1972	1350	1983	914	1994	1560	2005	1388.1
1962	1391.3	1973	929	1984	392	1995	752	2006	865.13
1963	775	1974	272	1985	515	1996	1100	2007	809.27
1964	540.1	1975	2250	1986	674	1997	875	2008	317
1965	473	1976	548	1987	438	1998	3080	2009	249
1966	1690.4	1977	913	1988	781	1999	976	2010	300
1967	262	1978	1060	1989	398	2000	835	2011	297
1968	3730	1979	2580	1990	1427	2001	629	2012	828
1969	555.99	1980	831	1991	509.8	2002	217.3	2013	607
1970	1044.2	1981	580	1992	631.3	2003	573.13	2014	301

Table 5.5 Modelling performance of different distributions on the AMAX flood series (see Table 5.4). This table is generated using the MATLAB script in Code 5.6.

Distribution	K–S test h	p	RMSE	AIC
LN	0.00	0.84	0.05	−318.90
Gumbel	0.00	0.47	0.18	−182.05
GEV	0.00	0.92	0.02	−411.40

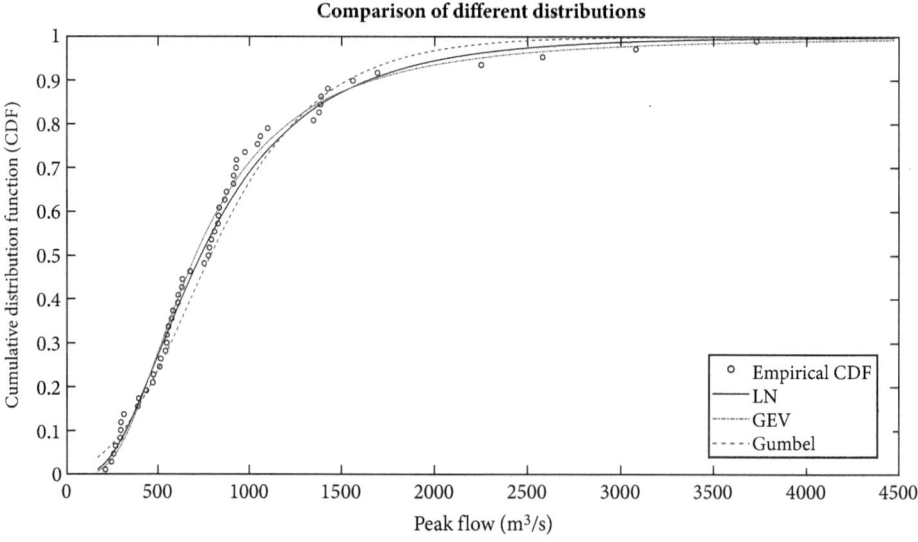

Figure 5.11 Comparison of different types of distributions (i.e., Gumbel, GEV, and LN) for modelling the probabilistic features of the AMAX flood series. This figure is generated using the MATLAB script in Code 5.6.

Table 5.6 Flood magnitudes under different return periods generated using the GEV distribution. This table is generated using the MATLAB script in Code 5.6.

Return period (Year)	Flood peak (m³/s)
10	1660.8
50	3161.2
100	4087.5
200	5249.8
500	7255.8

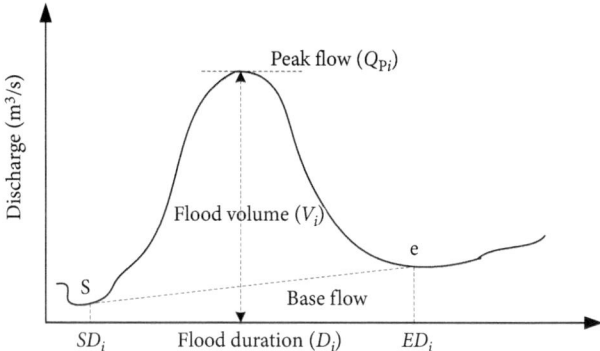

Figure 5.12 Typical flood hydrograph showing flood flow characteristics, adapted from Ganguli and Reddy (2012).

The flood duration can be given by

$$D_i = ED_i - SD_i, \qquad i = 1, 2, 3, \ldots, n \qquad (5.36c)$$

5.3.3.2 Modelling interdependence among flood variables

The copula function has been widely used for multivariate flood risk analyses when considering multiple flood variables (e.g., flood peak and volume). The main advantage of copula functions over classical bivariate frequency analyses is that the selection of marginal distributions and multivariate dependence modelling are two separate processes, giving additional flexibility to the practitioner in choosing different marginal and joint probability functions (Zhang and Singh 2006; Genest and Favre 2007; Karmakar and Simonovic 2009; Sraj et al. 2015).

The basic copula theorem was first introduced by Sklar in 1959, and defined the multivariate distribution function on $\boldsymbol{I}^d = [0, 1]^d$ with uniform marginals (Salvadori et al. 2011). Let F be the d-dimensional joint probability distribution for the random vector $\boldsymbol{X} = [X_1, X_2, \ldots, X_d]^T$. There exists a copula function such that

$$P\{X_1 \le x_1, \ldots, X_d \le x_d\} = F(x_1, \ldots, x_d) = C(u_1, \ldots, u_d) \qquad (5.37a)$$

where C is the copula function, $u_i = F_i(x_i)$ $(i = 1, 2, \ldots, d)$, and F_i is the marginal distribution for the X_i. The copula C is unique if the F_i $(i = 1, 2, \ldots, d)$ are continuous. The multivariate density function $f(x_1, x_2, \ldots, x_d)$ can be formulated as follows (Nelsen 2006):

$$f(x_1, x_2, \ldots, x_d) = c(u_1, u_2, \ldots, u_d) \prod_{i=1}^{d} f_i(x_i) \tag{5.37b}$$

where $c(u_1, u_2, \ldots, u_d)$ is the copula density function and $f_i(x_i)$ are the marginal density functions.

Several copula functions are available for multivariate flood risk analysis, and can be classified as Archimedean, elliptical, and extreme-value copulas. Among them, Archimedean copulas are quite attractive in hydrologic frequency analysis, because they can be easily generated and are capable of capturing a wide range of dependence struc- ture with several desirable properties, such as symmetry and associativity (Reddy and Ganguli 2012). In general, a bivariate Archimedean copula can be defined as follows (Nelsen 2006):

$$C_\theta(u_1, u_2) = \phi^{-1}(\phi(u_1) + \phi(u_2)) \tag{5.38}$$

where u_1 and u_2 are specific values of U_1 and U_2, respectively, with $U_1 = F_{X_1}(x_1)$ and $U_2 = F_{X_2}(x_2)$. F_{X_1} and F_{X_2} are the CDFs of the random variables X_1 and X_2, respectively. ϕ is the copula generator, a convex decreasing function with $\phi(1) = 0$ and $\phi^{-1}(\cdot) = 0$ when $u_2 \geq \phi(0)$. The subscript θ of the copula C is the parameter hidden in the generating function (i.e., ϕ). For a one-parameter copula model, the unknown parameter (i.e., θ) can be estimated using the method of moments using the Kendall correlation coefficient (Nelsen 2006). For copulas with two or more unknown param- eters, the maximum likelihood method can be selected (Zhang and Singh 2006; Sraj et al. 2015).

There are several models in the Archimedean copula family such as the Gumbel– Hougaard, Clayton, and Frank copulas. For a bivariate copula function $C_\theta = (u_1, u_2)$, considering two correlated variables x_1 and x_2 where $u_1 = F_1(x_1)$ and $u_2 = F_2(x_2)$, Table 5.7 presents some basic characteristics of the applied single-parameter bivari- ate Archimedean copulas, where τ is the Kendall coefficient between two correlated variables. Consequently, the copula function can be estimated through the following procedure:

(i) Determine Kendall's τ based on the observation data as follows:

$$\tau_N = \frac{2}{N(N-1)} \sum_{i<j} \text{sign}\left[(x_i - x_j)(y_i - y_j)\right]$$

Table 5.7 Basic properties of applied copulas.

Copula name	Gumbel	Clayton	Frank[*]
Function $[C_\theta(u_1, u_2)]$	$\exp\{-[(-\ln u_1)^\theta + (-\ln u_2)^\theta]^{1/\theta}\}$	$[u_1^{-\theta} + u_2^{-\theta} - 1]^{-1/\theta}$	$-\frac{1}{\theta} \ln\{1 + \frac{(e^{-\theta u}-1)(e^{-\theta v}-1)}{e^{-\theta}-1}\}$
Parameter (i.e., θ) range	$[-1, \infty)\setminus\{0\}$	$[1, \infty)$	$[-\infty, \infty)\setminus\{0\}$
Generator $[\phi(t)]$	$(-\ln t)^\theta$	$\frac{1}{\theta}(t^{-\theta} - 1)$	$-\ln\left[\frac{e^{-\theta t}-1}{e^{-\theta}-1}\right]$
$\tau = 1 + 4\int_0^1 \frac{\phi(t)}{\phi'(t)}\,dt$	$1 - \theta^{-1}$	$\frac{\theta}{\theta+2}$	$1 - \frac{4}{\theta}[D_1(-\theta) - 1]$

[*]D_1 is the first-order Debye function, and for any positive integer k, $D_k(x) = \frac{k}{x^k}\int_0^k \frac{t^k}{e^t-1}\,dt$.

where N is the member of observations, τ_N is the estimate of τ from the observations, and

$$\text{sign}\left[(x_i - x_j)(y_i - y_j)\right] = \begin{cases} 1 & \text{if } (x_i - x_j)(y_i - y_j) > 0 \\ -1 & \text{if } (x_i - x_j)(y_i - y_j) < 0 \end{cases}$$

(ii) Determine the parameter (i.e., θ) based on the correlation between τ and θ presented in Table 5.7. For instance, for the Gumbel copula, the relationship between Kendall's τ and the copula parameter θ is given by $\tau = 1 - \theta^{-1}$, and thus the unknown parameter $\theta = \frac{1}{1-\tau}$.

(iii) Obtain the copula function.

After obtaining the appropriate copula function, one can derive the conditional distribution of one random variable given the other based on the copula function. For instance, the conditional distribution function for U_1, when $U_2 = u_2$ is known, can be derived as

$$C_{U_1|U_2=u_2}(u_1) = C(U_1 \le u_1|U_2 = u_2) = \frac{\partial}{\partial u_2}C(u_1, u_2)|U_2 = u_2 \tag{5.39a}$$

Similarly, the conditional cumulative distribution for U_2, given $U_1 = u_1$, can be obtained. Moreover, the conditional cumulative distribution function of U_1, given $U_2 \le u_2$, can be expressed as

$$C_{U_1|U_2\le u_2}(u_1) = C(U_1 \le u_1|U_2 \le u_2) = \frac{C(u_1, u_2)}{u_2} \tag{5.39b}$$

Likewise, an equivalent formula for the conditional distribution function for U_2, given $U_1 \le u_1$, can be obtained.

The PDF of a copula function can be expressed as follows:

$$c(u_1, u_2) = \frac{\partial^2 C(u_1, u_2)}{\partial u_1 \partial u_2} \tag{5.39c}$$

And the joint PDF of two random variables can be obtained as

$$f(x_1, x_2) = \frac{\partial^2 C(u_1, u_2)}{\partial x_1 \partial x_2} = \frac{\partial^2 C(u_1, u_2)}{\partial u_1 \partial u_2} \frac{\partial u_1}{\partial x_1} \frac{\partial u_2}{\partial x_2} = f_{X_1}(x_1) f_{X_2}(x_2) c(u_1, u_2) \tag{5.39d}$$

where $u_1 = F_{X_1}(x_1)$ and $u_2 = F_{X_2}(x_2)$.

Consequently, the conditional PDF of X_1, given the value of X_2, can be formulated as

$$f(x_1|x_2) = \frac{f(x_1, x_2)}{f_{X_2}(x_2)} = f_{X_1}(x_1) c(u_1, u_2) \tag{5.40a}$$

and the conditional PDF of X_2, given the value of X_1, can be expressed as

$$f(x_2|x_1) = \frac{f(x_1, x_2)}{f_{X1}(x_1)} = f_{X_2}(x_2) c(u_1, u_2) \tag{5.40b}$$

5.3.3.3 Risk analyses for correlated flood variables within a multivariate context

When multiple flood variables are under consideration, the risk indices such as the return period for an individual flood variable, i.e., Eq (5.25), can be extended to the multivariate context, leading to a joint (or multivariate) return period. Moreover, there are differ-ent multivariate return periods corresponding to different multivariate hazard scenarios. Following previous studies (e.g., Salvadori et al. 2011, 2013, 2016; Fan et al. 2020a, 2020b), for a compound extreme with two attributes denoted as $X^* = (x_1^*, x_2^*)$ the concept of return period (RP) can be extended into the multivariate context, leading to the following multivariate return period types:

(i) 'OR' case, T^{OR}: This multivariate RP describes the occurrence time interval that at least one of the attributes of the compound extreme exceeds the prescribed threshold:

$$T^{OR} = \frac{\mu}{\Pr(x \in R_{OR}^2)} \tag{5.41a}$$

where

$$R_{OR}^2 = \{(x_1, x_2) \in R^2 : x_1 > x_1^* \vee x_2 > x_2^*\} \tag{5.41b}$$

Based on the copula function, the multivariate RP in the OR case can be formulated as

$$T^{OR} = \frac{\mu}{1 - C(u_1^*, u_2^*)} \tag{5.41c}$$

where μ indicates the time interval between two adjacent extremes, and $u_i^* = F_i(x_i^*)$.

(ii) 'AND' case, T^{AND}: Such a multivariate RP describes the occurrence time interval for all components of the compound extreme to exceed the prescribed thresholds:

$$T^{AND} = \frac{\mu}{Pr(x \in R^2_{AND})} \tag{5.42a}$$

$$R^2_{AND} = \{(x_1, x_2) \in R^2 : x_1 > x_1^* \wedge x_2 > x_2^*\} \tag{5.42b}$$

Also, the multivariate RP in the AND case can be reformulated based on the copula as

$$T^{AND} = \frac{\mu}{1 - u_1^* - u_2^* + C(u_1^*, u_2^*)} \tag{5.42c}$$

(iii) 'Kendall' case, $T^{Kendall}$: This multivariate RP describes the occurrence time interval when the compound extreme exceeds a critical layer:

$$T^{Kendall} = \frac{\mu}{Pr(x \in R^2_{Kendall})} \tag{5.43a}$$

$$R^2_{Kendall} = \{(x_1, x_2, \ldots, x_d) \in R^2 : F(x_1, x_2) > F(x_1^*, x_2^*), F(x_1^*, x_2^*) = t\} \tag{5.43b}$$

The multivariate RP in the Kendall case can be obtained based on the copula function as

$$T^{Kendall} = \frac{\mu}{1 - K_C(t)} \tag{5.43c}$$

where K_C is the Kendall distribution associated with the theoretical copula function C. For Archimedean copulas, K_C can be expressed as (Nelsen 2006)

$$K_C(t) = t - \frac{\phi(t)}{\phi'(t^+)} \tag{5.43d}$$

where $\phi'(t^+)$ is the right derivative of the copula generator function $\phi(t)$.

5.3.3.4 Numerical example

The records of flood peak and volume at the Zhangjiashan station in the Jing River basin (China) are employed to illustrate the applicability of the copula method for multivariate risk analyses. The detailed catchment information for the Jing River basin is described in Section 5.3.2.4 and the AMAX series of flood peaks are listed in Table 5.4. Also, the GEV distribution will be adopted to quantify the probabilistic feature of flood peaks as demonstrated in Table 5.5.

Table 5.8 presents the flood volumes corresponding to the AMAX flood peaks at the Zhangjiashan station. Similar to the estimation process for the probability distribution

Table 5.8 The flood volumes (in m^3/s day) corresponding to the AMAX flood peaks at the Zhangjiashan station.

Year	Volume (m^3/s day)	Year	Volume	Year	Volume	Year	Volume	Year	Volume
1960	1653.5	1971	584	1982	718.4	1993	1778.2	2004	534.86
1961	935.85	1972	3896.4	1983	2186	1994	5074.1	2005	4161.9
1962	1820.3	1973	1849.5	1984	1376.3	1995	1022.4	2006	1832.3
1963	598.5	1974	303.7	1985	595.1	1996	1804	2007	1156.5
1964	1571.9	1975	7576.1	1986	1128	1997	1627.3	2008	312.5
1965	1365.3	1976	650.2	1987	675	1998	5797.1	2009	659.45
1966	2528.3	1977	1464.1	1988	1330.1	1999	2863.8	2010	548.85
1967	369.7	1978	4033	1989	637.8	2000	1136.2	2011	508.15
1968	6069.8	1979	4768.2	1990	3316	2001	1391.5	2012	1720.7
1969	683.77	1980	2021	1991	815.07	2002	404.1	2013	1329.7
1970	1423	1981	1616	1992	750.2	2003	966.29	2014	553.1

Table 5.9 Modelling performance of different distributions on the flood volumes.

Distribution	K–S test h	p	RMSE	AIC
LN	0	0.73	0.07	−289.79
Gumbel	0	0.21	0.24	−152.14
GEV	0	0.58	0.04	−360.59

for flood peaks, the probability distribution for flood volume can also be calculated using the MATLAB script in Code 5.6, with the detailed results presented in Table 5.9. The results in Table 5.9 also show that the GEV distribution should be chosen for modelling the distribution of flood volume at this station as it provides the best performance.

The RMSE and AIC calculated using Eqs (5.34) and (5.35a) can also be employed to test the goodness of fit for the copula model. In a bivariate case, F_c in Eqs (5.34) and (5.35a) is the calculated CDF value from the copula function. In comparison, F_e is the corresponding empirical CDF which is calculated in a manner analogous to that for a univariate variable as in Eq (5.31).

Let $(x_1, y_1), (x_2, y_2), \ldots, (x_N, y_N)$ be the bivariate observations. These pairs were ranked in ascending order by the values of the variable X. For each pair, i.e., (x_i, y_i), by counting the number of pairs (x_j, y_j) such that $x_j \le x_i, y_j \le y_i, i, j = 1, 2, 3, \ldots, N$, the cumulative joint probability of (x_i, y_i) of flood peak and volume can be computed (Zhang and Singh 2006):

$$F_e(x_i, y_i) = P(X \le x_i, Y \le y_i) = \frac{No.of(x_j \le x_i \text{ and } y_j \le y_i) - 0.44}{N + 0.12} \tag{5.44}$$

A MATLAB function for producing the empirical probabilities in a bivariate case is presented in Code 5.7. Note that Code 5.7 is a MATLAB function which implies it cannot be run by itself, but needs to be called by other MATLAB programs.

Code 5.7 *The MATLAB function for the empirical probabilities for bivariate cases. This function is named 'EP2Var' and its inputs are the flood peak and flood volume; the output is the empirical probability. Save this as the file 'EP2Var.m' on your computer.*

```
function [ Pjoint ] = EP2Var(Peak, Vol)
%Empirical probability for bivariate cases
% Inputs, Flood Peak and Flood Volume
N = length(Vol);

for i = 1:N
    si = 0;
    for j = 1:N
        if((Peak(j)<=Peak(i))&&(Vol(j)<=Vol(i)))
            si = si + 1;
        else
            si = si +0;
        end
    end
    Pjoint(i) = (si - 0.44)/(N + 0.12);
end
end
```

Table 5.10 and Figure 5.13 present the performance of the copula functions in modelling the dependence between flood peaks and volumes at the chosen station. The results suggest that all three copula functions can perform well when compared with the empirical CDF values. However, the Gumbel copula seems to give the best performance. The MATLAB script to establish the copula models and do the relevant comparison work is presented in Code 5.8.

Table 5.10 Modelling performance of different copulas on the flood peak and volume. These calculations are performed using the MATLAB script in Code 5.8.

Copula	RMSE	AIC
Gumbel	0.008	−535.217
Clayton	0.030	−384.683
Frank	0.013	−476.964

Code 5.8 *The MATLAB script for establishment of the copula models. Note that the file containing the MATLAB function 'EP2Var.m' should be present in the same folder to run this code. This code generates Figure 5.13 and Table 5.10.*

```
clc;clear all;close all;set(0,'defaultaxesfontsize',16);
%Step 1 Load the flood data
flood_peak = [1380, 792, 1391.3, 775, 540.1, 473, 1690.4, 262, 3730, 555.99,...
    1044.2, 609, 1350, 929, 272, 2250, 548, 913, 1060, 2580, 831, 580, 547,...
    914, 392, 515, 674, 438, 781, 398, 1427, 509.8, 631.3, 927, 1560, 752,...
    1100, 875, 3080 , 976, 835, 629, 217.3, 573.13, 476.13, 1388.1, 865.13,...
    809.27, 317, 249, 300, 297, 828, 607, 301];
flood_vol = [1653.5, 935.85, 1820.3, 598.5, 1571.9, 1365.3, 2528.3, 369.7, 6069.8,...
    683.77, 1423, 584, 3896.4, 1849.5, 303.7, 7576.1, 650.2, 1464.1, 4033,...
    4768.2, 2021, 1616, 718.4, 2186, 1376.3, 595.1, 1128, 675, 1330.1, 637.8,...
    3316 , 815.07, 750.2, 1778.2, 5074.1, 1022.4, 1804, 1627.3, 5797.1,...
    2863.8, 1136.2, 1391.5, 404.1, 966.29, 534.86, 4161.9, 1832.3, 1156.5,...
    312.5, 659.45, 548.85, 508.15, 1720.7, 1329.7, 553.1];
%change the data to column vector
peak = flood_peak';  vol = flood_vol';
% Step 2: estimate the marginal distribution through GEV
Pgev_peak = mle(peak,'distribution','gev'); %Pgev includes the three model parameters
Pgev_vol = mle(vol,'distribution','gev'); %Pgev includes the three model parameters
% Step 3, estimate the parameter of the copula function
tau = corr(peak,vol,'type','Kendall');  % Kendall coefficient
theta_GH = 1/(1-tau); % parameter for Gumbel copula
theta_CJ=(2*tau)/(1-tau); % parameter for the Clayton copula
theta_Frank = copulaparam('Frank',tau); % parameter for Frank Copula
% Step 4: Comparison of different copula models
%--------------------define the copula functions-------------------
GHcdf = @(u,v,theta_GH)exp(-((-log(u))^theta_GH+(-log(v))^theta_GH)^(1/theta_GH));
%Gumbel copula
CJcdf = @(u,v,theta_CJ)((u^(-theta_CJ))+(v^(-theta_CJ))-1)^(-1/theta_CJ);
% Clayton Copula
Fcdf = @(u,v,theta_Frank)(-1/theta_Frank)*log(1+((exp(-theta_Frank*u)-1)*...
    (exp(-theta_Frank*v)-1)/(exp(-theta_Frank)-1)));%Frank Copula
%---------obtain the empirical probabililities-----------
Pjoint = EP2Var(peak,vol);
%---------obtain the theoretical probabilities through copula-------
N = length(peak);
for i = 1:N
    u = cdf('gev',peak(i),Pgev_peak(1), Pgev_peak(2),Pgev_peak(3));
        % cdf, flood peak
    v = cdf('gev',vol(i),Pgev_vol(1), Pgev_vol(2),Pgev_vol(3)); %
        % cdf, flood volume
    Pjoint_GH(i) =  GHcdf(u,v,theta_GH);
    Pjoint_CJ(i) =  CJcdf(u,v,theta_CJ);
    Pjoint_Frank(i) = Fcdf(u,v,theta_Frank);
end
TestMatrix = zeros(3,2); %define the matrix to store the test results;
%define the RMSE funtion
rmseFUN = @(n,k, Fc, Fe)sqrt((1/(n-k))*sum(Fc-Fe).^2);
%define the AIC function
aicFUN = @(n,k,MSE)n*log(MSE)+2*k;
TestMatrix(1,1) = rmseFUN(N,1,Pjoint_GH,Pjoint);
MSE = TestMatrix(1,1)^2;
TestMatrix(1,2) = aicFUN(N,1,MSE);
TestMatrix(2,1) = rmseFUN(N,1,Pjoint_CJ,Pjoint);
```

```
MSE = TestMatrix(2,1)^2;
TestMatrix(2,2) = aicFUN(N,1,MSE);
TestMatrix(3,1) = rmseFUN(N,1,Pjoint_Frank,Pjoint);
MSE = TestMatrix(3,1)^2;
TestMatrix(3,2) = aicFUN(N,1,MSE);
%%%%%%% PRINT the performances of different copulas (RMSE are AIC)
display('Performances of different copulas'); display(TestMatrix);
%-----------------Plot comparison---------------------------
x = 0:0.01:1; y = x;
figure
subplot('position',[0.1 0.75 0.4 0.21]); plot(x,y,'b-','linewidth', 1.5);
hold on; plot(Pjoint,Pjoint_GH,'ko','markersize',5,'linewidth', 1.0);
xlabel('Empirical CDF'); ylabel('Calculated CDF');
title('(a) Gumbel Copula vs. Empirical');set(gca,'linewidth', 1.5);
legend('45^o line','Gumble','location','southeast');
%
subplot('position',[0.1 0.42 0.4 0.21]); plot(x,y,'b-','linewidth', 1.5);
hold on; plot(Pjoint,Pjoint_CJ,'ko','markersize',5,'linewidth',1.0);
xlabel('Empirical CDF'); ylabel('Calculated CDF');
title('(b) Clayton Copula vs. Empirical');set(gca,'linewidth', 1.5);
legend('45^o line','Clayton Copula','location','southeast');
%
subplot('position',[0.1 0.09 0.4 0.21]); plot(x,y,'b-','linewidth', 1.5);
hold on; plot(Pjoint,Pjoint_Frank,'ko','markersize',5,'linewidth', 1.0);
xlabel('Empirical CDF'); ylabel('Calculated CDF');
title('(c) Frank Copula vs. Empirical');set(gca,'linewidth', 1.5);
legend('45^o line','Frank Copula','location','southeast');
```

Based on the results in Table 5.10 and Figure 5.13, it can be concluded that the Gumbel copula would perform better on modelling the dependence between flood peaks and volumes at the chosen station. Therefore, this copula will be adopted to derive the multivariate return periods in the AND, OR, and Kendall cases. Figure 5.14 presents contour plots of the three multivariate return periods, in which the blue circles show the historical observations. The comparison between univariate and multivariate return periods is presented in Table 5.11. It shows that the multivariate return periods in the AND and Kendall cases would generally be larger than the univariate return period, while the multivariate return period in the OR case is less than the corresponding univariate return period. The MATLAB script for deriving those multivariate return periods is presented in Code 5.9, which generates Figure 5.14 and Table 5.11.

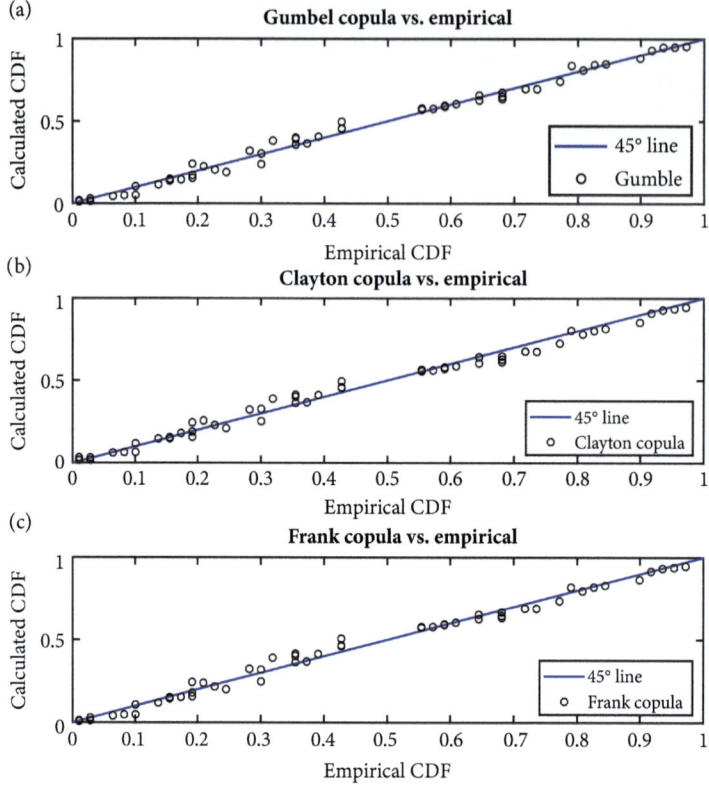

Figure 5.13 Comparison between empirical CDFs and calculated CDFs from different copulas. This figure is generated using the MATLAB script in Code 5.8.

Code 5.9 *The MATLAB script for generating the joint return periods based on a copula. Note that the file containing the MATLAB function 'EP2Var.m' should be present in the same folder to run this code. This code generates Figure 5.14 and Table 5.11. This code and other code in this chapter can be downloaded from the following website: http://www.oup.com/AdvancedMathematicalModelling.*

```
clc;clear all;close all;set(0,'defaultaxesfontsize',15);
%Step 1 Load the flood data
flood_peak = [1380, 792, 1391.3, 775, 540.1, 473, 1690.4, 262, 3730, 555.99,...
    1044.2, 609, 1350, 929, 272, 2250, 548, 913, 1060, 2580, 831, 580, 547,...
    914, 392, 515, 674, 438, 781, 398, 1427, 509.8, 631.3, 927, 1560, 752,...
    1100, 875, 3080 , 976, 835, 629, 217.3, 573.13, 476.13, 1388.1, 865.13,...
    809.27, 317, 249, 300, 297, 828, 607, 301];
flood_vol = [1653.5, 935.85, 1820.3, 598.5, 1571.9, 1365.3, 2528.3, 369.7, 6069.8,...
    683.77, 1423, 584, 3896.4, 1849.5, 303.7, 7576.1, 650.2, 1464.1, 4033,...
    4768.2, 2021, 1616, 718.4, 2186, 1376.3, 595.1, 1128, 675, 1330.1, 637.8,...
    3316 ,   815.07, 750.2, 1778.2, 5074.1, 1022.4, 1804, 1627.3, 5797.1,...
    2863.8, 1136.2, 1391.5, 404.1, 966.29, 534.86, 4161.9, 1832.3, 1156.5,...
    312.5, 659.45, 548.85, 508.15, 1720.7, 1329.7, 553.1];
peak = flood_peak';  vol = flood_vol';
% Step 2: estimate the marginal distribution through GEV
Pgev_peak = mle(peak,'distribution','gev'); %Pgev includes three model parameters
Pgev_vol = mle(vol,'distribution','gev'); %Pgev includes three model parameters
```

```
% Step 3, estimate the parameter of the copula function
tau = corr(peak,vol,'type','Kendall'); % Kendall coefficient
theta_GH = 1/(1-tau); % parameter for Gumbel copula
theta_CJ=(2*tau)/(1-tau); % parameter for the Clayton copula
theta_Frank = copulaparam('Frank',tau); % parameter for Frank Copula
% Step 4: Comparison of different copula models
%---------------------define the copula functions------------------
GHcdf = @(u,v,theta_GH)exp(-((-log(u))^theta_GH+(-log(v))^theta_GH)^(1/theta_GH));
%Gumbel copula
CJcdf = @(u,v,theta_CJ)((u^(-theta_CJ))+(v^(-theta_CJ))-1)^(-1/theta_CJ);
% Clayton Copula
Fcdf = @(u,v,theta_Frank)(-1/theta_Frank)*log(1+((exp(-theta_Frank*u)-1)*...
    (exp(-theta_Frank*v)-1)/(exp(-theta_Frank)-1)));%Frank Copula
%---------obtain the empirical probabililities-----------
Pjoint = EP2Var(peak,vol);
%---------obtain the theoretical probabilities through copula-------
N = length(peak);
for i = 1:N
    u = cdf('gev',peak(i),Pgev_peak(1), Pgev_peak(2),Pgev_peak(3));
        % cdf, flood peak
    v = cdf('gev',vol(i),Pgev_vol(1), Pgev_vol(2),Pgev_vol(3));
        % cdf, flood volume
    Pjoint_GH(i) =  GHcdf(u,v,theta_GH);
    Pjoint_CJ(i) =  CJcdf(u,v,theta_CJ);
    Pjoint_Frank(i) = Fcdf(u,v,theta_Frank);
end
% Step 5: Calculate the joint return period: % Based on Step 4, the Gumbel copula
% will be adopted
 %set the curve presence for the joint RP
v_or= [5 10 20 50 100];v_and= [5 10 20 50 100];v_kendall= [5 10 20 50 100];
% Secondary return period
fg = @(t)(-log(t))^theta_GH; % the generator function of Gumbel copula
delta = 1e-6;  Kc = @(t)t-fg(t)/((fg(t+delta)-fg(t))/delta);
dpeak = min(peak):50:1.2*max(peak);   dvol = min(vol):50:1.2*max(vol);
theta_GH = double(theta_GH);  N = length(peak);
for i =1:length(dpeak)
    for j = 1:length(dvol)
        u = cdf('gev',dpeak(i),Pgev_peak(1), Pgev_peak(2), Pgev_peak(3));
            % cdf for flood peak
        v = cdf('gev',dvol(j),Pgev_vol(1), Pgev_vol(2), Pgev_vol(3));
            % cdf for flood volume
        JCDF(i,j) = GHcdf(u,v,theta_GH);
        T_or(i,j) = 1/(1-JCDF(i,j));
        T_and(i,j) = 1/(1-u-v+JCDF(i,j));
        K_RP(i,j) =1/(1-Kc(JCDF(i,j)));
    end
end
figure; subplot('position',[0.1 0.75 0.42 0.20]) %--------subplot 1------
[C,h] = contour(dvol,dpeak,T_and,v_and);
```

Code 5.9 *Continued*

```
set(h,'color','k','LineWidth',1.5); clabel(C,h)
hold on; plot(vol,peak,'bo','markersize',5,'LineWidth',1.5);
ylabel('Peak (m^3/s)'); xlabel('Volume (m^3/s day)'); set(h,'color','k')
title('(a) T^A^N^D'); set(gca,'linewidth',1.5);
subplot('position',[0.1 0.42 0.42 0.20]) %----------------subplot 2-----
[C,h] = contour(dvol,dpeak,T_or,v_or);
set(h,'color','k','LineWidth',1.5); clabel(C,h)
hold on; plot(vol,peak,'bo','markersize',5,'LineWidth',1.5)
hold off; ylabel('Peak (m^3/s)'); xlabel('Volume (m^3/s day)')
set(h,'color','k','LineWidth',1.5);title('(b) T^O^R');set(gca, 'linewidth',1.5);
subplot('position',[0.1 0.09 0.42 0.20]) %----------------subplot 3----
[C,h] = contour(dvol,dpeak,K_RP,v_kendall);
set(h,'color','k','LineWidth',1.5); clabel(C,h)
hold on; plot(vol,peak,'bo','markersize',5,'LineWidth',1.5)
hold off; ylabel('Peak (m^3/s)'); xlabel('Volume (m^3/s day)')
set(h,'color','k'); title('(c) T^k^e^n^d^a^l^l'); set(gca, 'linewidth',1.5);
% ----Comparison between univariate and multivariate return periods---------
T = [10, 20, 50, 100, 200]; % set the univariate return periods
p = 1-1./T; %the corresponding CDF
OutputMatrix = zeros(length(T),6);
%colnames ='univariate RP','peak magnitude','vol magnitude','Tand','Tor','Tkendall'
for i = 1:length(T)
    OutputMatrix(i,1) =T(i);
    OutputMatrix(i,2) = icdf('gev',p(i),Pgev_peak(1), Pgev_peak(2),...
        Pgev_peak(3));
    OutputMatrix(i,3) = icdf('gev',p(i),Pgev_vol(1), Pgev_vol(2),...
        Pgev_vol(3));
    JointCDF = GHcdf(p(i),p(i),theta_GH);
    OutputMatrix(i,4) = 1/(1-p(i)-p(i)+JointCDF); %Tand
    OutputMatrix(i,5) = 1/(1-JointCDF); %Tor
    OutputMatrix(i,6) = 1/(1-Kc(JointCDF)); %Tkendall
end
```

5.4 Distribution and transport of pollutants

5.4.1 Pollutant transport in water

The water quality in a river or lake is significantly affected by pollutants in the discharges from urban, agricultural, and industrial sectors. There are numerous pollutants with various physical, chemical, and/or biological properties. Generally, these pollutants can be classified into two categories: nonconservative (nonpersistent) and conservative (persistent). The nonconservative pollutants such as bacteria or some chemical compounds can change in accordance with specific biological or chemical processes that are enhanced by contact with water, giving rise to other substances through reactions or mutations (Benedini and Tsakiris 2013). These pollutants include biochemical oxygen demand, ammonia, and certain other organic compounds. In comparison, the pollutants that do not undergo similar alterations are conservative (or persistent; Benedini and Tsakiris 2013). These include, but are not limited to, salts and metals. For the sake of clarity, only conservative pollutants will be considered in this chapter.

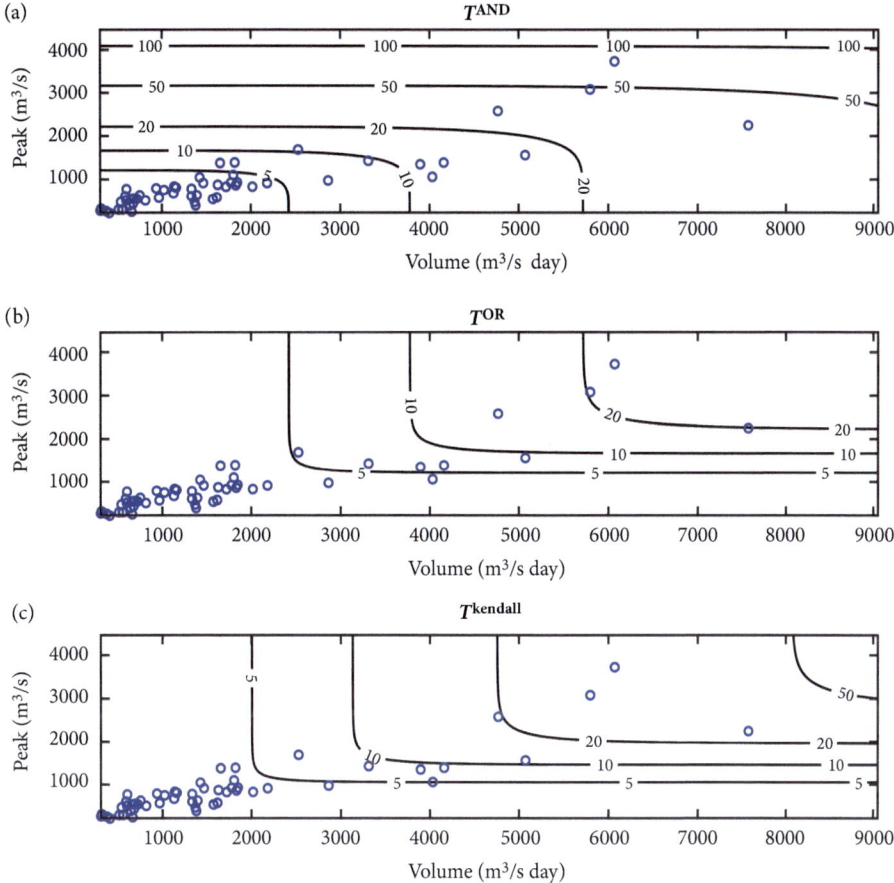

Figure 5.14 Contour plot for the joint return periods in the AND, OR, and Kendall cases. The blue circles indicate the real observations for flood peaks and volumes. The contour lines indicate the flood peak and volume combinations with return periods in the AND, OR, and Kendall cases. This figure is generated using Code 5.9.

Consider an elementary volume of water as shown in Figure 5.15. A pollutant in water is quantified by its concentration, expressed in terms of mass per unit volume, its dimensions being $[ML^{-3}]$. If V is the volume of the water body and M is the mass of the pollutant, the concentration of the pollutant (C) can be expressed as

$$C = \frac{M}{V} \tag{5.45}$$

Consider two contiguous points P and P' in the water volume as shown in Figure 5.16. The pollutant may be transported from the former point to the latter during a certain time interval, which will lead to the concentration decreasing at P and increasing at P'. Such a transport process is mainly conducted by two mechanisms: advection and dispersion.

Table 5.11 Comparison between univariate and multivariate return periods. The results listed in this table are generated using Code 5.9.

Univariate RP	Flood peak (m³/s)	Flood volume (m³/s day)	T^{AND}	T^{OR}	$T^{Kendall}$
10.0	1660.8	3777.2	12.5	8.3	11.3
20.0	2215.7	5717.9	25.2	16.6	22.7
50.0	3161.2	9683.8	63.2	41.3	56.9
100.0	4087.5	14305.8	126.7	82.6	114.0
200.0	5249.8	21051.1	253.6	165.1	228.2

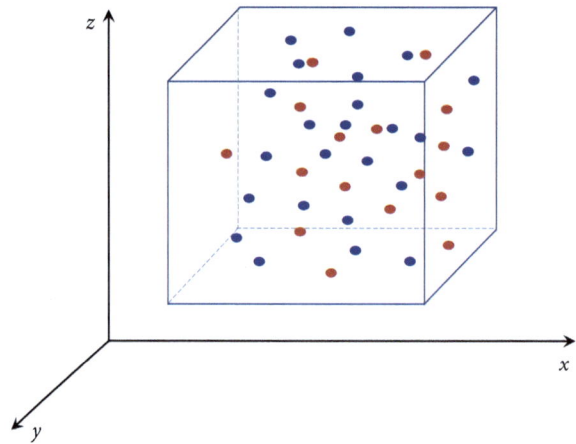

Figure 5.15 Pollutant particles (red dots) and water particles (blue dots) in the elementary volume of the water body.

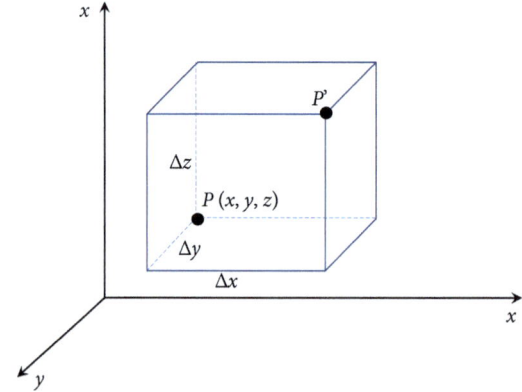

Figure 5.16 An elementary volume with dimensions $\Delta x \times \Delta y \times \Delta z$ around point P.

Advection refers to horizontal transport by flows that move patches of material around, but do not significantly distort or dilute them; it represents the primary pollutant transport process in the longitudinal direction in a river system (Ji 2008). Assume that the flow volume at point P in Figure 5.16 has a pollutant concentration of $C\left(x, y, z, t\right)$ and velocity $v\left(x, y, z, t\right)$ at time t. The advection transport depends on the water velocity $v\left(x, y, z, t\right)$ at the point P, which depends on the time t and has components v_x, v_y, and v_z in the x, y, and z directions, respectively. Along the x axis direction, the pollutant particles are transported with the moving water which has a velocity along the x axis of v_x. Consequently, the pollutant mass crossing the area of $\Delta y \times \Delta z$ of the elementary volume after a time interval of Δt would be

$$M_x = C \Delta y \Delta z \, v_x \, \Delta t \tag{5.46a}$$

The pollutant mass respectively crossing the areas of $\Delta y \times \Delta x$ (along the z axis direction) and $\Delta x \times \Delta z$ (along the y axis direction) can be similarly obtained:

$$M_y = C \Delta x \Delta z \, v_y \, \Delta t \tag{5.46b}$$

$$M_z = C \Delta x \Delta y \, v_z \, \Delta t \tag{5.46c}$$

Dispersion (sometimes also known as diffusion) is the horizontal spreading and mixing of water mass caused by turbulent mixing and molecular diffusion (Ji 2008). Mass diffusion in water can be described by Fick's law, which was developed by Adolph Fick in 1855 to quantify the movement of a substance from areas of high concentration to areas of low concentration. Fick's law states that the transfer rate of mass across an interface normal to the diffusion direction (e.g., the x direction) in a quiescent fluid varies directly with the coefficient of molecular diffusion E and the negative gradient of solute concentration (Chanson 2004). If $J\left(x, y, z, t\right)$ denotes the pollutant flux crossing the unit area in unit time, then we have (Benedini and Tsakiris 2013)

$$\vec{J} = -E \overrightarrow{grad}\left(C\right) \tag{5.47}$$

where E is the dispersion coefficient describing the characteristic of the liquid field, and $\overrightarrow{grad}\left(C\right)$ is the concentration gradient describing the difference of concentration in a given direction between two points. Along the x direction, the concentration gradient can be expressed as

$$grad(C)_x = \frac{\partial C}{\partial x} \tag{5.48}$$

and the corresponding pollutant flux can be obtained as

$$J_x = -E \frac{\partial C}{\partial x} \tag{5.49}$$

Therefore, the mass diffusion across the area of $\Delta y \times \Delta x$ of the elementary volume in a time interval of Δt can be obtained as (Benedini and Tsakiris 2013)

$$J_x \Delta x \Delta y \Delta t = -E \frac{\partial C}{\partial x} \Delta x \Delta y \Delta t \tag{5.50}$$

There are other processes that can change pollutant concentrations in water. Some examples include local injection (source) from outside the water volume, local subtraction (sink), leaching, or absorbing pollutants at the bottom, the wall, or at the free surface of the water body (Benedini and Tsakiris 2013). The effects of such processes can be expressed in a general form by means of a term $\mp S\,(x,\, y,\, z,\, t)$, known as the source term.

5.4.2 The general equation of pollutant transport in water

Based on the mechanisms for pollutant transport in water, it is now possible to formally develop the general mathematical equations to describe how the pollutant mass varies at a certain point of the water volume. Consider the elementary volume illustrated in Figure 5.16, assuming the water velocity along the x, y, and z directions is v_x, v_y, and v_z, respectively.

Due to the advection effect, the pollutant mass entering the face of $\Delta y \times \Delta z$ at the location of x in a time interval of Δt would be

$$M_x = v_x C_x \Delta y \Delta z \Delta t \tag{5.51a}$$

Similarly, the mass of pollutant going out of the face of $\Delta y \times \Delta z$ at the location of $x + \Delta x$ in a time interval of Δt would be

$$M_{x+\Delta x} = v_{x+\Delta x} C_{x+\Delta x} \Delta y \Delta z \Delta t \tag{5.51b}$$

The difference of Eqs (5.51a) and (5.51b) indicates the mass variation of pollutant in the time interval of Δt due to the advection:

$$\Delta M_{adv,x} = (v_{x+\Delta x} C_{x+\Delta x} - v_x C_x)\, \Delta y \Delta z \Delta t \tag{5.52a}$$

If $\Delta M_{adv} \geq 0$, the outflow of pollutant mass is greater than the inflow. Otherwise, the outflow of pollutant mass is less than the inflow. For the y and z directions, the mass variations of pollutant due to the advection can be similarly expressed:

$$\Delta M_{adv,y} = \left(v_{y+\Delta y} C_{y+\Delta y} - v_y C_y\right) \Delta x \Delta z \Delta t \tag{5.52b}$$

$$\Delta M_{adv,z} = \left(v_{z+\Delta z} C_{z+\Delta z} - v_z C_z\right) \Delta x \Delta y \Delta t \tag{5.52c}$$

In addition to advection, the elementary volume also undergoes the effect of dispersion. During the time interval of Δt, the mass inflow of pollutant entering the face of $\Delta y \times \Delta z$ at the location of x can be expressed as

$$D_x = -\left(E \frac{\partial C}{\partial x}\right)_x \Delta x \Delta y \Delta t \tag{5.53a}$$

The pollutant mass going out of the face of $\Delta y \times \Delta z$ at the location of $x + \Delta x$ in the time interval of Δt can be formulated as

$$D_{x+\Delta x} = -\left(E\frac{\partial C}{\partial x}\right)_{x+\Delta x} \Delta x \Delta y \Delta t \tag{5.53b}$$

Consequently, the difference of Eqs (5.53a) and (5.53b) denotes the mass change of pollutant due to dispersion in the elementary volume in a time interval of Δt:

$$\Delta D_{dis,x} = -\left[\left(E\frac{\partial C}{\partial x}\right)_{x+\Delta x} - \left(E\frac{\partial C}{\partial x}\right)_{x}\right]\Delta x \Delta y \Delta t \tag{5.54a}$$

The mass variations of pollutant resulting from the dispersion effect along the y and z directions can be expressed similarly:

$$\Delta D_{dis,y} = -\left[\left(E\frac{\partial C}{\partial y}\right)_{y+\Delta y} - \left(E\frac{\partial C}{\partial y}\right)_{y}\right]\Delta x \Delta z \Delta t \tag{5.54b}$$

$$\Delta D_{dis,z} = -\left[\left(E\frac{\partial C}{\partial z}\right)_{z+\Delta z} - \left(E\frac{\partial C}{\partial z}\right)_{z}\right]\Delta x \Delta y \Delta t \tag{5.54c}$$

Moreover, the mass variation of the pollutant due to local injection or subtraction can be considered as $S(x, y, z, t)\Delta x \Delta y \Delta z \Delta t$, where S is the concentration of the pollutant injected/subtracted.

The combined effects from advection, i.e., Eq (5.52), dispersion, i.e., Eq (5.54), and local injection/subtraction lead to the mass variation of the pollutant in the elementary volume, which will further cause the concentration variation of the pollutant. The mass flowing out of the elementary volume in the time interval Δt can be written as

$$\Delta M = -(C_{t+\Delta t} - C_t)\Delta x \Delta y \Delta z \tag{5.55}$$

Based on the principle of mass balance, the value of ΔM should be equal to the mass change resulting from advection, dispersion, and local injection/subtraction. Thus, we have

$$-(C_{t+\Delta t} - C_t)\Delta x \Delta y \Delta z =$$

$$(v_{x+\Delta x}C_{x+\Delta x} - v_x C_x)\Delta y \Delta z \Delta t + (v_{y+\Delta y}C_{y+\Delta y} - v_y C_y)\Delta x \Delta z \Delta t + (v_{z+\Delta z}C_{z+\Delta z} - v_z C_z)\Delta x \Delta y \Delta t$$

$$-\left[\left(E\frac{\partial C}{\partial x}\right)_{x+\Delta x} - \left(E\frac{\partial C}{\partial x}\right)_{x}\right]\Delta z \Delta y \Delta t - \left[\left(E\frac{\partial C}{\partial y}\right)_{y+\Delta y} - \left(E\frac{\partial C}{\partial y}\right)_{y}\right]\Delta x \Delta z \Delta t$$

$$-\left[\left(E\frac{\partial C}{\partial z}\right)_{z+\Delta z} - \left(E\frac{\partial C}{\partial z}\right)_{z}\right]\Delta x \Delta y \Delta t - S\Delta x \Delta y \Delta z \Delta t$$

$$\tag{5.56}$$

Dividing by $\Delta x\,\Delta y\,\Delta z\,\Delta t$ on both left- and right-hand sides, Eq (5.56) can be reformulated as

$$-\frac{C_{t+\Delta t}-C_t}{\Delta t}=\frac{v_{x+\Delta x}C_{x+\Delta x}-v_x C_x}{\Delta x}+\frac{v_{y+\Delta y}C_{y+\Delta y}-v_y C_y}{\Delta y}+\frac{v_{z+\Delta z}C_{z+\Delta z}-v_z C_z}{\Delta z}$$

$$-\frac{\left(E\frac{\partial C}{\partial x}\right)_{x+\Delta x}-\left(E\frac{\partial C}{\partial x}\right)_x}{\Delta x}-\frac{\left(E\frac{\partial C}{\partial y}\right)_{y+\Delta y}-\left(E\frac{\partial C}{\partial y}\right)_y}{\Delta y}-\frac{\left(E\frac{\partial C}{\partial z}\right)_{z+\Delta z}-\left(E\frac{\partial C}{\partial z}\right)_z}{\Delta z}-S$$

$$(5.57)$$

Let Δx, Δy, Δz, and Δt tend to 0. As a result, Eq (5.57) takes the following form:

$$\frac{\partial C}{\partial t}+\frac{\partial\left(v_x C\right)}{\partial x}+\frac{\partial\left(v_y C\right)}{\partial y}+\frac{\partial\left(v_z C\right)}{\partial z}=E\left(\frac{\partial^2 C}{\partial x^2}+\frac{\partial^2 C}{\partial y^2}+\frac{\partial^2 C}{\partial z^2}\right)+S \qquad (5.58)$$

For a one-dimensional pollutant transportation problem along the x direction in the water body, Eq (5.58) can be simplified as follows:

$$\frac{\partial C}{\partial t}=-\frac{\partial\left(v_x C\right)}{\partial x}+E\frac{\partial^2 C}{\partial x^2}+S \qquad (5.58)$$

If the fluid moves with velocity v in the x direction, Eq (5.58) can be solved analytically under the assumptions that (i) the sink/source term is neglected, (ii) the flow is uniform, and (iii) the solute is fully mixed with a constant E in the channel (Holzbecher 2012; Ramezani et al. 2019), leading to the following equation for the concentration of the pollutant (C):

$$C(x,t)=\frac{M}{\sqrt{4\pi t\sqrt{E}}}\exp\left(-\frac{(x-vt)^2}{4tE}\right) \qquad (5.59)$$

where M denotes the total mass per unit area in the fluid system.

5.4.3 Finite difference method

5.4.3.1 Finite difference

In many cases, pollutant transport in a water body is generally expressed as a PDE, which can seldom be solved analytically and thus are generally solved by numerical methods. The finite difference method (FDM) is one of the major numerical methods for solving PDEs. The FDM is developed based on numerical differentiation for the partial derivatives in PDEs, in which the infinitesimal terms are replaced with finite steps of convenient size. For example, if f is a function of x (Figure 5.17), the partial derivative of f at the point x_0

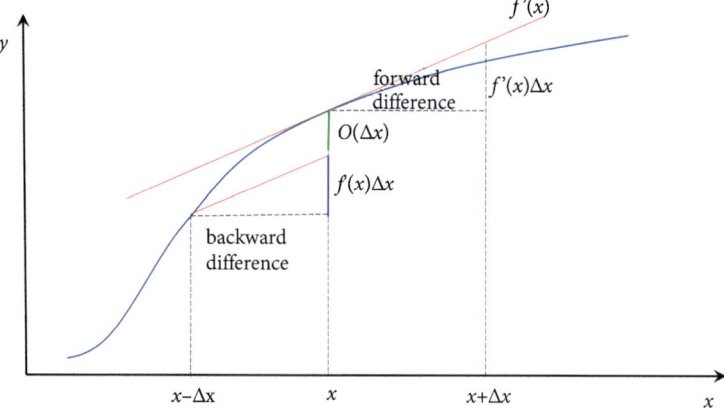

Figure 5.17 Forward and backward difference approximations to a first derivative as used in the finite difference method.

can be written as

$$\frac{\partial f}{\partial x} = \lim_{\Delta x \to 0} \frac{f(x_0 + \Delta x) - f(x)}{\Delta x} \tag{5.60}$$

According to Taylor's series, the value of f at the point $x_0 + \Delta x$ can be approximated as

$$f(x_0 + \Delta x) = f(x_0) + \Delta x \frac{\partial f}{\partial x} + \frac{\Delta x^2}{2} \frac{\partial^2 f}{\partial x^2} + \dots \tag{5.61a}$$

Similarly, the value of f at the point $x_0 - \Delta x$ can be approximated as

$$f(x_0 - \Delta x) = f(x_0) + (-\Delta x) \frac{\partial f}{\partial x} + \frac{(-\Delta x)^2}{2} \frac{\partial^2 f}{\partial x^2} + \dots \tag{5.61b}$$

By truncating the Taylor series after the first-derivative term, Eq (5.61) can be reformulated as

$$f(x_0 \pm \Delta x) = f(x_0) \pm \Delta x \frac{\partial f}{\partial x} + O(\Delta x)^2 \tag{5.62}$$

By rearranging Eq (5.62), the partial derivative of f can be approximated as follows:

$$\frac{\partial f}{\partial x} = \frac{f(x_0 + \Delta x) - f(x)}{\Delta x} + O(\Delta x) \tag{5.63a}$$

$$\frac{\partial f}{\partial x} = \frac{f(x_0) - f(x - \Delta x)}{\Delta x} + O(\Delta x) \tag{5.63b}$$

Equation (5.63a) is called a forward difference and Eq (5.63b) a backward difference (Figure 5.17). Both expressions have a truncation error of $O(\Delta x)$.

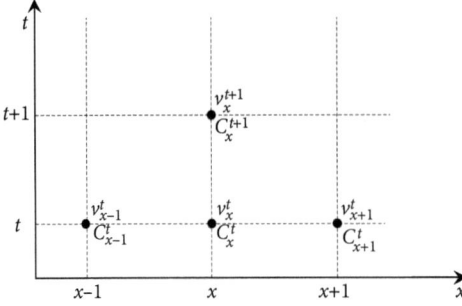

Figure 5.18 Discretization for the explicit FDM scheme (adapted from Benedini and Tsakiris 2013).

The derivative of f can be approximated with a smaller error through using the first three terms of the Taylor series in Eq (5.61). Let Eq (5.61b) be subtracted from Eq (5.61a), which gives the following equation (Hornberger and Wiberg 2013):

$$f(x_0 + \Delta x) - f(x_0 - \Delta x) = 2\Delta x \frac{\partial f}{\partial x} + O\left((\Delta x)^3\right) \tag{5.64}$$

By rearranging Eq (5.64), another approximation for the first-order derivative of f is obtained:

$$\frac{\partial f}{\partial x} = \frac{f(x_0 + \Delta x) - f(x - \Delta x)}{2\Delta x} + O\left((\Delta x)^2\right) \tag{5.65}$$

This is referred to as a central difference approximation, which has an error of order $O((\Delta x)^2)$.

5.4.3.2 Explicit numerical schemes

To understand the basic concepts of the FDM, a one-dimensional pollutant transport problem consisting of both advection and dispersion is employed. First, the domain of x and t is discretized in Figure 5.18. The six points in the grid represent the situation of three cross sections along the x direction (i.e., at the $x - 1$, x, and $x + 1$ coordinates) and at two time instants (i.e., t and $t + 1$), with each point characterized by a specific value of velocity, v, in the x direction, and a pollutant concentration C.

Consider the one-dimensional advection–dispersion equation

$$\frac{\partial C}{\partial t} = -\frac{\partial (vC)}{\partial x} + E\frac{\partial^2 C}{\partial x^2} \tag{5.66}$$

For the four points available based on the grid presented in Figure 5.18, the velocity v and the concentration C at time t are supposed to be known. The explicit numerical scheme called forward time / centred space (FTCS) can be developed with the following discrete terms (Benedini and Tsakiris 2013):

$$\frac{\partial C}{\partial t} = \frac{C_x^{t+1} - C_x^t}{\Delta t} \tag{5.67a}$$

$$\frac{\partial C}{\partial x} = \frac{C_{x+1}^t - C_{x-1}^t}{2\Delta x} \tag{5.67b}$$

$$\frac{\partial^2 C}{\partial x^2} = \frac{C_{x+1}^t - 2C_x^t + C_{x-1}^t}{\Delta x^2} \tag{5.67c}$$

Equation (5.67) can be converted to the following form:

$$\frac{C_x^{t+1} - C_x^t}{\Delta t} = -\frac{v}{2\Delta x}\left(C_{x+1}^t - C_{x-1}^t\right) + \frac{E}{\Delta x^2}\left(C_{x+1}^t - 2C_x^t + C_{x-1}^t\right) \tag{5.68}$$

Thus, we have

$$C_x^{t+1} = \beta_1 C_{x-1}^t + \beta_2 C_x^t + \beta_3 C_{x+1}^t \tag{5.69a}$$

where

$$\beta_1 = \Delta t\left(\frac{v}{2\Delta x} + \frac{E}{\Delta x^2}\right) \tag{5.69b}$$

$$\beta_2 = \Delta t\left(\frac{1}{\Delta t} - \frac{2E}{\Delta x^2}\right) \tag{5.69c}$$

$$\beta_3 = \Delta t\left(-\frac{v}{2\Delta x} + \frac{E}{\Delta x^2}\right) \tag{5.69d}$$

Equation (5.69) is a linear equation for every space and time step, with only one variable to be calculated (i.e., C_x^{t+1}). Such a numerical scheme is considered as explicit, which is stable only under certain conditions. Fletcher (1990) stated that the FTCS is stable if the following condition is met:

$$0 \le \left(\frac{v\Delta t}{\Delta x}\right)^2 \le \frac{2E\Delta t}{\Delta x^2} \le 1 \tag{5.70}$$

Fletcher (1990) also recommends the following condition for achieving sufficient accuracy:

$$\frac{v\Delta x}{E} \ll \frac{2\Delta x}{v\Delta t} \tag{5.71}$$

5.4.3.3 Implicit numerical schemes

Compared with the explicit numerical schemes, implicit numerical schemes are performed at the end of the iteration to formulate a system of linear equations. These linear equations are solved through numerical methods such as Gaussian elimination or the Gauss–Seidel

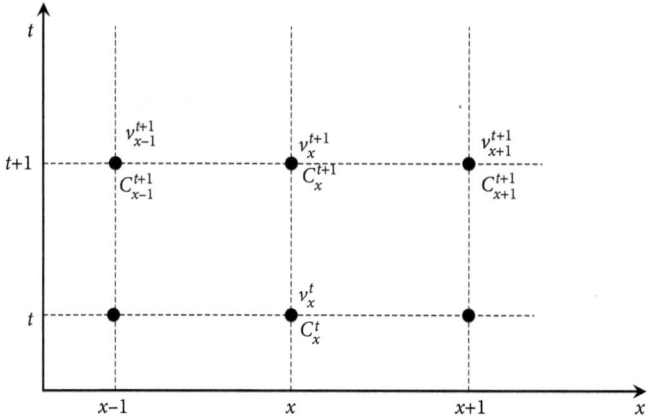

Figure 5.19 Discretization for the backward time / centred space scheme as used in FDM (adapted from Benedini and Tsakiris 2013).

method to obtain the variables for the next step. One major advantage of implicit schemes is that they are stable in all cases. Backward time / centred space (BTCS) is one of the most popular implicit numerical schemes.

Figure 5.19 presents the discretization for the BTCS scheme, in which the velocity and concentration at time t are supposed to be known. The first- and second-order derivatives in Eq (5.66) can be approximated as

$$\frac{\partial C}{\partial t} = \frac{C_x^{t+1} - C_x^t}{\Delta t} \tag{5.72a}$$

$$\frac{\partial C}{\partial x} = \frac{C_{x+1}^{t+1} - C_{x-1}^{t+1}}{2\Delta x} \tag{5.72b}$$

$$\frac{\partial^2 C}{\partial x^2} = \frac{C_{x+1}^{t+1} - 2C_x^{t+1} + C_{x-1}^{t+1}}{\Delta x^2} \tag{5.72c}$$

Thus, Eq (5.66) can be reformulated as

$$\frac{C_x^{t+1} - C_x^t}{\Delta t} = -\frac{v}{2\Delta x}\left(C_{x+1}^{t+1} - C_{x-1}^{t+1}\right) + \frac{E}{\Delta x^2}\left(C_{x+1}^{t+1} - 2C_x^{t+1} + C_{x-1}^{t+1}\right) \tag{5.73}$$

Equation (5.73) can be further converted to

$$\alpha_1 C_{x-1}^{t+1} + \alpha_2 C_x^{t+1} + \alpha_3 C_{x+1}^{t+1} + \frac{C_x^t}{\Delta t} = 0 \tag{5.74a}$$

where

$$\alpha_1 = \left(\frac{E}{\Delta x^2} + \frac{v}{2\Delta x}\right) \tag{5.74b}$$

$$\alpha_2 = -\left(\frac{2E}{\Delta x^2} + \frac{1}{\Delta t}\right) \tag{5.74c}$$

$$\alpha_3 = \left(\frac{E}{\Delta x^2} - \frac{v}{2\Delta x}\right) \tag{5.74d}$$

5.4.3.4 Boundary and initial conditions

In order to solve the PDEs for pollutant transport in the water body, appropriate boundary and initial conditions are required. The initial conditions reflect the values of variables at $t = 0$ in the computational field. In terms of the advection–dispersion equation Eq (5.66), the initial value of the pollutant concentration, $C(x, 0) = C_0$, is assumed to be known. Also, the water velocity v at $t = 0$ is also assumed to be predetermined, namely $v(x, 0) = v_0$.

The boundary conditions describe the status of the variables (e.g., pollutant concentration) at the boundary of the computational field. There are different alternatives for considering the boundary conditions. The first type, a Dirichlet boundary condition, indicates the fixed constants for the variables at a boundary, such as the following:

$$C(0, t) = C_L, \qquad\qquad C(L, t) = C_R \qquad\qquad (5.75a)$$

where $C(L, t) = C_R$ is the right-hand boundary condition at $x = L$ in the computational field. The secondary boundary condition reflects the case where a stable concentration is finally achieved but cannot be determined a priori. This is a Neumann boundary condition, which is generally expressed as

$$\frac{\partial C}{\partial x}\bigg|_{x=L} = 0 \qquad\qquad (5.75b)$$

5.2.4 Numerical example

A numerical example, which is adapted from the case discussed by Benedini and Tsakiris (2013), is employed here to demonstrate the application of different numerical methods for solving the pollutant transport problem in a river stream.

Consider the river stream shown in Figure 5.20. One conservative pollutant with a constant concentration of C_0 (assume $C_0 = 100$ ppm) is continuously injected into the river at the point $x = 0$. This means that at the boundary $x = 0$, the constant of pollutant injection $C = 100$. At the cross section $x = 0$ the pollutant is injected continuously, and the concentration C_0 remains indefinitely constant. The water velocity, constant all over the stream, is $v = 0.3$ m/s. The dispersion coefficient, also constant, is $E = 5.0$ m^2/s. The problem considers a space discretization $\Delta x = 25\,m$ and time steps $\Delta t = 20\,s$.

5.4.4.1 Explicit numerical method

As stated in the example, $\Delta x = 25$ m and $\Delta t = 20$ s. Thus, according to Eq (5.70), we can derive the stability condition as

$$\left(\frac{v\Delta t}{\Delta x}\right)^2 = 0.0576 \qquad\qquad (5.76a)$$

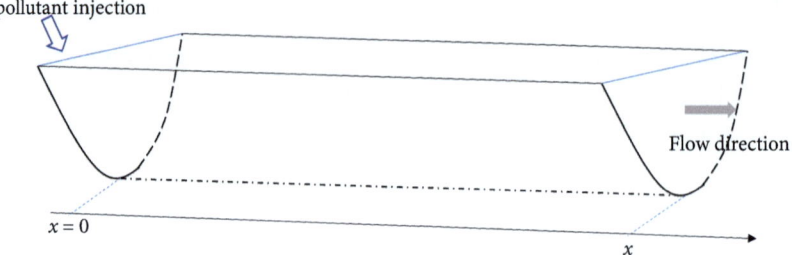

pollutant injection

Flow direction

$x = 0$

x

Figure 5.20 Schematic plot of pollutant injection in a one-dimensional stream (adapted from Benedini and Tsakiris 2013).

$$\frac{2E\Delta t}{\Delta x^2} = 0.32 \tag{5.76b}$$

$$0 \le \left(\frac{v\Delta t}{\Delta x}\right)^2 \le \frac{2E\Delta t}{\Delta x^2} \le 1 \tag{5.76c}$$

Therefore, the proposed discretization scheme is sufficiently accurate to ensure the stability of the explicit numerical scheme.

Based on Eq (5.69), the pollutant concentration at different locations and time steps can be expressed as follows:

$$C_x^{t+1} = \beta_1 C_{x-1}^t + \beta_2 C_x^t + \beta_3 C_{x+1}^t \tag{5.77a}$$

where

$$\beta_1 = \Delta t \left(\frac{v}{2\Delta x} + \frac{E}{\Delta x^2}\right) = 0.28 \tag{5.77b}$$

$$\beta_2 = \Delta t \left(\frac{1}{\Delta t} - \frac{2E}{\Delta x^2}\right) = 0.68 \tag{5.77c}$$

$$\beta_3 = \Delta t \left(-\frac{v}{2\Delta x} + \frac{E}{\Delta x^2}\right) = 0.04 \tag{5.77d}$$

The MATLAB script for the explicit scheme of FDM for the one-dimensional river stream example is presented in Code 5.10, and the resulting solutions are presented in Table 5.12 and Figure 5.21.

Code 5.10 *The MATLAB script used for solving the one-dimensional river stream example through explicit FDM. This code generates Figure 5.21 and the results in Table 5.12.*

```
clc;clear all;close all;set(0,'defaultaxesfontsize',17);
%explicit numerical method
% water velocity v = 0.3 m/s, % dispersion coefficient E = 5.0
% m^2/s
% dx = 25 m, % dt = 20 s
```

```
dt = 20; % time step for finite difference method
dx = 25; % grid spacing for finite difference method
E = 5; % dispersion coefficient
v = 0.3; % water velocity
beta1 = dt*(v/(2*dx)+E/(dx^2));
beta2 = dt*(1/dt-2*E/(dx^2));
beta3 = dt*(-v/(2*dx)+E/(dx^2));
C = zeros(50, 10); % row is time t, column is distance;
C(:,1) = 100; % at t = 0; the pollutant concentration is 100;
[nrow, ncol] = size(C);

for t = 1:(nrow-1)
    for x = 2:(ncol-1)   %x = 1 means t = 0
        C((t+1),x) = beta1*C(t,(x-1))+beta2*C(t,x)+beta3*C(t,(x+1));
    end
end
%%%%%%%%%%%%%%%%%% PLOT %%%%%%%%%%%%%%%%%%%%%%%%%%%%%%%%%%%%%%%%%%%%%
%it = 1:10;
T = 0:20:20*49
plot(T,C(:,2),'k','linewidth',1.5); % x = 25;
hold on; plot(T,C(:,3),'k--','linewidth',1.5); %x = 50
hold on; plot(T,C(:,4),'b-.','linewidth',1.5); %x = 75;
hold on ; plot(T,C(:,5),'mo','markersize',5,'linewidth',1.5); %x = 100;
xlabel('Time (s)'); ylabel('Concentration (ppm)');
legend('Location x=25','Location x=50','Location x=75','Location x=100',...
    'location','southeast');   set(gca,'linewidth',1.5);
```

5.4.4.2 Implicit numerical method using BTCS

In addition to the explicit numerical scheme, the implicit scheme for the above example, based on Eq (5.74), can be formulated as

$$\alpha_1 C_{x-1}^{t+1} + \alpha_2 C_x^{t+1} + \alpha_3 C_{x+1}^{t+1} + \frac{C_x^t}{\Delta t} = 0 \qquad (5.78a)$$

Table 5.12 Pollutant concentrations for different locations (x) at different times (t) obtained from the explicit numerical scheme. This table is generated using Code 5.10.

	$x=0$	$x=25$	$x=50$	$x=50$	$x=75$	$x=100$	$x=120$	$x=150$	$x=175$
$t=0$	100	0	0	0	0	0	0	0	0
$t=20$	100	28.00	0.00	0.00	0.00	0.00	0.00	0.00	0.00
$t=40$	100	47.04	7.84	0.00	0.00	0.00	0.00	0.00	0.00
$t=60$	100	60.30	18.50	2.20	0.00	0.00	0.00	0.00	0.00
$t=80$	100	69.74	29.55	6.67	0.61	0.00	0.00	0.00	0.00
$t=100$	100	76.61	39.89	12.84	2.29	0.17	0.00	0.00	0.00
$t=120$	100	81.69	49.09	19.99	5.16	0.76	0.05	0.00	0.00
$t=140$	100	85.51	57.05	27.55	9.13	1.96	0.24	0.01	0.00
$t=160$	100	88.43	63.84	35.07	14.00	3.90	0.72	0.08	0.00
$t=180$	100	90.69	69.58	42.28	19.50	6.60	1.58	0.25	0.02
$t=200$	100	92.45	74.40	49.01	25.36	10.01	2.93	0.62	0.09

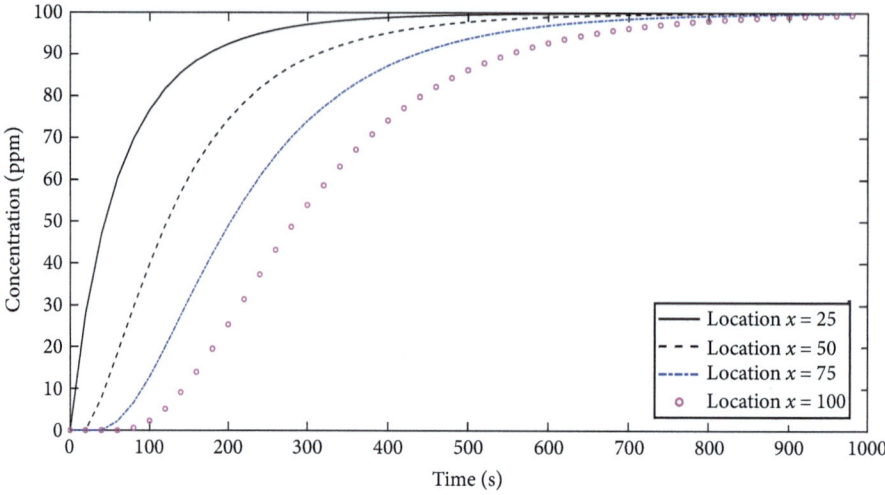

Figure 5.21 The calculated concentrations for the river stream at different locations (x) and different times (t) from the explicit numerical scheme. This figure is generated using Code 5.10.

where

$$\alpha_1 = \left(\frac{E}{\Delta x^2} + \frac{v}{2\Delta x}\right) = 0.014 \tag{5.78b}$$

$$\alpha_2 = -\left(\frac{2E}{\Delta x^2} + \frac{1}{\Delta t}\right) = -0.066 \tag{5.78c}$$

$$\alpha_3 = \left(\frac{E}{\Delta x^2} - \frac{v}{2\Delta x}\right) = 0.002 \tag{5.78d}$$

For the implicit scheme of FDM, the pollutant concentrations along all the locations of the stream would be solved simultaneously at time $t + 1$ based on the results at time t. If the river stream is discretized into five grid points (i.e., $x_0, x_1, \ldots x_5$), a linear equation system can be formulated at each time step. For instance, for $t = 0$, the linear system can be formulated as

$$\begin{cases} \alpha_1 C_0^1 + \alpha_2 C_1^1 + \alpha_3 C_2^1 & = -\dfrac{C_1^0}{\Delta t} \\[2mm] \alpha_1 C_1^1 + \alpha_2 C_2^1 + \alpha_3 C_3^1 & = -\dfrac{C_2^0}{\Delta t} \\[2mm] \alpha_1 C_2^1 + \alpha_2 C_3^1 + \alpha_3 C_4^1 & = -\dfrac{C_3^0}{\Delta t} \\[2mm] \alpha_1 C_3^1 + \alpha_2 C_4^1 + \alpha_3 C_5^1 & = -\dfrac{C_4^0}{\Delta t} \end{cases} \tag{5.79a}$$

From the initial condition we know that $C_0^1 = 100$ ppm. However, here we can apply the Neumann boundary condition for the right boundary, which leads to the following condition:

$$C_4^1 = C_5^1 \tag{5.79b}$$

Therefore, the linear system can be obtained based on Eq (5.79), as follows:

$$
\begin{bmatrix}
\alpha_2 & \alpha_3 & 0 & 0 & 0 \\
\alpha_1 & \alpha_2 & \alpha_3 & 0 & 0 \\
0 & \alpha_1 & \alpha_2 & \alpha_3 & 0 \\
0 & 0 & \alpha_1 & \alpha_2 & \alpha_3 \\
0 & 0 & 0 & 1 & -1
\end{bmatrix}
\begin{bmatrix}
C_1^1 \\
C_2^1 \\
C_3^1 \\
C_4^1 \\
C_5^1
\end{bmatrix}
=
\begin{bmatrix}
-\dfrac{C_1^0}{\Delta t} - \alpha_1 C_0^1 \\
-\dfrac{C_2^0}{\Delta t} \\
-\dfrac{C_3^0}{\Delta t} \\
-\dfrac{C_4^0}{\Delta t} \\
0
\end{bmatrix}
\tag{5.80}
$$

The MATLAB script for the implicit scheme of FDM for the one-dimensional river stream example is presented in Code 5.12. Note that to run Code 5.12, the MATLAB function named 'GaussianSolver.m' is required from Code 5.11. Therefore, save Code 5.11 with the filename 'GaussianSolver.m' in the same folder as Code 5.12. The resulting solutions are presented in Table 5.13.

Code 5.11 *The MATLAB function used for solving the one-dimensional river stream example through implicit FDM by applying the Gaussian elimination method for solving a linear equation system. Note that this is a MATLAB function and therefore does not run by itself. It can be called by other MATLAB programs, such as in Code 5.12.*

```
function x = GaussianSolver(A,b)
%A = the coefficient matrix for the linear equation system
% B = right-hand parameters
[n,m] = size(A);
x=zeros(n,1);

for j = 1:n-1
    for i = j+1:n
        rul = A(i,j)/A(j,j);
        A(i,:)= A(i,:)-rul*A(j,:);
        b(i)= b(i)-rul*b(j);
    end
end
for i=n:-1:1
    sum=0;
    for j=n:-1:i+1
        sum=sum+x(j)*A(i,j);
    end
    x(i)=(b(i)-sum)/A(i,i);
end
end
```

Code 5.12 *The MATLAB script used for solving the one-dimensional river stream example through implicit FDM. Note that this code requires the function from Code 5.11 in the same folder.*

```matlab
clc;clear all;close all;set(0,'defaultaxesfontsize',17);
% implicit numerical method-BTCS
% water velocity v = 0.3 m/s; % dispersion coefficient E = 5.0 m^2/s;
% dx = 25 m; % dt = 20 s; %E = 5.0;
dt = 20; dx = 25;  E = 5;   v = 0.3;
alpha1 = E/(dx^2)+v/(2*dx);
alpha2 = -(2*E/dx^2+1/dt);
alpha3 = E/dx^2 - v/(2*dx);
% set initial conditions, at x = 0, C = 100; C = 0 for x >0
C = zeros (6, 6);
% C(i,j) i = 1, 2 ..., 6 corresponds to t = 0, 20, 40, ...;
% j = 1, 2, ..., 6 corresponds to x = 0, 25, 50, 75, 100, 125
% Input initial conditions to the Matrix
C(:,1) = 100;   [nrow, ncol] = size(C);
for i = 1: (nrow-1) % start when t = 20
    %set the coefficient matrix for the linear system
    ParMatrix = zeros(5,5);
    ParMatrix(1, 1:2) = [alpha2, alpha3];
    for k = 2:4
        ParMatrix(k,(k-1):(k+1))= [alpha1, alpha2, alpha3];
    end
    ParMatrix(5, 4:5) = [1, -1];
    %Set the right-hand values for the linear system
    b = zeros(5,1);
    b(1,1) = -C(i,2)/dt - alpha1*C((i+1),1);  % C()
    for k = 2:4
        b(k,1) = -C(i,(k+1))/dt;
    end
    b(5,1) = 0;
    %solve the linear system by Gaussian Elimination Method
    X = GaussianSolver(ParMatrix,b);
    C((i+1),2:ncol) = X';
end
%%%%%%%%%%%%%%%%%% PRINT  %%%%%%%%%%%%%%%%%%%%%%%%%%%%%%%%%%%%%%%%
display (C);
%%%%%%%%%%%%%%%%%%% PLOT %%%%%%%%%%%%%%%%%%%%%%%%%%%%%%%%%%%%%%%%%%
plot(C(:,2),'k','linewidth',1.5); % x = 25;
hold on; plot(C(:,3),'k--','linewidth',1.5); %x = 50
hold on; plot(C(:,4),'b-.','linewidth',1.5); %x = 75;
hold on ; plot(C(:,5),'mo','markersize',5,'linewidth',1.5); %x =
100;
xlabel('Time (s)'); ylabel('Concentration (ppm)');
legend('Location x=25','Location x=50','Location x=75','Location x=100',...
'location','northeast');   set(gca,'linewidth',1.5);
ylim([0 100]);
%
```

Table 5.13 Pollutant concentrations for different locations at different times obtained from the implicit numerical scheme. These values are generated using Code 5.12.

	$x = 0$	$x = 25$	$x = 50$	$x = 50$	$x = 75$	$x = 100$
$t = 0$	100	0	0	0	0	0
$t = 20$	100	21.35	4.56	0.97	0.21	0.21
$t = 40$	100	37.74	11.56	3.21	0.87	0.87
$t = 60$	100	50.40	19.65	6.67	2.14	2.14
$t = 80$	100	60.24	28.00	11.11	4.10	4.10
$t = 100$	100	67.94	36.12	16.29	6.77	6.77

5.5 Problems for further study

Problem 5.1

In Section 5.2, the streamflow rates were predicted via multivariate linear regression (MLR) and ANN considering the independent variables current rainfall (i.e., P_t) and PET (i.e., PET_t), and one-day-lagged streamflow (i.e., Q_{t-1}), rainfall (i.e., P_{t-1}), and PET (i.e., PET_{t-1}). This problem entails extending the feature set of both the MLR and ANN models to include two-day-lagged variables (i.e., Q_{t-2}, P_{t-2}, PET_{t-2}). The objective is to assess whether incorporating these additional lagged variables enhances the models' predictive accuracy.

Problem 5.2

In Section 5.3, the GEV distribution was employed to calculate flood peaks associated with various return periods. This problem involves recalculating these flood peaks using the Gumbel and lognormal distributions. Compare the results from these distributions with those obtained using the GEV distribution to evaluate their relative performance and suitability for modelling flood peak data.

Problem 5.3

Assume that in a watershed, the flood peaks and volumes are correlated, with a Kendall τ of 0.8. For a flood peak and volume each having a return period of 50 years, calculate their joint return periods in the AND, OR, and Kendall cases via Gumbel and Clayton copulas, and discuss their differences and implications.

Problem 5.4

In Section 5.2.4.1, the explicit FDM scheme was established with $\Delta x = 25\ m$ and $\Delta t = 20\ s$. Now, consider adjusting the scheme to $\Delta x = 15\ m$ and $\Delta t = 10\ s$, and recalculate the pollutant concentrations at various locations. Compare these results with those presented in Table 5.12. Discuss the impact of using different discretization schemes on the accuracy of the FDM.

References

Adamowski, J., Chan, H.F., Prasher, S.O., Ozga-Zielinski, B., and Sliusarieva, A. (2012). Comparison of multiple linear and nonlinear regression, autoregressive integrated moving average, artificial neural network, and wavelet artificial neural network methods for urban water demand forecasting in Montreal, Canada. *Water Resources Research*, 48, W01528.

Adamowski, J., and Karapataki, C. (2008). Comparison of multivariate regression and artificial neural networks for peak urban water demand forecasting: The evaluation of different ANN learning algorithms. *Journal of Hydrologic Engineering (ASCE)*, 15(10), 729–743.

Araghinejad, S. (2014). *Data-Driven Modeling: Using MATLAB in Water Resources and Environmental Engineering*. Springer.

Benedini, M., and Tsakiris, G. (2013). *Water Quality Modelling for Rivers and Streams*. Springer.

Chanson, H. (2004). *Environmental Hydraulics of Open Channel Flows*. Elsevier-Butterworth-Heinemann.

Chen L., Singh, V.P., Guo, S.L., Mishar, A.K., and Guo, J. (2013). Drought analysis using copulas. *Journal of Hydrologic Engineering*, 18(7), 797–808.

Cigizoglu, H.K. (2005). Generalized regression neural network in monthly flow forecasting. *Civil Engineering and Environmental Systems*, 22(2), 71–84.

Estrela T., and Vargas E. (2012). Drought management plans in the European Union. The case of Spain. *Water Resources Management*, 26(6), 1537–1553.

Fan, Y., Huang, K., Huang, G., and Li, Y. (2020a). A factorial Bayesian copula framework for partitioning uncertainties in multivariate risk inference. *Environmental Research*, 183, 109215.

Fan, Y., Huang, K., Huang, G., Li, Y., and Wang, F. (2020b). An uncertainty partition approach for inferring interactive hydrologic risks. *Hydrology and Earth System Sciences*, 24(9), 4601–4624.

Fan, Y.R., Huang, W., Huang, G.H., Li, Z., Li, Y.P., Wang, X.Q, Cheng, G.H., and Jin, L. (2015). A stepwise-cluster forecasting approach for monthly streamflows based on climate teleconnections. *Stochastic Environmental Research and Risk Assessment*, 29, 1557–1569.

Fletcher, C.A.J. (1990). *Computational Techniques for Fluid Dynamics 1 – Fundamental and General Techniques*, 2nd edn. Springer.

Gabriel, R.K., and Fan, Y. (2022). Multivariate hydrologic risk analysis for River Thames. *Water*, 14, 384. https://doi.org/10.3390/w14030384

Gebregiorgis, A.S., and Hossain F. (2012). Hydrological risk assessment of old dams: Case study on Wilson dam of Tennessee River basin. *Journal of Hydrologic Engineering*, 17, 201–212.

Genest, C., and Favre, A.C. (2007). Everything you always wanted to know about copula modeling but were afraid to ask. *Journal of Hydrologic Engineering*, 12(4), 347–368.

Holzbecher, E. (2012). *Environmental Modelling: Using MATLAB*, 2nd edn. Springer.

Intergovernmental Panel on Climate Change (IPCC) (2007). *Climate Change 2007 – Impacts, Adaptation and Vulnerability: Working Group II Contribution to the Fourth Assessment Report of the IPCC*. Cambridge University Press.

Hornberger, G., and Wiberg, P. (2013). *Numerical Methods in the Hydrological Sciences*. American Geophysical Union.

Ji, Z.-G. (2008). *Hydrodynamics and Water Quality: Modeling Rivers, Lakes, and Estuaries*. John Wiley & Sons, Inc.

Karmakar, S., and Simonovic, S.P. (2009). Bivariate flood frequency analysis. Part 2: A copula-based approach with mixed marginal distributions. *Journal of Flood Risk Management*, 2, 32–44.

Nelsen, R.B. (2006). *An Introduction to Copulas*. Springer.

Okkan, U., and Serbes, Z.A. (2012). Rainfall-runoff modeling using least squares support vector machines. *Environmetrics*, 23, 549–564.

Parajuli, P.B. (2007). SWAT bacteria sub-model evaluation and application. PhD dissertation, Kansas State University.

Qi, W., Zhang, C., Fu, G., and Zhou, H. (2016). Imprecise probabilistic estimation of design floods with epistemic uncertainties. *Water Resources Research*, 52, 4823–4844. doi:10.1002/2015WR017663

Ramezani, M., Noori, R., Hooshyaripor, F., Deng Z., and Sarang, A. (2019). Numerical modelling-based comparison of longitudinal dispersion coefficient formulas for solute transport in rivers. *Hydrological Sciences Journal*, 64(7), 808–819.

Reddy, M.J., and Ganguli, P. (2012). Bivariate flood frequency analysis of Upper Godavari River flows using Archimedean copulas. *Water Resources Management*, 26, 3995–4018.

Requena, A.I., Mediero, L., and Garrote, L. (2013). A bivariate return period based on copulas for hydrologic dam design: Accounting for reservoir routing in risk estimation. *Hydrology and Earth System Sciences*, 17, 3023–3038.

Sachindra, D.A., Huang, F., Barton, A., and Perera, B.J.C. (2013). Least square support vector and multi-linear regression for statistically downscaling general circulation model outputs to catchment streamflows. *International Journal of Climatology*, 33, 1087–1106.

Sadri, S., and Burn, D.H. (2012). Nonparametric methods for drought severity estimation at ungauged sites. *Water Resources Research*, 48, W12505.

Salvadori, G., De Michele, C., and Durante, F. (2011). On the return period and design in a multivariate framework. *Hydrology and Earth System Sciences*, 15, 3293–3305.

Salvadori, G., Durante, F., and De Michele, C. (2013). Multivariate return period calculation via survival functions. *Water Resources Research*, 49, 2308–2311. doi:10.1002/wrcr.20204

Salvadori, G., Durante, F., De Michele, C., Bernardi, M., and Petrella, L. (2016). A multivariate copula-based framework for dealing with hazard scenarios and failure probabilities, *Water Resources Research*, 52, 3701–3721. doi:10.1002/2015WR017225

Sraj, M., Bezak, N., and Brilly, M. (2015). Bivariate flood frequency analysis using the copula function: A case study of the Litija station on the Sava River. *Hydrological Processes*, 29(2), 225–238.

Sujay, R.N., and Deka, P.C. (2014). Support vector machine application in the field of hydrology: A review. *Applied Soft Computing*, 19, 371–386.

Sun, C.X., Huang, G.H., Fan, Y., Zhou, X., Lu, C., and Wang, X.Q. (2019). Drought occurring with hot extremes: Changes under future climate change on Loess Plateau, China. *Earth's Future*, 7(6), 587–604.

Tanguy, M., Prudhomme, C., Smith, K., and Hannaford, J. (2017). *Historic gridded potential evapotranspiration (PET) based on temperature-based equation McGuinness–Bordne calibrated for the UK (1891–2015)*. NERC Environmental Information Data Centre. https://doi.org/10.5285/17b9c4f7-1c30-4b6f-b2fe-f7780159939c

Turan, E.M., and Yurdusev, A.M. (2009). River flow estimation from upstream flow records by artificial intelligence methods. *Journal of Hydrology*, 369, 71–77.

Wang, W.C., Chau, K.W., Cheng, C.T., and Qiu, L. (2009). A comparison of performance of several artificial intelligence methods for forecasting monthly discharge time series. *Journal of Hydrology*, 374, 294–306.

Wang, X., Huang, G., Lin, Q., Nie, X., Cheng, G., Fan, Y., Li, Z., Yao, Y., and Suo, M. (2013). A stepwise cluster analysis approach for downscaled climate projection: A Canadian case study. *Environmental Modelling & Software*, 49, 141–151.

Xu, Z.X., Ito, K., and Li, J.Y. (2001). Risk estimation for flood and drought: Case studies. *IAHS-AISH publication*, 272, 333–340.

Yue, S. (2000). The bivariate lognormal distribution to model a multivariate flood episode. *Hydrological Processes*, 14(14), 2575–2588.

Yue, S. (2001). A bivariate gamma distribution for use in multivariate flood frequency analysis. *Hydrological Processes*, 15(6), 1033–1045.

Zhang, L., and Singh, V.P. (2006). Bivariate flood frequency analysis using the copula method. *Journal of Hydrologic Engineering*, 11(2), 150–164.

Zhou, V. (2019). Machine Learning for Beginners: An Introduction to Neural Networks. https://victorzhou.com/blog/intro-to-neural-networks/

Chapter 6
Transportation Engineering

6.1 Introduction

Transportation engineering is a multidisciplinary subject within civil engineering involving the planning, design, operation, and maintenance of transportation systems and associated infrastructure such as highways, roads, and railways (Figure 6.1). The primary aim of transportation engineering is to guarantee efficient, safe, and sustainable movement of people and goods from one place to another. In transportation engineering, several pressing issues demand attention, including optimizing vehicle flow, enhancing transport network efficiency, and accurately predicting traffic patterns.

Queue theory, a fundamental concept in operations research and systems analysis, plays a pivotal role in comprehending and effectively managing the intricate flow dynamics of vehicles, pedestrians, and information within complex transportation systems. The principles of queue theory come to the fore in scenarios where vehicles patiently line up at critical points such as intersections, toll booths, and traffic signals. As transportation systems evolve and face new challenges, the principles of queue theory remain a cornerstone, continually adapting to provide innovative solutions for the complex dynamics of modern mobility.

Efficiently optimizing traffic flow within transportation networks is a paramount objective that significantly contributes to enhancing overall system efficiency and minimizing congestion. This optimization process involves the strategic use of mathematical

Motorway M4, UK

Figure 6.1 A typical transportation system showing the M4 motorway in the west of the United Kingdom.

A Practical Approach to Advanced Mathematical Modelling in Civil Engineering. Mohammad Heidarzadeh et al., Oxford University Press. © Mohammad Heidarzadeh, Theodosios K. Papathanasiou, Yurui Fan, Hamid Bahai (2025). DOI: 10.1093/9780191888656.003.0006

A traffic signal in Bath, UK

Figure 6.2 A traffic signal in Bath, United Kingdom.

programming techniques, specifically leveraging the capabilities of linear programming and dynamic programming. These powerful tools play a pivotal role in systematically addressing complex challenges associated with network optimization, allowing transportation engineers and planners to design, manage, and improve traffic systems effectively.

As transportation systems become more complex, the integration of advanced technologies becomes imperative. Specifically, machine learning and deep learning techniques have been applied to predicting traffic flow patterns. By leveraging historical data and real-time inputs, these methodologies offer predictive capabilities for anticipating congestion, optimizing traffic signal timings, and improving overall traffic management. These technologies not only facilitate the anticipation of congestion but also enable the optimization of traffic signal timings (Figure 6.2), adaptive route planning, and the implementation of proactive strategies for mitigating potential disruptions. As transportation systems continue to evolve, the integration of these advanced technologies becomes paramount for creating smarter, more responsive, and efficient networks.

This chapter presents three key aspects of transportation engineering. Basic concepts and applications of queue theory will be introduced in detail. Mathematical programming techniques, including linear programming and dynamic programming methods, will then be illustrated for optimizing a traffic flow network. Finally, the application of machine learning and deep learning to traffic flow prediction will be described.

6.2 Queuing theory

6.2.1 Fundamentals of queuing models

Queuing theory is a mathematical discipline that studies the behaviour and performance of waiting lines or queues. In the context of transportation engineering, queuing theory finds applications in analysing traffic flow, wait times, and resource utilization efficiency.

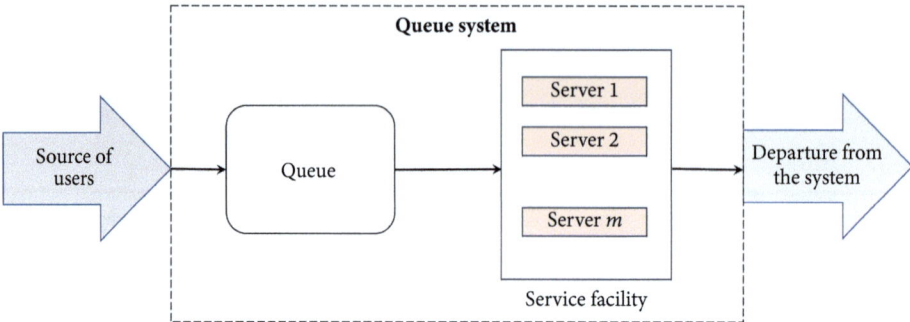

Figure 6.3 Sketch showing a generic queuing system, consisting of a user source, a queue, and a service facility.

Queuing theory plays a central role in the analysis of and planning for urban services since almost all urban service systems can be viewed as queuing systems (Larson and Odoni 1981). Queuing models have many applications in transportation engineering such as traffic signal control, toll booths and checkpoints, public transportation systems, airport security and check-in, parking facilities, freight terminals, and emergency evacuation planning. Queuing models provide a quantitative framework for understanding and optimizing the performance of transportation systems, making them invaluable tools for transportation engineers and planners. By considering factors such as arrival rates, service times, and system capacities, these models help in designing more efficient and resilient transportation networks.

A generic model for a queuing system consists of three elements, as shown in Figure 6.3: (i) a user source from which the population (or vehicles) arrive with certain time intervals at the queuing system, (ii) a queue formed by the arriving vehicles, and (iii) a service facility that accommodates one or more identical servers, potentially extending to an infinite number, operating in parallel. Thus, customers requiring service are generated over time by an input source, pass through the queue where they may remain for a nonnegative period of time (including possibly zero time), and then a member of the queue is processed by a single server because of the parallel arrangement of servers. Once a vehicle has left any of the servers in the system, after obtaining service there, the user is considered to have left the queuing system as well.

A typical queue system is characterized by some model parameters such as the number of parallel servers (m), arrival rate (λ), and the service rate of a server (μ). Also, the arrival distribution, service distribution, and queue discipline are required to determine a queue system. Kendall's (1953) notation is generally used to classify queues as follows:

$$A/S/m/K/P/QD \qquad (6.1)$$

where A denotes the interarrival distribution, referring to the probability distribution that describes the time between consecutive arrivals; S indicates the service time distribution;

m describes the number of parallel servers; K is the total system size or capacity; P is the population size; and QD describes the queue discipline.

Some standard code letters are used to represent probability distributions for interarrival times and service times in queuing theory (Larson and Odoni 1981):

- M = Poissonian or random phenomena (i.e., negative exponential PDF for user interarrival times or for service times. Here, M stands for 'memoryless')
- D = deterministic or regular phenomena (i.e., interarrival or service times are constant)
- E_k = kth-order Erlang distribution
- H_k = kth-order hyperexponential distribution
- G = a 'general' distribution (i.e., any type of distribution).

The queue discipline, denoted as QD in Eq (6.1), determines the order in which members of the queue are selected for service and can take one of the following forms:

- FIFO indicates the 'first in, first out' queuing arrangement, also known as FCFS (first come, first served). Typical FIFO queues include a single road lane, and airport check-in counters.
- LIFO (i.e., last in, first out) or LCFS (last come, first served) indicates the situation in which the last user to join the queue becomes the next in line for entering service (Larson and Odoni 1981), such as disembarking from the rear of a bus.
- SIRO is 'service in random order'.

Among these queue disciplines, FIFO is frequently encountered in the analysis of transport systems.

The fundamental elements of a queuing model include the arrival rate (λ), service rate (μ), and queue length, involving entities within the system, such as vehicles or pedestrians:

- Arrival rate (λ): the rate at which entities arrive at the system.
- Service rate (μ): The rate at which entities are serviced or processed.
- Queue length: The number of entities waiting in the queue.

6.2.2 D/D/1 model

The D/D/1 queuing model, also known as the deterministic/deterministic/single server queuing model, is a simple yet versatile model used to analyse queuing systems with specific characteristics. It assumes that:

- The arrival process in a D/D/1 queue is characterized by deterministic interarrival times (i.e., the time interval between successive arrivals in a queue). This means that the time between consecutive arrivals is constant and predictable.

- The service process is also deterministic in the D/D/1 model. Each customer receives service for a fixed and constant amount of time.
- The '1' in D/D/1 denotes a single server in the system. Only one customer can be served at a time.

In order to elaborate the D/D/1 queuing model, an example is illustrated to investigate its intuitive graphic or mathematical solution.

Example 6.1

This example is from Mannering and Washburn (2020). Vehicles arrive at an entrance to a park. There is a single gate where all vehicles should stop, and a park attendant distributes a free brochure. The park opens at 8:00 a.m., at which time vehicles begin to arrive at a rate of 480 vehicles/hr. After 20 min, the arrival flow rate declines to 120 vehicles/hr, and it continues at that level for the remainder of the day. If the time required to distribute the brochure is 15 s, and assuming a D/D/1 queuing model, describe the operational characteristics of the queue.

Solutions

From the problem description, the arrival rate in units of vehicles per minute (i.e., vehicles/min) can be obtained as follows:

$$\lambda = \frac{480 \text{ veh/hr}}{60 \text{ min/hr}} = 8 \text{ veh/min for } t \le 20 \text{ min} \tag{6.2a}$$

$$\lambda = \frac{120 \text{ veh/hr}}{60 \text{ min/hr}} = 2 \text{ veh/min for } t > 20 \text{ min} \tag{6.2b}$$

Consequently, the total vehicles arriving ($D_{arrival}$) at any time t is generated as

$$D_{arrival} = \begin{cases} 8t & t \le 20 \text{ min} \\ 160 + 2(t-20) & t > 20 \text{ min} \end{cases} \tag{6.3}$$

Similarly, the service rate (μ) in one minute can be obtained as

$$\mu = \frac{60 \text{ s/min}}{15 \text{ s/veh}} = 4 \text{ veh/min} \tag{6.4}$$

The number of vehicles departing the queue ($D_{departure}$) is given by the following equation:

$$D_{departure} = 4t \text{ for all } t \tag{6.5}$$

The number of the vehicles in the queuing system (D_S) becomes

$$D_S = \begin{cases} 8t - 4t & t \le 20 \text{ min} \\ 160 + 2(t-20) - 4t & t > 20 \text{ min} \end{cases} \tag{6.6a}$$

The queue length (denoted as D_L) is determined by subtracting the vehicles being served from the total number of vehicles in the queue system. In this case, only one server is available (i.e., vehicles being served = 1). Thus, the queue length is formulated as

$$D_L = D_S - 1 \tag{6.6b}$$

From Eqs (6.2)–(6.6), it can be observed that the arrival rate (i.e., λ) exceeds the service rate (i.e., μ) in the first 20 min, resulting in the formation of a queue. Subsequently, the service rate surpasses the arrival rate, causing the queue to dissipate. Figure 6.4 shows the arrival and departure vehicles and the corresponding total and queue lengths for Example 6.1. It indicates that the queue begins to form at 8:00 a.m. and dissipates 60 min later (9:00 a.m.). A total number of 240 vehicles will have arrived and departed (4 veh/min × 60 min) during this period. Also, the maximum vehicles would occur at $t = 20$ min with 80 vehicles in the queue system (see Figure 6.4b), and 79 vehicles in the queue. The MATLAB script for the D/D/1 queue model is presented in Code 6.1 and can be used to plot Figure 6.4. Note that this MATLAB code and all the other code presented in this chapter is available at the following website and can be downloaded: http://www.oup.com/AdvancedMathematicalModelling.

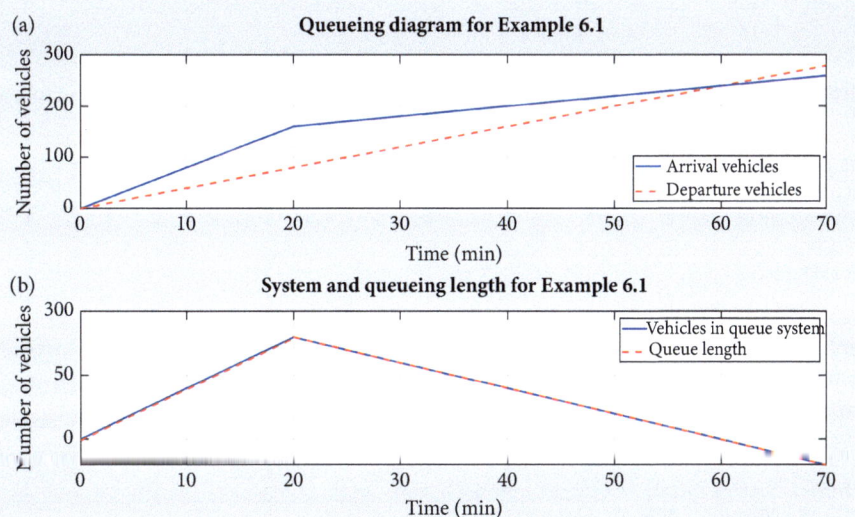

Figure 6.4 Simulations of a D/D/1 queuing system presented in Example 6.1: (a) arrival and departure of vehicles, and (b) queue length.

Code 6.1 *The MATLAB script used for solving the D/D/1 queue system described in Example 6.1.*
This code produces Figure 6.4. This MATLAB code can be downloaded at the following website:
http://www.oup.com/AdvancedMathematicalModelling.

```
clc; clear all; close all;
set(0,'defaultaxesfontsize',18);
% create the time intervals
t1 = 0:0.1:20; %  t <= 20
t2 = 20:0.1:70; % for t > 20
t = 0:0.1:70;
%  vehicle arrival
R1 = 8 * t1;   % for t <= 20
R2 = 160 + 2 * (t2 - 20);% for t >20
% vehicle departure
D = 4*t;
% arrival
figure; subplot(2,1,1);
plot(t1, R1, 'b-', 'LineWidth', 2); % arrival vehicles t <= 20
hold on;
plot(t, D, 'r--', 'LineWidth', 2);% departure vehicles for t >20
hold on;
plot(t2, R2, 'b-', 'LineWidth', 2); % arrival vehicles for t >20
%
xlabel('Time (min)');
ylabel('Number of vehicles');
title('(a) Queueing diagram for example 6.1');
legend('Arrival vehicles','Departure vehicles','Location', 'southeast');
grid on; hold off; set(gca,'linewidth', 1.5);
%%%% =============Queue length ===========
Dqs1 = 4*t1; % for t <= 20;
Dq1 = Dqs1 - 1;
Dqs2 = 160-2*t2 - 40;
Dq2 = Dqs2 - 1;
subplot(2,1,2)
plot(t1, Dqs1, 'b-', 'LineWidth', 2); % vehicles in queue system
hold on;
plot(t1, Dq1, 'r--', 'LineWidth', 2);% queue length
hold on;
plot(t2, Dqs2, 'b-', 'LineWidth', 2); % vehicles in queue system
hold on
plot(t2, Dqs2, 'r--', 'LineWidth', 2); % vehicles in queue system
xlabel('Time (min)');
ylabel('Number of vehicles');
title('(b) System and queueing length for example 6.1');
legend('Vehicles in queue system','Queue length');
grid on; hold off; set(gca,'linewidth', 1.5);
```

6.2.3 M/M/1 model

In the previous section, the D/D/1 model was described which represents the simplest
queuing model, where both interarrival time and service time are constant. However, in
most traffic applications the interarrival and service times exhibit random characteristics

and necessitate description by probability distributions. The exponential distribution is often used to describe the randomness of interarrival time and service time. When only one server is available in the queue system, an M/M/1 model will be formulated, in which 'M' signifies that both the interarrival and service times follow memoryless exponential probability distributions, and '1' denotes a single server in the system. The M/M/1 queuing model is particularly valuable for traffic systems where the random nature of arrivals and service times is a more accurate representation of reality.

The exponential distribution

The PDF, denoted $f_T(t)$, and CDF, represented by $F_T(t)$, for an exponential distributed variable are

$$f_T(t) = \begin{cases} \lambda e^{-\lambda t} & t \geq 0 \\ 0 & t < 0 \end{cases} \tag{6.7a}$$

$$F_T(t) = \begin{cases} 1 - e^{-\lambda t} & t \geq 0 \\ 0 & t < 0 \end{cases} \tag{6.7b}$$

The mean, $E(T)$, and variance, $\sigma(T)$, of the exponential distribution described by Eqs (6.7a) and (6.7b) can be obtained as follows:

$$E(T) = \frac{1}{\lambda} \tag{6.7c}$$

$$\sigma(T) = \frac{1}{\lambda} \tag{6.7d}$$

There are several key properties of the exponential distribution that illustrate its applicability in queuing models.

Property 1 The PDF function $f_T(t)$ is a strictly decreasing function of t ($t \geq 0$).

Property 2 Lack of memory ('memoryless'). This means that the probability of an event occurring in the next infinitesimally small interval of time is constant, irrespective of the past. Mathematically, for any $s, t \geq 0$, the memoryless property is expressed as $P(T > s + t | T > s) = P(T > t)$, where T is a random variable representing the time until the next event. The exponential distribution is the only continuous distribution with the lack of memory property.

Property 3 The minimum of exponential variables is still an exponential random variable. Let $T_1, T_2, \ldots, T_n$ be independent exponential variables with parameters of $\lambda_1, \lambda_2, \ldots, \lambda_n$. If $U = \min\{T_1, T_2, \ldots, T_n\}$, U is an exponential variable with $\lambda = \lambda_1 + \lambda_2 + \cdots + \lambda_n$ (i.e., $U \sim \exp\left(\sum_{i=1}^{n} \lambda_i\right)$).

Property 4 Relationship to the Poisson process. The interarrival (waiting) times between events in a Poisson process occurring at the rate of λ per unit of time exhibit the following characteristics: (i) they are mutually independent, and (ii) they are described by an exponential PDF with a parameter λ.

The Poisson process

If successive interevent (or interarrival) times are independent and identically distributed as exp(λ), then $N(t)$, denoting the number of arrivals during the interval (0, t], forms a Poisson process with rate λ.

The probability that n events occur during the time interval is expressed as

$$P[N(t) = n] = \frac{(\lambda t)^n e^{-\lambda t}}{n!} \tag{6.8a}$$

where t is the time interval length, and λ indicates the arrival rate (vehicle/unit time). Also, the mean and variance of $N(t)$ can be obtained as follows:

$$E[N(t)] = \lambda t \tag{6.8b}$$

$$\sigma[N(t)] = \lambda t \tag{6.8c}$$

Based on Eq (6.8a), the probability that at least one event would occur in (0, t] in a Poisson process is given by

$$P[N(t) \geq 1] = 1 - P[N(t) = 0] = 1 - e^{-\lambda t} \tag{6.9}$$

In addition, the probability that n events will occur in the next Δt in a Poisson process is approximated as

$$P[N(t + \Delta t) - N(t) = n] \simeq \begin{cases} 1 - \lambda \Delta t & n = 0 \\ \lambda \Delta t & n = 1 \\ 0 & n \geq 2 \end{cases} \tag{6.10}$$

A queuing system with $m = 1, 2, 3, \ldots$ parallel identical servers and infinite system capacity is operated in the following fashion (Larson and Odoni 1981):

(i) Whenever there are n users in the system (in queue plus in service), new users arrive at the system according to a Poisson process with a mean arrival rate of λ_n expected arrivals per unit time and service completions also occur in a Poisson manner with a mean service rate of μ_n per unit time.

(ii) The queuing system operates under an FCFS queue discipline. If the queuing system contains n users at time t, then at time $t + \Delta t$ it will contain either $n + 1$ users with probability $\lambda_n \Delta t$ or $n - 1$ users with probability $\mu_n \Delta t$, or n users with probability $1 - (\lambda_n + \mu_n)\Delta t$, which can be derived from Eq (6.10) and is illustrated in Figure 6.5.

Let $P_n(t) = P[N(t) = n]$ describe the probability that there are exactly n vehicles in a queuing system at a specific time t. Based on Eq (6.10), $P_n(t + \Delta t)$ can be derived as follows (Larson and Odoni 1981):

$$P_n(t + \Delta t) = P_{n+1}(t) \mu_{n+1}\Delta t + P_n(t)\left[1 - (\lambda_n + \mu_n)\Delta t\right] + P_{n-1}(t)\lambda_{n-1}\Delta t \tag{6.11a}$$

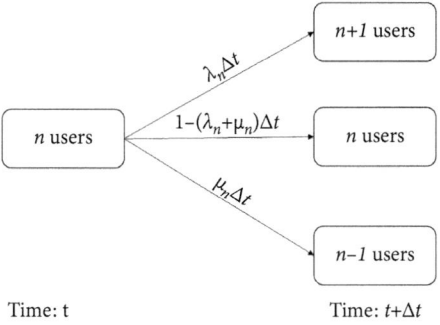

Figure 6.5 The probabilities of transition in the next Δt.

By rearranging Eq (6.11a) and dividing by Δt, we have

$$\frac{P_n(t + \Delta t) - P_n(t)}{\Delta t} = P_{n+1}(t)\mu_{n+1} - (\lambda_n + \mu_n)P_n(t) + P_{n-1}(t)\lambda_{n-1} \qquad (6.11b)$$

By letting $\Delta t > 0$ in Eq. (6.11b), we can obtain the following differential equation:

$$\frac{dP_n(t)}{dt} = P_{n+1}(t)\mu_{n+1} - (\lambda_n + \mu_n)P_n(t) + P_{n-1}(t)\lambda_{n-1}, \ n = 1, \ 2, \ 3, \dots \qquad (6.11c)$$

When $n = 0$ (i.e., no vehicle in the queuing system), the differential equation can be similarly derived as follows:

$$\frac{dP_0(t)}{dt} = P_1(t)\mu_1 - \lambda_0 P_0(t) \qquad (6.11d)$$

In the application of queuing models, the steady-state condition is mainly considered, which indicates that the state of the system (i.e., the number of customers in the system) is independent of the initial state and time elapsed. The steady-state probability that n customers are in the queuing system is defined as

$$P_n = \lim_{t \to \infty} P_n(t) \qquad (6.12)$$

Also, under steady-state conditions, we have $dP_n(t)/dt = 0$. Therefore, Eqs (6.11c) and (6.11d) can be reformulated as

$$\lambda_0 P_0 = \mu_1 P_1 \qquad (6.13a)$$

and

$$(\lambda_n + \mu_n)P_n = \lambda_{n-1}P_{n-1} + \mu_{n+1}P_{n+1}, \ n = 1, 2, 3, \dots \qquad (6.13b)$$

M/M/1: One operator, infinite number of lines

In a queuing system of the M/M/1 type with FCFS service and an infinite capacity in the queue, the interarrival time is assumed to be an exponential variable with mean $1/\lambda$ and the service time is also exponentially distributed with a mean of $1/\mu$. Under steady-state conditions, we can have

$$\lambda_0 = \lambda_1 = \lambda_2 = \cdots = \lambda \qquad (6.14a)$$

and

$$\mu_0 = \mu_1 = \mu_2 = \cdots = \mu \qquad (6.14b)$$

Let us define the traffic intensity (also called the utilization factor) as $\rho = \lambda/\mu$ for the M/M/1 queuing model. The queue is stable if and only if $\rho < 1$, meaning the queue will not grow indefinitely. Based on Eqs (6.13) and (6.14), we can have the steady-state probabilities as follows:

$$P_1 = \frac{\lambda}{\mu} P_0 \qquad (6.15a)$$

$$P_2 = \left(\frac{\lambda}{\mu}\right)^2 P_0 \qquad (6.15b)$$

$$P_n = \left(\frac{\lambda}{\mu}\right)^n P_0 \qquad (6.15c)$$

Since $\sum_{n=0}^{\infty} P_n = 1$, we have the following:

$$\sum_{n=0}^{\infty} \left(\frac{\lambda}{\mu}\right)^n P_0 = 1 \Leftrightarrow P_0 = \frac{1}{\sum_{n=0}^{\infty} \left(\frac{\lambda}{\mu}\right)^n} \qquad (6.16)$$

Also, $\rho = \lambda/\mu$ and when $\rho < 1$, Eqs (6.15) and (6.16) can be reformulated as

$$P_0 = \frac{1}{\sum_{n=0}^{\infty} \rho^n} = 1 - \rho \qquad (6.17a)$$

$$P_n = \rho^n(1 - \rho) \qquad (6.17b)$$

Based on Eq (6.17), the expected number of vehicles in the system (L_s) can be derived as (Larson and Odoni 1981)

$$L_s = \sum_{n=0}^{\infty} nP_n = \sum_{n=0}^{\infty} n\rho^n(1-\rho) = (1-\rho)\sum_{n=0}^{\infty} n\rho^n = (1-\rho)\rho\sum_{n=0}^{\infty} n\rho^{n-1}$$

$$= (1-\rho)\rho\frac{d}{d\rho}\sum_{n=0}^{\infty} \rho^n = (1-\rho)\rho\frac{d}{d\rho}\left(\frac{1}{1-\rho}\right) = (1-\rho)\rho\left(\frac{1}{(1-\rho)^2}\right) = \frac{\rho}{1-\rho} = \frac{\lambda}{\mu-\lambda}$$

$$(6.18a)$$

The expected queue length (L_q) can be obtained as $L_q = L_s -$ expected number of vehicles being served.

In the M/M/1 queuing model, only one server is available and thus the expected queue length (L_q) can be generated as

$$L_q = \sum_{n=1}^{\infty} (n-1)P_n = \sum_{n=0}^{\infty} nP_n - (P_1 + P_2 + \cdots) = \sum_{n=0}^{\infty} nP_n - (1 - P_0)$$

$$= L_s - (1 - P_0) = \frac{\rho^2}{1 - \rho} = \frac{\lambda^2}{\mu(\mu - \lambda)} \tag{6.18b}$$

Similarly, the expected waiting time (W_s) in the system can be obtained as (Larson and Odoni 1981)

$$W_s = \frac{1}{\mu} P_0 + \frac{2}{\mu} P_1 + \frac{3}{\mu} P_2 + \cdots = \sum_{n=0}^{\infty} \frac{n+1}{\mu} P_n = \frac{1}{\mu(1 - \rho)} = \frac{1}{\mu - \lambda} \tag{6.19a}$$

The expected waiting time in the queue (W_q) can be calculated as $W_q = W_s -$ expected service time, which can be formulated as

$$W_q = \sum_{n=0}^{\infty} \frac{n}{\mu} P_n = \frac{\rho}{\mu(1 - \rho)} = \frac{\lambda}{\mu(\mu - \lambda)} \tag{6.19b}$$

An application of the M/M/1 queuing model is illustrated in the following example.

Example 6.2

Consider a toll booth at a highway entrance, where vehicles arrive and queue up for service. The arrival of vehicles follows a Poisson process, with an average rate of $\lambda = 3$ vehicle/min. The toll booth operates as a single-server system, and the service times for each vehicle follow an exponential distribution with a mean service rate of $\mu = 4$ vehicle/min. The objective is to analyse the average queue length and waiting time, providing insights for optimizing toll booth operations.

Solution

From the description, the system can be described by the M/M/1 queuing model with the following parameters:

- arrival rate (λ): 3 veh/min
- service rate (μ): 4 veh/min

Then, the utilization factor is $\rho = \lambda/\mu = 0.75$. Consequently, the average vehicles in the queuing system and the queue can be obtained as

$$L_s = \frac{\lambda}{\mu - \lambda} = 3 \text{ vehicles}$$

$$L_q = \frac{\lambda^2}{\mu(\mu - \lambda)} = 2.25 \text{ vehicles}$$

Example 6.2 *Continued*

$$W_s = \frac{1}{\mu - \lambda} = 1 \text{ min}$$

$$W_q = \frac{\lambda}{\mu(\mu - \lambda)} = 0.75 \text{ min}$$

The MATLAB code for solving Example 6.2 is presented in Codes 6.2 and 6.3, where Code 6.2 calculates the parameters and Code 6.3 produces the simulations and generates Figure 6.6.

Code 6.2 *The MATLAB script used for calculating the parameters for Example 6.2. This MATLAB code and other code from this chapter can be downloaded at the following website: http:// www.oup.com/AdvancedMathematicalModelling.*

```
clc; clear all;
% Initialize parameters
lambda = 3; % Arrival rate of 3 vehicles per minute
mu = 4;     % Service rate of 4 vehicles per minute
% Calculate utilization, average number of vehicles, and average time in system
rho = lambda / mu;
Ls = lambda / (mu - lambda);
Lq = lambda^2/(mu*(mu-lambda));
Ws = 1 / (mu - lambda);
Wq = lambda/(mu*(mu-lambda));
% print the obtained results
fprintf('Utilization (rho): %0.2f \n', rho);
fprintf('Average Number of Vehicles in System (Ls): %0.2f \n', Ls);
fprintf('Average Number of Vehicles in Queue (Lq): %0.2f \n', Lq);
fprintf('Average Time a Vehicle Spends in System (W): %0.2f minutes \n', Ws);
fprintf('Average Time a Vehicle Spends in Queue (W): %0.2f minutes \n', Wq);
```

Code 6.3 *The MATLAB script for the M/M/1 queue model simulation in Example 6.2 which produces Figure 6.6.*

```
clc; clear all; close all;
set(0,'defaultaxesfontsize',18);
Total_time = 60; % minutes, simulation time
N=10000000000; %,maximum vehicles in the queue
lambda=3, mu=4; %arrival and service rate
arr_mean = 1/lambda; %average arrival time
ser_mean = 1/mu; % average service time
arr_num =  Total_time*lambda*2; %The number of arrival vehicles
events = [];
events(1,:)=exprnd(arr_mean,1,arr_num); %generate interarrival times
events(1,:)=cumsum(events(1,:)); %Arrival time of each car=sum interarrival times;
events(2,:)=exprnd(ser_mean,1,arr_num); %generate service times for vehicles
```

```
len_sim = sum(events(1,:)<= Total_time); %total cars arrived in simulation time;
%-------------information for the first vehicle ------------
events(3,1) = 0; % first vehicle is served directly without any waiting
events(4,1)=events(1,1)+events(2,1); %Departure time=sum arrival & service time;
%first car is accepted, and only 1 car in the system. Assign 1 for 1 vehicle
events(5,1) = 1;
member = [1]; %1st vehicle enter the system, 1 member in the system
for i = 2:arr_num
    if events(1,i)>Total_time % break the loop if arrival time > simulation time
        break;
    else
        number = sum(events(4,member) > events(1,i));
        %if the system is full, the system would reject service and assign 0
        if number >= N+1
            events(5,i) = 0;
        else
            if number == 0  % if system is empty, no waiting time
                events(3,i) = 0;
                %departure time = arrival time + service time
                events(4,i) = events(1,i)+events(2,i);
                %assign 1 at its position
                events(5,i) = 1;
                member = [member,i];
                % if system have vehicles but is not full, vehicle i enters
            else
                len_mem = length(member);
% waiting time=departure time of preceding customer minus its arrival time
                events(3,i)=events(4,member(len_mem))-events(1,i);
%departure time is equal to departure time of preceding customer + service time
                events(4,i)=events(4,member(len_mem))+events(2,i);
    %indicates the total number of customers in the system after it enters
                events(5,i) = number+1;
                member = [member,i];
            end
        end
    end
end

len_mem = length(member); % total vehicles entered in the simulation period
% Ploat the results
figure
subplot(2,1,1)
stairs([0 events(1,member)],0:len_mem,'b','linewidth',1.5);
hold on;
stairs([0 events(4,member)],0:len_mem,'r--','linewidth',1.5);
legend('Arrival time ','Departure time','Location', 'southeast');
xlabel('Time (min)');
ylabel('Number of vehicles');
title('(a) Arrival and departure time in M/M/1');
hold off; grid on; set(gca,'linewidth', 1.5);
%figure;
subplot(2,1,2)
plot(1:len_mem,events(3,member),'r-*',...
1: len_mem,events(2,member)+events(3,member),'b-', 'linewidth', 1.5);
legend('Waiting time in queure ','Waiting time in system ');
xlabel('Number of vehicles'); ylabel('Time (min)');
title('(b) Waiting time in queue and system in M/M/1');
grid on; set(gca,'linewidth', 1.5);
```

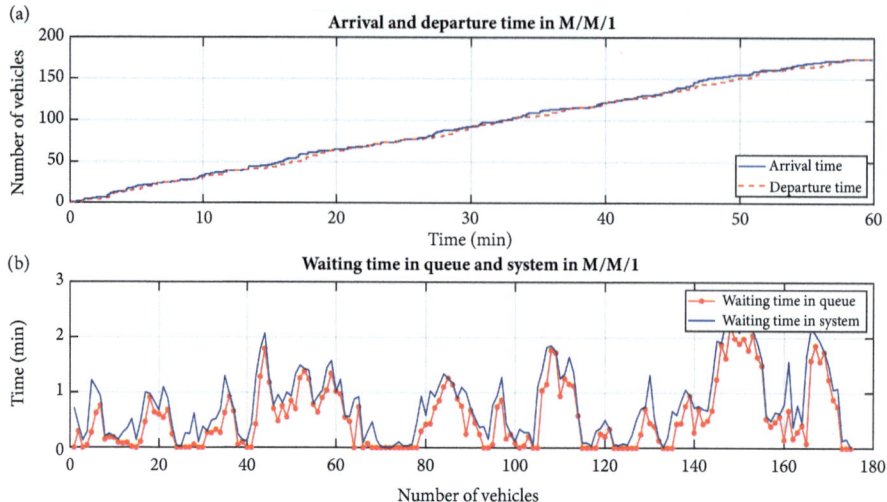

Figure 6.6 The M/M/1 queuing model simulations for Example 6.2: (a) vehicle arrival and departure time, and (b) vehicle waiting times in the queue and the system. This figure is created using the MATLAB script in Code 6.3.

6.3 Optimization of transport networks

Optimization methods play a crucial role in enhancing the efficiency and functionality of traffic network planning. Optimizing transportation networks involves maximizing or minimizing objectives to enhance efficiency, reduce congestion, or lower costs. A number of optimization methods are applied in transport network optimization, such as the linear programming method for optimizing resource allocation and minimizing costs, and dynamic programming for finding optimal solutions through sequential decision-making.

6.3.1 Linear programming

Linear programming (LP) is a powerful mathematical technique and has been widely used in transportation engineering for tasks such as traffic signal optimization, vehicle routing and scheduling, and network flow optimization (e.g., Revelle and Whitlach 1996). The benefits of using LP in transportation engineering include: (i) LP models provide efficient solutions to complex optimization problems; (ii) by optimizing routes, schedules, and resource allocation, LP can lead to cost reductions in transportation operations; (iii) LP helps in improving the overall quality of transportation services by minimizing delays and congestion; and (iv) LP models can be adapted to different transportation scenarios, making them versatile tools.

Formulation of an LP model

In a mathematical programming model, the n related quantifiable decisions are represented as decision variables $x_1, x_2, \ldots, x_n$ whose values are to be determined. Next, the measure of performance (e.g., profit) is expressed as a mathematical function of these decision variables, termed the objective function. Any constraints on the values that can be assigned to the decision variables are also expressed mathematically, typically by means of inequalities (Hillier and Lieberman 2010). In general, a mathematical programming model can be formulated as

$$\max f(\boldsymbol{X}) \tag{6.20a}$$

subject to

$$g_i(\boldsymbol{X}) \leq b_i, \ i = 1, \ 2, \ ..., \ m \tag{6.20b}$$

$$\boldsymbol{X} \geq 0 \tag{6.20d}$$

where $\boldsymbol{X} = (x_1, x_2, ..., x_n)^{\mathrm{T}}$ are decision variables and f is a function of $\boldsymbol{X}$, representing the objective function. The functions g_i are also functions of $\boldsymbol{X}$, representing the inequality constraints. Note that when equality constraints are included, additional functions like $h_j(\boldsymbol{X}) = b'_j, \ j = 1, \ 2, \ ..., \ m'$, can be added to the model. The constants (namely, the coefficients and right-hand sides, e.g., b_i, b'_j) in the constraints and the objective function are called the parameters of the model. The mathematical model of Eq (6.20) articulates that the problem is to choose the values of the decision variables that maximize the objective function, subject to the specified constraints.

In the model represented by Eq (6.20), when both $f(\boldsymbol{X})$ and $g_i(\boldsymbol{X})$ are linear functions of the decision variables (e.g., $f(\boldsymbol{X}) = c_1x_1 + c_2x_2 + \cdots + c_nx_n$), Eq (6.20) is formulated as an LP model. Following Hillier and Lieberman (2010), the standard form of the LP model can be formulated as follows:

$$\max Z = c_1x_1 + c_2x_2 + \cdots + c_nx_n \tag{6.21a}$$

subject to

$$a_{11}x_1 + a_{12}x_2 + \cdots + a_{1n}x_n \leq b_1 \tag{6.21b}$$
$$a_{21}x_1 + a_{22}x_2 + \cdots + a_{2n}x_n \leq b_2 \tag{6.21c}$$

$$\cdots$$

$$a_{m1}x_1 + a_{m2}x_2 + \cdots + a_{mn}x_n \leq b_m \tag{6.21d}$$

$$x_1 \geq 0, \ x_2 \geq 0, ..., x_n \geq 0 \tag{6.21e}$$

Also, the LP model represented by Eq (6.21) can be reformulated in matrix notation as

$$\max Z = \boldsymbol{C}^{\mathrm{T}}\boldsymbol{X} \tag{6.22a}$$

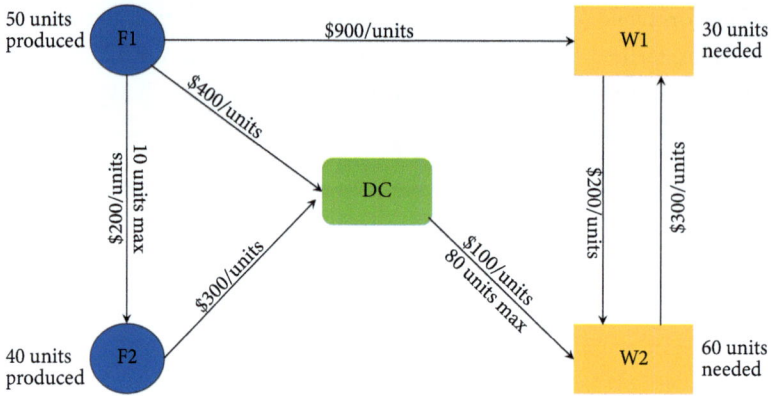

Figure 6.7 The distribution network for Example 6.3. Abbreviations are: F, factory; DC, distribution centre; and W, warehouse (modified from Hillier and Lieberman 2010).

subject to

$$AX \leq b \tag{6.22b}$$

$$X \geq 0 \tag{6.22c}$$

where $C = [c_1, c_2, ..., c_n]^T$, $X = [x_1, x_2, ..., x_n]^T$, $A = [a_{ij}]_{m \times n}$, $A \in R^{m \times n}$, and $b = [b_1, b_2, ..., b_m]^T$.

In the context of LP and its extensions, any set of values assigned to the decision variables $(x_1, x_2, ..., x_n)$ is referred to as a **solution**.

For a solution to be considered **feasible**, it must simultaneously satisfy all four of these constraint conditions, namely $X \in \{X \mid AX \leq b\}$. The **feasible region** is the collection of all feasible solutions. In comparison, an **infeasible solution** is a solution for which at least one constraint is violated.

Applications of LP to transport network optimization

An example, modified from Hillier and Lieberman (2010), is provided to describe how the LP method can be used in transport network optimization.

Example 6.3

A company is going to produce the same products from different factories (F1 and F2 in Figure 6.7), which will be further shipped to two warehouses (W1 and W2). It is assumed that either factory can supply either warehouse. Figure 6.7 presents the distribution network available for shipping this product in which the two factories are denoted as F1 and F2, and the two warehouses are represented as W1 and W2. Additionally, a distribution centre, denoted as DC in Figure 6.7, is also available in the distribution network.

As shown in Figure 6.7, there are 50 units from F1 and 40 units from F2 which will be shipped to the two warehouses (i.e., W1 and W2), while W1 and W2 will respectively receive 30 and 60 units of the product. The feasible shipping lines are denoted as the arrow lines. Therefore, F1 can ship directly to W1 and would have three possible routes for shipping to W2:

(i) F1 → DC → W2
(ii) F1 → F2 → DC →W2
(iii) F1 → W1 → W2.

F2 has one shipping route to W2 (i.e., F2 → DC → W2) and also one route to W1 (i.e., F2 → DC → W2 → W1). The unit cost for shipping through each route is also shown in Figure 6.7. It is assumed that the routes F1 → F2 and DC → W2 have maximum shipping amounts (i.e., 10 units max for the route F1 → F2 and 80 units max for the route DC → W2). The other lanes have sufficient shipping capacity to handle everything the factories can send.

The decision to be made here concerns how much to ship through each shipping lane. The objective is to minimize the total shipping cost.

Solution

The first step is formulation as a linear programming problem. There are seven shipping lanes, and thus seven decision variables are set up ($x_{F1-F2}, x_{F1-DC}, x_{F1-W1}, x_{F2-DC}, x_{DC-W2}, x_{W1-W2}, x_{W2-W1}$) to represent the product shipping amounts through the respective lanes. There are several restrictions on the values of these variables. First, all decision variables are to be nonnegative. In addition, as shown in Figure 6.7, there are two upper-bound constraints, $x_{F1-F2} \leq 10$ and $x_{DC-W2} \leq 80$, imposed by the limited shipping capacities for the two routes F1 → F2 and DC → W2. All the other restrictions arise from five net flow constraints, one for each of the five locations. These constraints have the following form:

Amount shipped out − amount shipped in = required amount of shipping out

From Figure 6.7, the specified quantities to be shipped out for F1 and F2 are 50 and 40, respectively. For W1 and W2, the specified quantities to be shipped out would be −30 and −60, respectively, since they are to receive the product. Concerning the distribution centre (DC), as all units produced at the factories are ultimately needed at the warehouses, any units shipped from the factories to the distribution centre should be forwarded to the warehouses. Therefore, the total amount shipped from the distribution centre to the warehouses should equal the total amount shipped from the factories to the distribution centre. In other words, the difference between these two shipping amounts (representing the required net flow constraint) should be zero.

Example 6.3 *Continued*

Since the objective is to minimize the total shipping cost, the coefficients for the objective function come directly from the unit shipping costs given in Figure 6.7. Therefore, using money units of **hundreds of dollars** in this objective function, the complete linear programming model becomes:

$$Minimize \ Z = 2x_{F1-F2} + 4x_{F1-DC} + 9x_{F1-W1} + 3x_{F2-DC} + x_{DC-W2} + 3x_{W1-W2} + 2x_{W2-W1}$$

$$(6.22a)$$

subject to:

(1) Net flow constraints:

$$x_{F1-F2} + x_{F1-DC} + x_{F1-W1} = 50 \ \text{(factory 1)} \qquad (6.22b)$$

$$-x_{F1-F2} + x_{F2-DC} = 40 \ \text{(factory 2)} \qquad (6.22c)$$

$$-x_{F1-DC} - x_{F2-DC} + x_{DC-W2} = 0 \ \text{(distribution centre)} \qquad (6.22d)$$

$$-x_{F1-W1} + x_{W1-W2} - x_{W2-W1} = -30 \ \text{(warehouse 1)} \qquad (6.22e)$$

$$-x_{DC-W2} - x_{W1-W2} + x_{W2-W1} = -60 \ \text{(warehouse 2)} \qquad (6.22f)$$

(2) Upper-bound constraints:

$$x_{F1-F2} \leq 10 \qquad (6.22g)$$

$$x_{DC-W2} \leq 80 \qquad (6.22h)$$

(3) Nonnegativity constraints:

$$x_{F1-F2} \geq 0$$
$$x_{F1-DC} \geq 0$$
$$x_{F1-W1} \geq 0$$
$$x_{F2-DC} \geq 0 \qquad (6.22i)$$
$$x_{DC-W2} \geq 0$$
$$x_{W1-W2} \geq 0$$
$$x_{W2-W1} \geq 0$$

This typical linear program can be solved by the simplex method (for details, see Hillier and Lieberman 2010). The MATLAB script for solving the linear programming model in Eq (6.22) is presented in Code 6.4. Finally, the optimal shipping schemewould be:

$$x_{F1-F2} = 0$$

$$x_{F1-DC} = 40$$

$$x_{F1-W1} = 10$$

$$x_{F2-DC} = 40$$

$$x_{DC\text{-}W2} = 80$$

$$x_{W1\text{-}W2} = 0$$

$$x_{W2\text{-}W1} = 20$$

Such a shipping scheme will lead to a minimum cost of \$49,000. Figure 6.8 shows the optimal shipping strategy obtained by the linear programming model.

Code 6.4 *The MATLAB script for the linear programming model presented in Eq (6.22) for solving Example 6.3.*

```
clc; clear all;
% coefficients of the objective function
f = [2; 4; 9; 3; 1; 3; 2];
% Coefficients for equal constraints
%  F1-F2, F1-DC, F1-W1, F2-DC, DC-W2, W1-W2, W2-W1
Aeq = [1      1       1      0      0      0      0;...
       -1     0       0      1      0      0      0;...
       0      -1      0      -1     1      0      0;...
       0      0       -1     0      0      1      -1;...
       0      0       0      0      -1     -1     1];
%Right-hand side for equal constraints
beq = [50; 40; 0; -30; -60];
% Coefficients for inqulity constraints
%  F1-F2, F1-DC, F1-W1, F2-DC, DC-W2, W1-W2, W2-W1
A = [1      0       0      0      0      0      0;...
     0      0       0      0      1      0      0];
%Right-hand side for inequality constraints
b = [10; 80];
% define the lower bound of decision variables
lb = [0; 0; 0; 0; 0; 0; 0];

% solve the linear programming
[x, fval, exitflag, output] = linprog(f, A, b, Aeq, beq, lb, []);

% show the results
disp('optimal solutions:');
disp(x);
disp('optimal objective values:');
disp(fval);
```

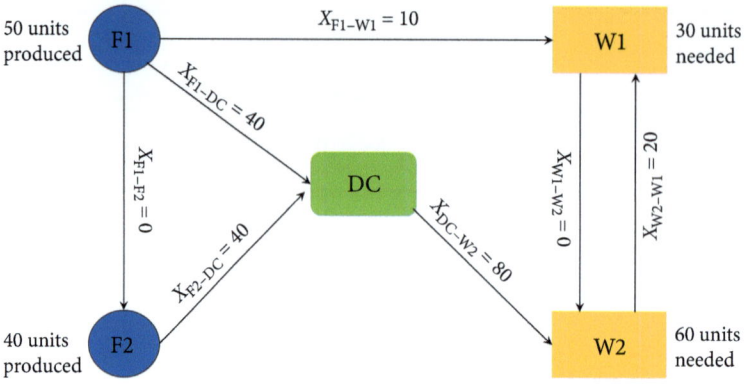

Figure 6.8 The optimal distribution network for Example 6.3 obtained through the linear programming model.

6.3.2 Dynamic programming method

Dynamic programming is a valuable mathematical technique employed to make a sequence of interconnected decisions. It offers a systematic procedure for identifying the optimal combination of decisions. Unlike linear programming, there is no standardized mathematical formulation for dynamic programming problems. Instead, dynamic programming represents a general approach to problem-solving, and the specific equations employed should be tailored to suit each unique situation. The main concept of dynamic programming is to divide a problem into smaller nested subproblems, and then combine the solutions to reach an overall solution. Dynamic programming plays a pivotal role in addressing various challenges within traffic engineering, particularly in optimizing traffic flow, signal control, and route planning.

Optimality and characteristics of dynamic programming problems

Dynamic programming's ability to handle problems with sequential and interdependent decisions makes it particularly valuable in the transportation domain in areas such as route optimization, traffic signal timing, traffic flow control, vehicle routing, and scheduling. Consider the following multistage decision process (Moura 2014):

$$\min_{x_k, u_k} J = \sum_{k=0}^{N-1} g_k(x_k, u_k) + g_N(x_N) \tag{6.23a}$$

subject to:

$$x_{k+1} = f(x_k, u_k), \quad k = 0, 1, \ldots, N-1 \tag{6.23b}$$

$$x_0 = x_{\text{init}} \tag{6.23c}$$

where k is the discrete time index, x_k is the state at time k, u_k is the control decision applied at time k, N is the time horizon, g_k is the instantaneous cost, and g_N is the final or terminal cost.

In principle, the optimality principle can be described as follows. At time step k, assuming that all future optimal decisions, $u^*(k + 1), u^*(k + 2), ..., u^*(N - 1)$, are known, the decision-maker can calculate the best solution for the current time step by considering the pairing with future decisions (Moura 2014). This recursive process starts from the end at time $N - 1$ and progresses backward.

Mathematically, the principle of optimality can be precisely expressed as follows. Define $V_k(x_k)$ as the optimal 'cost to go' (also known as the 'value function') from time step k to the end of the time horizon N, given the current state x_k. The principle of optimality can then be formulated in recursive form as follows (Moura 2014):

$$V_k(x_k) = \min_{u_k} \{g(x_k, u_k) + V_{k+1}(x_{k+1})\} \tag{6.24a}$$

with the boundary condition

$$V_N(x_N) = g_N(x_N) \tag{6.24b}$$

As stated in Hillier and Lieberman (2010), there are some basic features that characterize dynamic programming problems:

(i) The problem can be divided into stages, with a policy decision required at each stage. Dynamic programming problems require making a sequence of interrelated decisions, where each decision corresponds to one stage of the problem.

(ii) Each stage has a number of states associated with the beginning of that stage. In general, the states are the various possible conditions the system might be in at that stage of the problem. The number of states may be either finite, as in the stagecoach problem, or infinite, as in some subsequent examples.

(iii) The effect of the policy decision at each stage is to transform the current state to a state associated with the beginning of the next stage, possibly according to a probability distribution.

(iv) The solution procedure is designed to find an optimal policy for the overall problem, i.e., a prescription of the optimal policy decision at each stage for each of the possible states. For any problem, dynamic programming provides this kind of policy prescription of what to do under every possible circumstance, which is why the actual decision made upon reaching a particular state at a given stage is referred to as a policy decision. Providing this additional information beyond simply specifying an optimal solution (i.e., optimal sequence of decisions) can be helpful in a variety of ways, including sensitivity analysis.

(v) Given the current state, an optimal policy for the remaining stages is independent of the policy decisions adopted in previous stages. Therefore, the optimal immediate decision depends on only the current state and not on how you got there. This is the principle of optimality for dynamic programming.

(vi) The solution procedure begins by finding the optimal policy for the last stage. The optimal policy for the last stage prescribes the optimal policy decision for

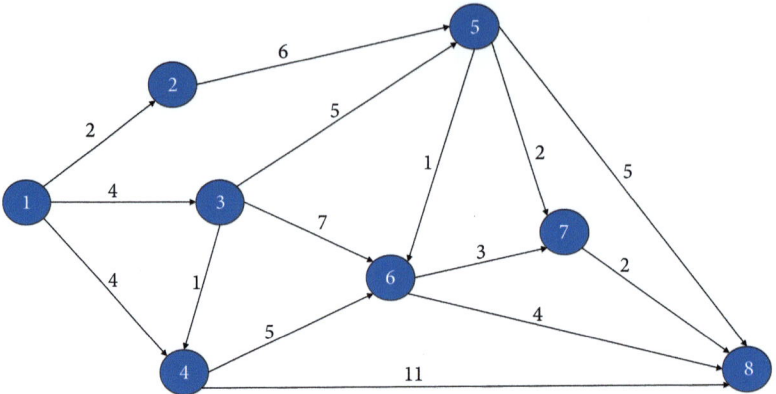

Figure 6.9 Traffic network for the shortest path problem (adapted from Moura 2014).

each of the possible states at that stage. The solution of this one-stage problem is usually trivial.

(vii) A recursive relationship that identifies the optimal policy for stage k, given the optimal policy for stage $k + 1$, is expressed as Eq (6.24a). Therefore, finding the optimal policy decision when you start in state x at stage k requires finding the minimizing value of u_k. For this particular problem, the corresponding minimum cost is achieved by using this value of u_k and then following the optimal policy when you start in state u_k at stage $k + 1$.

(viii) When we use this recursive relationship, the solution procedure starts at the end and moves backward stage by stage—each time finding the optimal policy for that stage—until it finds the optimal policy starting at the initial stage. This optimal policy immediately yields an optimal solution for the entire problem, namely, u_1^* for the initial state x_1, then u_2^* for the resulting state x_2, then u_3^* for the resulting state x_3, and so forth to u_N^* for the final stage x_N.

Application of dynamic programming in transportation

The following shortest path problem (Figure 6.9) is used to illustrate the application of dynamic programming methods in transportation engineering.

Example 6.4

This example is modified from Moura (2014). A traffic network is depicted in Figure 6.9, featuring eight cities indicated by circles with the distances between two adjacent cities shown on the connecting lines. The objective is to find the shortest route from city 1 to city 8.

Solutions

We can now define $V(i)$ as the shortest path length from city i to city 8. For instance, $V(8) = 0$. Let $d(i,j)$ denote the distance from city i to city j. For example, $d(3,5) = 5$.

Then $d(i, j) + V(j)$ represents the distance from city i to city j and then from city j to city 8 along the shortest path. This allows us to formulate the principle of optimality equation and establish boundary conditions.

$$V(i) = \min_{j \in N_i^d} \{d(i, j) + V(j)\}$$

(6.25a)

$$V(8) = 0$$

(6.25b)

where the set N_i^d represents the cities that descend from city i. For example, $N_3^d = \{4, 5, 6\}$. Now Eqs (6.25) can be solved recursively, starting from city 8 and backward to city 1, as follows:

$$V(7) = d(7, 8) + V(8) = 2 + 0 = 2$$

(6.26a)

$$V(6) = \min\{d(6, 7) + V(7), \ d(6, 8) + V(8)\} = \min\{3 + 2, \ 4 + 0\} = 4$$

(6.26b)

$$V(5) = \min\{d(5, 7) + V(7), \ d(5, 8) + V(8), \ d(5, 6) + V(6)\}$$

(6.26c)

$$= \min\{2 + 2, \ 5 + 0, \ 1 + 4\} = 4$$

$$V(4) = \min\{d(4, 6) + V(6), \ d(4, 8) + V(8)\} = \min\{5 + 4, \ 11 + 0\} = 9$$

(6.26d)

$$V(3) = \min\{d(3, 4) + V(4), d(3, 5) + V(5), d(3, 6) + V(6)\}$$

(6.26e)

$$= \min\{1 + 9, \ 5 + 4, \ 7 + 4\} = 9$$

$$V(2) = \min\{d(2, 5) + V(5)\} = 6 + 4 = 10$$

(6.26f)

$$V(1) = \min\{d(1, 2) + V(2), d(1, 3) + V(3), d(1, 4) + V(4)\}$$

(6.26g)

$$= \min\{2 + 10, \ 4 + 9, \ 4 + 9\} = 12$$

Consequently, the shortest path can be identified as $1 \rightarrow 2 \rightarrow 5 \rightarrow 7 \rightarrow 8$, shown in thick arrows in Figure 6.10. The MATLAB script for solving the dynamic programming model illustrated in Eq (6.26) is presented in Code 6.5.

Code 6.5 *The MATLAB script for the dynamic programming model in Eqs (6.25) and (6.26) described in Example 6.4, where the shortest path from city 1 to city 8 is desired.*

```
clc; clear all; close all;
%city 1 2 3 4 5 6 7 8
now=[0,2,4,4,0,0,0,0
     0,0,0,0,6,0,0,0
     0,0,0,1,5,7,0,0
     0,0,0,0,0,5,0,11
     0,0,0,0,0,1,2,5
     0,0,0,0,0,0,3,4
     0 0 0 0 0 0 0 2];
now(now==0)=inf;
BestRouteID = zeros(7,8);
```

Code 6.5 *Continued*

```
V = zeros(1,8);
for h = 7:-1:1
   allRoutes = find(now(h,:)~=inf);
   [V(h),minIndex] = min(now(h,allRoutes)+V(allRoutes));
   BestRouteID(h,allRoutes(minIndex)) = 1;
   % show the shortest pat
end

% show the shortest path
h = 1;
while(h < 8)
%for h = 1:7
   cityID = find(BestRouteID(h,:) == 1);
   disp(['shortest path is from City ' num2str(h) ' to City ' num2str(cityID)]);
   h = cityID;
end
```

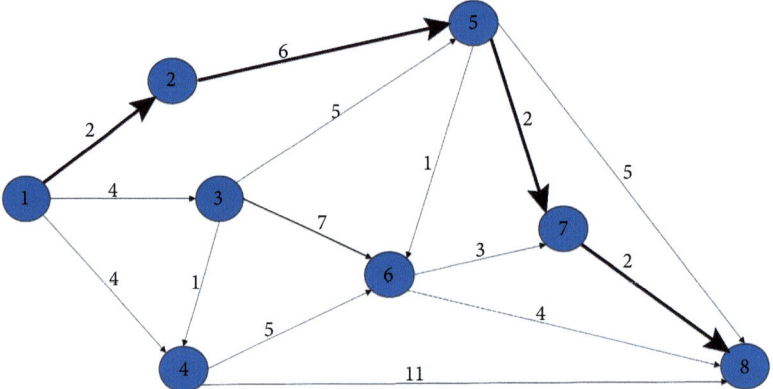

Figure 6.10 The shortest path from city 1 to city 8 is marked by the thick arrows. This solution is achieved using the MATLAB script in Code 6.5.

6.4 Traffic flow predictions

6.4.1 Background

Predicting traffic flow has emerged as a pivotal and intricate area of research in the development of intelligent traffic systems. Intelligent traffic systems are primarily designed to optimize traffic operations, alleviate congestion, and play a pivotal role in guiding vehicle route planning (Kim and Lee 2023). Traffic flow prediction forms a cornerstone for establishing traffic demand strategies within intelligent transportation systems, enabling the anticipation of traffic patterns in congested regions. This, in turn, facilitates the judicious utilization of traffic resources and contributes to sustainable traffic development on an economic scale (Afrin and Yodo 2020).

Traffic flow prediction models can be broadly categorized into traditional physics-based models and modern machine/deep learning models. Traditional physics-based models have long been utilized in the field of traffic flow prediction to understand and simulate the dynamics of vehicular movement. These models rely on fundamental principles of physics and empirical observations to capture the complex interactions within traffic systems. Some typical physics-based models include the Lighthill–Whitham–Richards model (Lighthill and Whitham 1955), cellular automaton models (Nagel and Schreckenberg 1992), the Greenshields model (Greenshields 1935), and the Daganzo–Newell model (Newell 1982; Daganzo 1994). Traditional physics-based models lay the foundation for understanding traffic flow dynamics. While these models have their merits, they often exhibit limitations in capturing the intricacies of real-world traffic scenarios. As traffic systems become increasingly complex, researchers are turning toward integrating advanced modelling approaches, including machine learning and data-driven techniques, for more comprehensive and accurate traffic flow predictions and management.

ML and deep learning (DL) have emerged as powerful tools in traffic flow prediction, revolutionizing how we model and forecast traffic dynamics. Typical ML or DL techniques in traffic flow predictions include ANNs, long short-term memory (LSTM), convolutional neural networks, and hybrid models (e.g., Hou et al. 2019; Redhu and Kumar 2023). ML and DL have significantly advanced traffic flow prediction, offering more accurate and adaptable models for managing and optimizing transportation systems.

6.4.2 Long short-term memory method

The LSTM method attempts to preserve long-term useful information and skip short-term irrelevant information. To solve the problem of long-term dependencies, Hochreiter and Schmidhuber (1997) developed the LSTM method. Subsequently, it has been applied to many different areas, including speech recognition (He and Droppo 2016), dynamic trajectory prediction and time series forecasting (Altché and Fortelle 2017), correlation analysis (Mallinar and Rosset 2018), and traffic flow prediction (e.g., Medina-Salgado et al. 2022). LSTM has become one of the focuses of deep learning.

The architecture of LSTM is shown in Figure 6.11. It is inspired by the logic gates of a computer. Gated memory cells are major components—a number of gates are used to control a memory cell. LSTM can be expressed succinctly in the following discrete dynamic and vector equations (Xiong 2022):

$$f_t = \sigma\left(W_{fx}x_t + W_{fh}h_{t-1} + b_f\right) \tag{6.27a}$$

$$i_t = \sigma\left(W_{ix}x_t + W_{ih}h_{t-1} + b_i\right) \tag{6.27b}$$

$$\tilde{c}_t = \tanh\left(W_{cx}x_t + W_{ch}h_{t-1} + b_c\right) \tag{6.27c}$$

$$c_t = f_t \odot c_{t-1} + i_t \odot \tilde{c}_t \tag{6.27d}$$

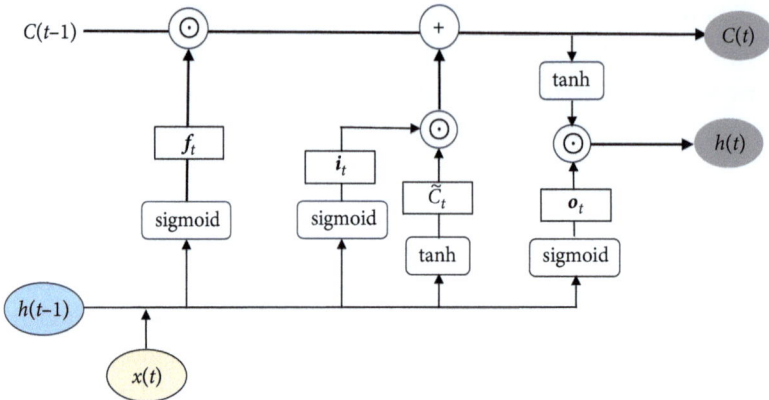

Figure 6.11 Architecture of the LSTM model. Here, $x(t)$ is the input at time step t; f_t represents the output vector of the forget gate; $h(t-1)$ and $h(t)$ respectively indicate the hidden state at $t-1$ and t; $C(t-1)$ and $C(t)$ represent the cell memory at $t-1$ and t, respectively; $\tilde{c}_t$ is the candidate cell memory at time step t; i_t is the input gate; and o_t represents the output gate.

$$o_t = \sigma\left(W_{ox}x_t + W_{oh}h_{t-1} + b_o\right) \tag{6.27e}$$

$$h_t = o_t \odot \tanh\left(c_t\right) \tag{6.27f}$$

$$z_t = \text{softmax}\left(W_{zh}h_t + b_z\right) \tag{6.27g}$$

Equation (6.27a) signifies the forget gate, determining which information is discarded from the cell state. Here, f_t represents the output vector of the forget gate; the inputs to the forget gate include the current input vector x_t and the hidden state h_{t-1}. The parameters W_{fx} and W_{fh} are weight matrices and b_f is a bias vector. σ is the sigmoid activation function. Equation (6.27b) indicates that the input gate determines which new information is stored in the cell state. This gate functions as a fully connected layer with a sigmoid activation function. Here, W_{ix} and W_{ih} are weight matrices and b_i is a bias vector.

Equation (6.27c) represents the candidate memory cell, determining the extent to which new data should be incorporated, in which W_{cx} and W_{ch} are weight matrices and b_c is a bias vector. The tanh activation function is employed in this context. Equation (6.27d) represents the memory cell, determining the balance between memorizing and retaining information from the old state and incorporating new information into the current hidden state within the cell. In this context, the symbol $\odot$ denotes element-wise multiplication.

Equation (6.27e) defines the output gate. This gate, functioning as a fully connected layer with a sigmoid activation function, determines which portions of the cell state will be output. The hidden state h_t can then be updated through Eq (6.27f) using information from the output gate o_t and cell state c_t. Equation (6.27g) will ultimately generate the prediction at time step t.

6.4.3 Traffic flow predictions through LSTM

In the realm of transportation and urban planning, predicting traffic flow is a critical task for optimizing traffic management and improving overall efficiency. LSTM networks offer a powerful solution for modelling complex temporal patterns in traffic data. We apply the LSTM method to a traffic problem in the following example.

Example 6.5

Table 6.1 shows part of the traffic flow monitoring at M25 motorway in London, in which the total traffic flow and different vehicle types are measured every 15 minutes. Data monitored in the period August 1–31 2018 were used to test the applicability of LSTM for traffic flow predictions. All data used in this case are collected from UK National Highways (https://webtris.highwaysengland.co.uk/). The full dataset can be downloaded from the book website at: http://www.oup.com/ AdvancedMathematicalModelling. Note that Table 6.1 shows only a small part of this large dataset.

Solution

In this scenario, the prediction of the total traffic carriageway flow will be based on the total flows observed in the past five measurement time steps. In simpler terms, the traffic flow prediction model can be expressed as $Q_t = f(Q_{t-1}, Q_{t-2}, Q_{t-3}, Q_{t-4}, Q_{t-5})$. The LSTM model will be used to construct the traffic flow prediction model. Figure 6.12 illustrates traffic flow predictions at different time steps during the training and testing periods. In this case, 70% of the measurements were used to train the LSTM model, while the remaining 30% were used to test the developed LSTM model. The results in Figure 6.12 indicate that the LSTM model performed exceptionally well for traffic flow predictions, achieving R^2 values of 0.97 in the training period and 0.96 in the testing period. This case demonstrates the effectiveness of LSTM in traffic flow predictions.

The MATLAB script for the LSTM-based traffic flow prediction model is presented in Code 6.6, which produces Figure 6.12. Note that to run this code, the input data file named 'LSTM-traffic-flow-data.xlsx' should be placed in the same folder. This input file can be downloaded from the following website: http://www.oup.com/ AdvancedMathematicalModelling.

Table 6.1 Some sample data for traffic flow monitoring at M25/4919A (longitude −0.5089, latitude 51.4676). The full dataset can be downloaded from the book website at: http://www.oup.com/AdvancedMathematicalModelling.

Local date	Local time (from midnight)	Total car-riageway flow	Vehicles less than 5.2 m	Vehicles 5.21–6.6 m	Vehicles 6.61–11.6 m	Vehicles above 11.6 m	Speed value (km/h)
2018/8/1	0:14:00	246	179	20	17	30	101.87
2018/8/1	0:29:00	246	181	20	13	32	81.37
2018/8/1	0:44:00	247	179	23	14	31	73.24
2018/8/1	0:59:00	193	132	17	15	29	77.25
2018/8/1	1:14:00	189	117	12	21	39	72.71
2018/8/1	1:29:00	152	96	21	10	25	75.28
2018/8/1	1:44:00	157	100	4	14	39	80.23
2018/8/1	1:59:00	166	112	18	17	19	106.44
2018/8/1	2:14:00	141	82	10	17	32	102.28
2018/8/1	2:29:00	125	75	13	12	25	104.88

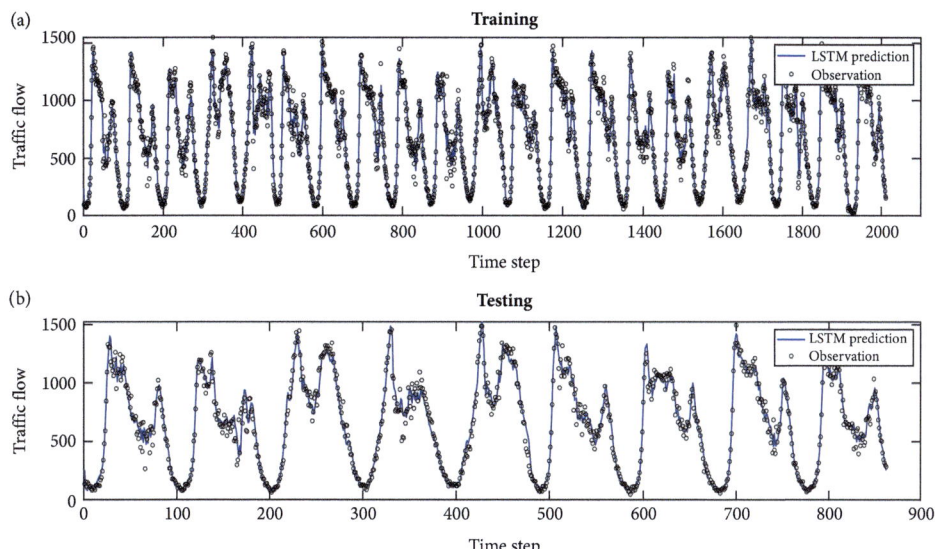

Figure 6.12 Comparison between LSTM predictions and actual measurements (i.e., observations) in the training and testing period for traffic flow in M25 motorway in London (Example 6.5). This figure is produced using the MATLAB script presented in Code 6.6.

Code 6.6 *The MATLAB script for traffic flow prediction through LSTM. This code produces Figure 6.12. Note that to run this code, the input data file named 'LSTM-traffic-flow-data.xlsx' should be placed in the same folder. This input file can be downloaded from the website: http://www. oup.com/AdvancedMathematicalModelling.*

```
clc; clear all; close all;
set(0,'defaultaxesfontsize',17);
%  load the traffic flow data from supplementary materials
[NumData,TxtData,CellData] = xlsread('LSTM-traffic-flow-data.xlsx');
% assigen data to different observation variables
TotalTrafficFlow = NumData(:,3);
Nsam = length(TotalTrafficFlow);
% using the past 5 time steps to predict current traffic flow rate
X_All(:,1) = TotalTrafficFlow(1:(Nsam-5));
X_All(:,2) = TotalTrafficFlow(2:(Nsam-4));
X_All(:,3) = TotalTrafficFlow(3:(Nsam-3));
X_All(:,4) = TotalTrafficFlow(4:(Nsam-2));
X_All(:,5) = TotalTrafficFlow(5:(Nsam-1));
Y_All =  TotalTrafficFlow(6:Nsam);
%
ntraining = floor(length(Y_All)*0.7)
Xmatrix_train = X_All(1:ntraining,:);
Yt_training = Y_All(1:ntraining);
Xmatrix_test = X_All((ntraining+1):2875,:);
Yt_test = Y_All((ntraining+1):2875);
inputs = Xmatrix_train';
targets = Yt_training';
% standardization
[XTrain, PSx] = mapminmax(inputs);
```

Code 6.6 *Continued*

```matlab
[YTrain, PSy] = mapminmax(targets);
%XTrain = inputs;
%YTrain = targets;
XTest = mapminmax('apply', Xmatrix_test', PSx);
% --------------LSTM model ----------
%  this part should be run in the lastest version of matlab
numFeatures = size(Xmatrix_train, 2);
% LSTM units in the hidden layer h(t)
numHiddenUnits = 10;
% dimension of output yt
numResponses = 1;
% setting regressionLayer
layers = [sequenceInputLayer(numFeatures)
          lstmLayer(numHiddenUnits)
          fullyConnectedLayer(numResponses)
          regressionLayer];
miniBatchSize = 64;
options = trainingOptions('adam', 'ExecutionEnvironment', 'cpu', 'MaxEpochs',...
    1000, 'MiniBatchSize', miniBatchSize, 'GradientThreshold', 1,...
    'InitialLearnRate', 0.01, 'LearnRateSchedule', 'piecewise', ...
  'LearnRateDropPeriod', 250, 'LearnRateDropFactor', 0.2,  'Verbose',...
  false,  'Plots', 'training-progress');
% training
net = trainNetwork(XTrain, YTrain, layers, options);
% prediction in training
yTrain_pre = predict(net, XTrain, 'MiniBatchSize', miniBatchSize,...
    'SequenceLength', 'longest');
% inverse from standardization
yTrain_pre = mapminmax('reverse', yTrain_pre, PSy);
yTrain_pre = yTrain_pre';
filename = 'output_training.txt';
delimiter = '\t';  % You can change the delimiter as needed
% Write data to the file
dlmwrite(filename, yTrain_pre, 'delimiter', delimiter);
% --------------Testing -------------
yTtest_pre = predict(net, XTest, 'MiniBatchSize', miniBatchSize,...
    'SequenceLength', 'longest');
% reverse from standardization
yTtest_pre = mapminmax('reverse', yTtest_pre, PSy);
yTtest_pre = yTtest_pre';
filename = 'output_testing.txt';
delimiter = '\t';  % You can change the delimiter as needed
% Write data to the file
dlmwrite(filename, yTtest_pre, 'delimiter', delimiter);
% plot comparison %%%%%%%%%%%%%%%%%%%%%%%%%%%%%%%%%%%%%%
figure
subplot(2,1,1)
days = 1:length(Yt_training);
plot(days,yTrain_pre,'b-','linewidth',1.5)
hold on
plot(days,Yt_training,'ko','markersize',4,'linewidth',0.8)
xlabel('Time step'); title('(a) Training')
xlim([0,2100]); set(gca,'linewidth', 1.5);
ylabel('Traffic flow'); legend('LSTM prediction','Observation')
%%%%%%%%%%%%%%%%%%%%%%%%%%%%%%%%%%%%%%%%%%%%%%%%%%%%%%%%
subplot(2,1,2)
```

```
days2 = 1:length(Yt_test);
plot(days2,yTtest_pre,'b-','linewidth',1.5)
hold on
plot(days2,Yt_test,'ko','markersize',4,'linewidth',0.8)
xlabel('Time step')
ylabel('Traffic flow')
legend('LSTM prediction','Observation')
title('(b) Testing')
% setup fontSize
set(gca, 'FontName', 'Arial'); set(gca,'linewidth', 1.5);
%save the plot
saveas(gcf, 'myfigure.jpg');
```

6.5 Problems for further study

Problem 6.1

A single-lane tunnel is undergoing maintenance, which restricts vehicles to passing through it one by one in a deterministic manner. The tunnel opens for vehicle passage at precisely 6:00 a.m. At opening, vehicles start to arrive at a fixed rate of one vehicle every 10 seconds, maintaining this pace for the first 30 minutes. After this period, at 6:30 a.m., the arrival rate changes to one vehicle every 40 seconds, which continues for the rest of the day. Given that each vehicle requires 10 seconds to pass through the tunnel, and considering this scenario under a D/D/1 queue model, explore the operational characteristics of the queue.

Problem 6.2

Consider a single-lane bridge that needs to be crossed by vehicles coming from both directions. Due to maintenance work, traffic lights have been installed at both ends of the bridge, allowing vehicles to pass from one direction at a time. The arrival of vehicles at one end of the bridge follows a Poisson process, with an average arrival rate of $\lambda = 2$ vehicle/min. The bridge acts as a single-server system, where the crossing time for each vehicle follows an exponential distribution with a mean rate of $\mu = 3$ vehicle/min. Analyse the average queue length and waiting time for vehicles waiting to cross the bridge. Additionally, assess the impact on the average queue length and waiting time if the crossing time were to be doubled.

Problem 6.3

There are two farms, one managed by Jessica, producing 3 tonnes of oranges per day, and another managed by Alex, producing 5 tonnes of oranges daily. Two factories require these oranges for their juice production processes. The first factory, managed by Lily, needs 4 tonnes of oranges per day, and the second factory, managed by Tom, requires 3 tonnes daily. Henry, who owns both farms and factories, is tasked with managing the logistics and costs associated with shipping the oranges from the farms to the factories. The shipping costs per tonne are outlined as follows:

From Jessica's farm to Lily's factory: £800 per tonne.

From Alex's farm to Lily's factory: £900 per tonne.

From Jessica's farm to Tom's factory: £950 per tonne.

From Alex's farm to Tom's factory: £850 per tonne.

The objective is to determine the most cost-effective distribution strategy for the oranges from the farms to the factories, ensuring that the supply meets the factories' demand for orange juice production.

Problem 6.4

In Example 6.5 in this chapter, the total traffic flow at location M25/4919A was predicted using an LSTM model that utilized the past five measurements. That model may be further refined and potentially improved through experiment with different lengths of historical input data. Specifically, test models that use the past three measurements (creating a general model $Q_t = f(Q_{t-1}, Q_{t-2}, Q_{t-3})$) and the past ten measurements ($Q_t = f(Q_{t-1}, Q_{t-2}, ..., Q_{t-10})$). The goal of this investigation is to determine whether including a greater number of past measurements can enhance the model's performance in predicting traffic flow.

References

Afrin, T., and Yodo, N. (2020). A survey of road traffic congestion measures towards a sustainable and resilient transportation system. *Sustainability*, 12(11), 4660.

Altché, F., and Fortelle, A.D.L. (2017). An LSTM network for highway trajectory prediction. In *Proceedings of the IEEE 20th International Conference on Intelligent Transportation Systems*.

Daganzo, C. (1994). The cell transmission model, part II: Network traffic. *Transportation Research Part B: Methodological*, 28(2), 279–293.

Greenshields, B.D. (1935). A study of traffic capacity. *Highway Research Board Proceedings*, 14, 448–477.

He, T., and Droppo, J. (2016). Exploiting LSTM structure in deep neural networks for speech recognition. In *Proceedings of the IEEE International Conference on Acoustics, Speech and Signal Processing*, 5445–5449.

Hillier, F.S., and Lieberman, G.J. (2010). *Introduction to Operations Research.*, 9th edn. McGraw-Hill.

Hou, Q., Leng, J., Ma, G., Liu, W., and Cheng, Y. (2019). An adaptive hybrid model for short-term urban traffic flow prediction. *Physica A: Statistical Mechanics and its Applications*, 527, 121065.

Hochreiter, S., and Schmidhuber, J. (1997). Long short-term memory. *Neural Computation*, 9(8), 1735–1780.

Kendall, D.G. (1953). Stochastic processes occurring in the theory of queues and their analysis by the method of the imbedded Markov chain. *The Annals of Mathematical Statistics*, 24(3), 338–354. doi: 10.1214/aoms/1177728975

Kim, M., and Lee, D. (2023). Why uncertainty in deep learning for traffic flow prediction is needed. *Sustainability*, 15(23), 16204.

Larson, R.C., and Odoni, A.R. (1981). *Urban Operations Research*. Prentice-Hall.

Lighthill, M.J., and Whitham, G.B. (1955). On kinematic waves II. A theory of traffic flow on long crowded roads. *Proceedings of the Royal Society of London, Series A*, 229(1178), 317–345.

Mallinar, N., and Rosset, C. (2018). Deep canonically correlated LSTMs. arXiv:1801.05407.

Mannering, F.L., and Washburn, S.S. (2020). *Principles of Highway Engineering and Traffic Analysis*, 7th edn. Wiley.

Medina-Salgado B., Sánchez-Delacruz E., Pozos-Parra P., and Sierra J.E. (2022). Urban traffic flow prediction techniques: A review. *Sustainable Computing: Informatics and Systems*, 35, 100739.

Moura, S., (2014). Dynamic Programming. Available at: https://ecal.berkeley.edu/ce191.html.

Nagel, K., and Schreckenberg, M. (1992). A cellular automaton model for freeway traffic. *Journal de Physique I*, 2(12), 2221–2229.

Newell, G. (1982). *Applications of Queueing Theory*, 2nd edn. Chapman and Hall.

Redhu, P., and Kumar, K. (2023). Short-term traffic flow prediction based on optimized deep learning neural network: PSO-Bi-LSTM. *Physica A: Statistical Mechanics and its Applications*, 625, 129001.

Revelle, C.A., and Whitlach, E.E. (1996). *Civil and Environmental Systems Engineering*. Prentice Hall.

Xiong, M. (2022). *Artificial Intelligence and Causal Inference*. Chapman and Hall/CRC.

Selected Topics in Engineering Mathematics

A.1 Sets and numbers

The concept of a set is fundamental in mathematics. The following definition of a set (Stoll 1963) is sufficient for the purposes of this chapter:

> A **set** is a gathering together into a whole of definite, distinct objects of our perception or of our thought, which are called elements of the set.

The elements of a set typically share common characteristics and properties. Sets are usually denoted with capital letters, e.g., A, B, V, U, while their elements are denoted with lowercase letters inside curly brackets, such as $A = \{a, b, c, d\}$. A set that has no elements is called the **null set** and is denoted as $\emptyset$ or $\{\}$.

The following sets are **sets of numbers**:

- **Natural numbers** $\mathbb{N}$: $\{1, 2, 3, 4, \ldots\}$
- **Integers** $\mathbb{Z}$: $\{\ldots, -4, -3, -2, -1, 0, 1, 2, 3, 4\}$
- **Rational numbers** $\mathbb{Q}$: numbers of the form $\frac{p}{q}$, where p, q are integers and $q \neq 0$. Rational numbers can therefore be written as fractions. Any number that cannot be written in fractional form is called irrational. Irrational numbers have infinite decimal expansions. Examples of irrational numbers are $\sqrt{2}, \pi, e$.
- **Real numbers** $\mathbb{R}$: the set of all rational and irrational numbers put together.
- **Complex numbers** $\mathbb{C}$: numbers of the form $a + ib$ where a, b are real numbers and $i = \sqrt{-1}$ is the imaginary unit.

Sets and operations with sets can conveniently be depicted with the use of **Venn diagrams**. In a Venn diagram a set is depicted by a closed curve in the plane (e.g., a circle; see Figure A.1).

Some basic operations with sets are listed below:

> The **union** of A and B is a set we denote as $A \cup B$ that contains all the elements in A and all the elements in B. It can be expressed as the operation 'A or B', which implies that the set has all the elements that are either in A or in B or in both at the same time (see Figure A.1a).
>
> The **intersection** of A and B is a set we denote as $A \cap B$ that contains all the common elements of A and B. It can be expressed as the operation 'A and B', which implies that the set has all the elements that belong to A and B at the same time (see Figure A.1b).
>
> The **absolute complement** of a set A in another set U is the set of all elements that belong in U but do not belong in A. It is denoted as $A^c = U \backslash A$. It can be expressed as the operation 'U and not A' (see Figure A.1c).

Consider the set $A = \{\ldots, -12, -10 - 8, -6 - 4, -2, 0, 2, 4, 6, 8, 10, 12, \ldots\}$, which is called the set of **even numbers**. Even numbers can be divided exactly by 2.

The set $B = \mathbb{Z} \backslash A = \{-11, -9, -7, -5, -3, -1, 1, 3, 5, 7, 9, 11, \ldots\}$ is called the set of **odd numbers**. Odd numbers cannot be divided exactly by 2.

(a)   (b)   (c) U

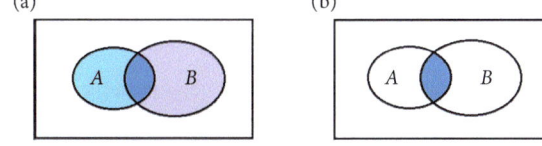
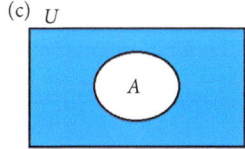

Figure A.1 Venn diagrams of some basic operations with sets: (a) union, (b) intersection, and (c) complement.

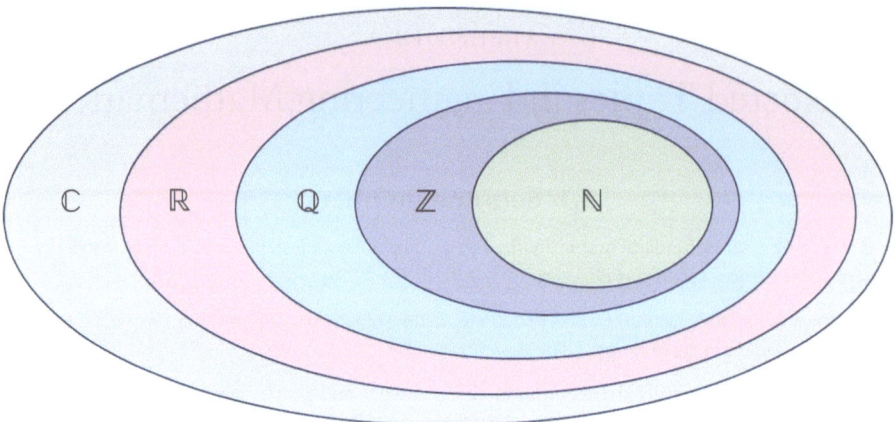

Figure A.2 Sets of numbers in the form of a Venn diagram.

Two sets are equal if and only if they have the same elements. The equality of two sets is denoted as $A = B$. If all the elements of a set A are also elements of a larger set B then we say that A is a **subset** of B.

The sets of numbers that were introduced above can also be represented in a Venn diagram—see Figure A.2.

Using the basic operations with sets and Venn diagrams, it is easy to prove **De Morgan's laws**, which state:

- The complement of the union of two sets is the intersection of their complements:

$$(A \cup B)^c = A^c \cap B^c \tag{A.1}$$

- The complement of the intersection of two sets is the union of their complements:

$$(A \cap B)^c = A^c \cup B^c \tag{A.2}$$

Sets of real numbers are denoted using parentheses if the end points are not included in the set and square brackets if the end points are included. For example, for any two real numbers a, b where $a < b$, the following sets are possible:

- (a, b) contains all real numbers between a and b, but not a, b themselves. This is called an **open set**.
- $[a, b]$ contains all real numbers between a and b, as well as a, b themselves. This is called a **closed set**.
- $[a, b)$ contains all real numbers between a and b, as well as a, but not b.
- $(a, b]$ contains all real numbers between a and b, as well as b, but not a.

The basic operations with numbers are **addition** and **multiplication**.

- Addition is denoted by $+$ and provides the sum of two numbers.
- Multiplication is denoted by $\cdot$ or $\times$ and provides the product of two numbers.
- The product of a number a by itself is denoted using the exponent 2: $a \cdot a = a^2$.
- This form can be generalized: $\underbrace{a \cdot a \cdot a \cdots a}_{n \text{ times}} = a^n$.
- Addition of a negative number is termed subtraction, while multiplication by the inverse of a number $1/a$ is termed division. The inverse of a number is conveniently denoted as a^{-1}.

- Exponents that are not integers produce radicals, such as the square root of a number: $\sqrt{2} = 2^{1/2}$.
- Addition and multiplication are **associative** and **commutative** operations. This means that, for any real numbers a, b, and c, the following properties hold true:

Associative: $(a + b) + c = a + (b + c)$ and $(a \cdot b) \cdot c = a \cdot (b \cdot c)$
Commutative: $a + b = b + a$ and $a \cdot b = b \cdot a$

A.1.1 Complex numbers

In the two-dimensional setting, there is a correspondence between points in the two-dimensional plane and complex numbers. Any point P in the plane with coordinates (x_p, y_p) can be represented using the complex number $p = x_p + iy_p$. The x_p coordinate is then called the **real part** of complex number p and is denoted as $\mathrm{Re}\,(p) = x_p$. The y_p coordinate is called the **imaginary part** of complex number p and is denoted as $\mathrm{Im}\,(p) = y_p$. In other words, the horizontal axis with the x coordinates corresponds to the real part of a complex number p and the vertical axis with the y coordinates corresponds to the imaginary part. For that reason, the vertical axis in two dimensions is sometimes called the **imaginary axis**. A complex number that has zero imaginary part is a real number. A complex number that has zero real part is called a **purely imaginary number**.

If $p = x_p + iy_p$, we define the conjugate complex number of p as follows:

$$\bar{p} = x_p - iy_p \tag{A.3}$$

The **measure** of complex number p is the length of the line connecting the point with coordinates $(0, 0)$ to the point (x_p, y_p); at the same time it is the square root of the product of p and its conjugate $\bar{p}$. That is,

$$\sqrt{p \cdot \bar{p}} = \sqrt{(x_p + iy_p) \cdot (x_p - iy_p)} = \sqrt{x_p^2 + y_p^2} = |p| \tag{A.4}$$

The sum of a complex number and its conjugate is always twice the real part of the complex number:

$$p + \bar{p} = x_p + iy_p + x_p - iy_p = 2 \cdot x_p = 2 \cdot \mathrm{Re}\,(p) \tag{A.5}$$

The difference between a complex number and its conjugate is always twice the imaginary part of the complex number times the imaginary unit:

$$p - \bar{p} = x_p + iy_p - (x_p - iy_p) = 2 \cdot i \cdot y_p = 2 \cdot \mathrm{Im}\,(p) \cdot i \tag{A.6}$$

If we want to calculate the result of the division of two complex numbers, we multiply the numerator and denominator of the division fraction by the conjugate number of the denominator. In that manner, the denominator becomes real. This means:

$$\frac{a + ib}{c + id} = \frac{(a + ib)(c - id)}{(c + id)(c - id)} = \frac{ac - iad + ibc + bd}{c^2 - idc + idc + d^2} = \frac{ac + bd}{c^2 + d^2} + i\frac{bc - ad}{c^2 + d^2} \tag{A.7}$$

Of major importance is **Euler's formula**, which relates the exponential function with an imaginary exponent to the cosine and sine functions:

$$e^{i\theta} = \cos(\theta) + i \cdot \sin(\theta) \tag{A.8}$$

A.1.2 Further reading

The interested reader can find more information on relevant topics in Stroud and Booth (2007), Yang (2018), and Singh (2011).

A.2 Functions

A.2.1 Introduction

Functions are used to correlate members of one set to members of another set. This means:

> A function f is an operation (mapping) from one set X (called the domain) to another set Y (called the codomain) such that every element of X is mapped to one and only one element of Y.

Assuming a set of numbers $X = \{x_1, x_2, x_3, x_4, \ldots, x_m\}$ and another set of numbers $Y = \{y_1, y_2, y_3, y_4, \ldots, y_n\}$, a function $y = f(x)$ with domain X and codomain Y is a mapping from X to Y, in the sense that **every element of X is mapped to one and only one element of** Y. Note that two or more different elements of X can be mapped into the same element of Y. In this case those elements of X have the same **image** in Y. As an example, it could happen that $y_4 = f(x_1)$ and $y_4 = f(x_3)$. However, a mapping that would produce $y_2 = f(x_1)$ and $y_4 = f(x_1)$ is not a function.

Functions are typically denoted using lowercase letters (e.g. f, g, h) followed by parentheses that contain the input of the function. An input of a function is termed an **argument** or **independent variable**. The output of a function (e.g., y) is called the **dependent variable**. The mapping $y = g(x)$ in Figure A.3a is a function because each element of the domain has a unique image in the codomain. The mapping $y = h(x)$ in Figure A.3b is not a function since x_3 is mapped to y_3 but also to y_5.

An efficient way to study functions is through their **graphs**. Graphs of single-variable functions are plots in a Cartesian system Oxy, where the horizontal axis has values of the independent variable x and the vertical axis is reserved for the dependent variable y. For every pair $(x, y) = (x, f(x))$ the corresponding point in the Oxy plane is depicted. Figure A.4a shows the graph of a mapping that is a function, while Figure A.4b shows a mapping that is not a function.

Let us now consider three sets, namely X, Y, and Z, and two functions g and f. Function g maps elements of X into Z. This means function g is an operation that, for each element x from set X, assigns one and only one element z from set Z. As a result, it can be written

$$z = g(x) \tag{A.9}$$

Function f is an operation that, for each element z from set Z, assigns one and only one element y from set Y and we write

$$y = f(z) \tag{A.10}$$

(a) $y = g(x)$

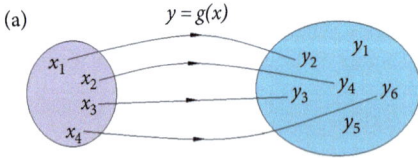

(b) $y = h(x)$

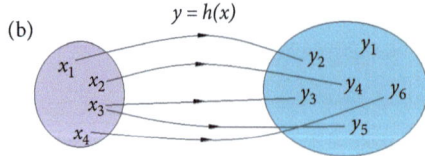

Figure A.3 Diagram of two mappings. The mapping in (a) is a function, while the mapping in (b) is not a function.

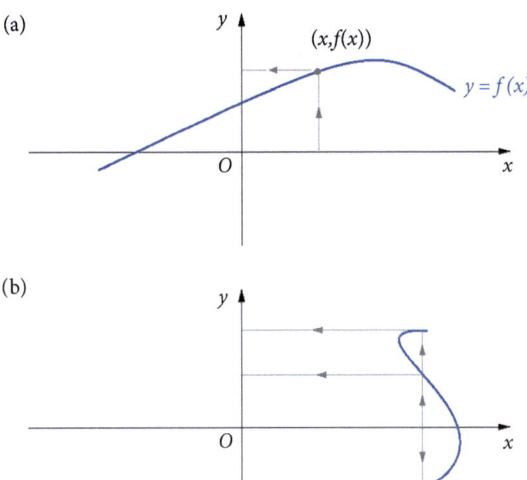

Figure A.4 Graphs of two mappings. The mapping in (a) is a function, while the mapping in (b) is not a function.

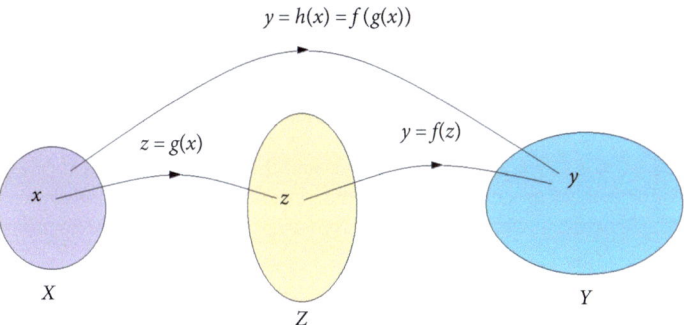

Figure A.5 Schematic representation of function composition.

Then, the **composition** of these two functions is a function h that maps elements from set X directly to set Y according to the rule $h = f(g(x))$, that is,

$$y = h(x) = f(g(x)) \tag{A.11}$$

A schematic representation of function composition is depicted in Figure A.5. Note that for function composition to be employed, the codomain of g must be the domain of f. If the composition of two functions f and g reproduces the original independent variable, then we say that f is the **inverse** of g and we write $f = g^{-1}$. In this case, it is written as follows:

$$g^{-1}(g(x)) = x \tag{A.12}$$

A schematic representation of inverse function composition is shown in Figure A.6. In this case, the codomain of $f = g^{-1}$ (set Y) is the same as the domain of g, which is X.

A function is one-to-one, often denoted as '1-1', if each element in the codomain corresponds to only one element from the domain. For the inverse function to exist, the initial function must be one-to-one.

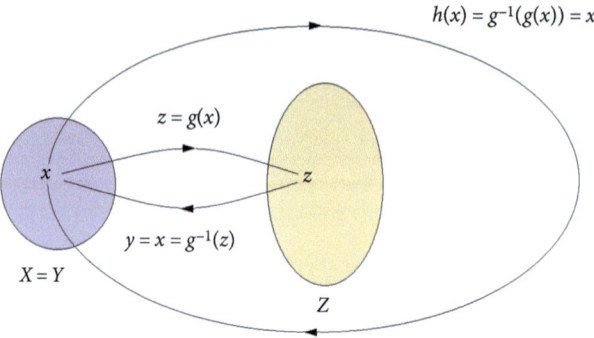

Figure A.6 Schematic representation of the inverse function.

This rule can be validated using Figure A.6 and considering the fact that a function never produces two different images for the same argument.

A.2.2 Important function categories

Polynomial functions

Polynomial functions are typically denoted as $y = p(x)$. However, the general notation $y = f(x)$ is also used. Polynomial functions take the following general form:

$$y = p(x) = c_0 + c_1 x + c_2 x^2 + c_3 x^3 + \cdots + c_{n-1} x^{n-1} + c_n x^n \tag{A.13}$$

where $c_0, c_1, c_2, c_3, \ldots, c_{n-1}, c_n$ are constants (numbers) and the maximum exponent n is the **degree** of the polynomial. The number c_0 is the **constant term**. Some examples of polynomials are:

$$p(x) = 4 \text{ (this is a constant function)}$$

$$p(x) = 5x \text{ (this is a linear function)}$$

$$p(x) = 1 - 2.5x + 9x^2 \text{ (this a quadratic function or quadratic polynomial)}$$

Note that the constant term can be written as $c_0 = c_0 \cdot 1 = c_0 x^0$. In other words, in a polynomial, all constants are multiplied by the independent variable at a power equal to the subscript of the constant. The simple functions $1, x, x^2, x^3, x^4, \ldots, x^{n-1}, x^n$ are called **monomials**. The graphs of some monomials are illustrated in Figure A.7. The MATLAB script in Code A.1 can be used to plot them. This MATLAB code and other scripts from this book can be downloaded from the following website: http://www.oup.com/AdvancedMathematicalModelling.

Code A.1 *MATLAB code to plot the graphs of the first four monomials. This code produces Figure A.7.*

```
clear all; clc; close all;

set(0,'defaultaxesfontsize',16);

x=-2:0.01:2;  % Domain
y=-4:0.01:4;

p0=ones(size(x)); % constant
p1=x;             % linear
```

```
p2=x.^2;            % quadratic
p3=x.^3;            % qubic
figure(1)
subplot(2,2,1)
plot(x,p0,'b-','linewidth',1.5)
hold on
plot(x,zeros(size(x)),'k-','linewidth',0.2)
plot(zeros(size(y)),y,'k-','linewidth',0.2)
hold off
ylim([-4 4])
xlabel('x','FontSize',18,'FontWeight','bold')
ylabel('y','FontSize',18,'FontWeight','bold')
title('p(x) = 1','FontSize',18,'FontWeight','bold')
subplot(2,2,2)
plot(x,p1,'b-','linewidth',1.5)
hold on
plot(x,zeros(size(x)),'k-','linewidth',0.2)
plot(zeros(size(y)),y,'k-','linewidth',0.2)
hold off
ylim([-4 4])
xlabel('x','FontSize',18,'FontWeight','bold')
ylabel('y','FontSize',18,'FontWeight','bold')
title('p(x) = x','FontSize',18,'FontWeight','bold')
subplot(2,2,3)
plot(x,p2,'b-','linewidth',1.5)
hold on
plot(x,zeros(size(x)),'k-','linewidth',0.2)
plot(zeros(size(y)),y,'k-','linewidth',0.2)
hold off
ylim([-4 4])
xlabel('x','FontSize',18,'FontWeight','bold')
ylabel('y','FontSize',18,'FontWeight','bold')
title('p(x) = x^2','FontSize',18,'FontWeight','bold')
subplot(2,2,4)
plot(x,p3,'b-','linewidth',1.5)
hold on
plot(x,zeros(size(x)),'k-','linewidth',0.2)
plot(zeros(size(y)),y,'k-','linewidth',0.2)
hold off
ylim([-4 4])
xlabel('x','FontSize',18,'FontWeight','bold')
ylabel('y','FontSize',18,'FontWeight','bold')
title('p(x) = x^3','FontSize',18,'FontWeight','bold')
```

Trigonometric functions

The basic trigonometric functions are **sine, cosine, tangent,** and **cotangent**, denoted respectively as $\sin(x)$, $\cos(x)$, $\tan(x)$, and $\cot(x)$. The definition of all the basic trigonometric functions can be depicted using the trigonometric circle (Figure A.8). Based on Figure A.8., we have

$$\tan(x) = \frac{\sin(x)}{\cos(x)} \text{ and } \cot(x) = \frac{\cos(x)}{\sin(x)} = \frac{1}{\tan(x)} \tag{A.14}$$

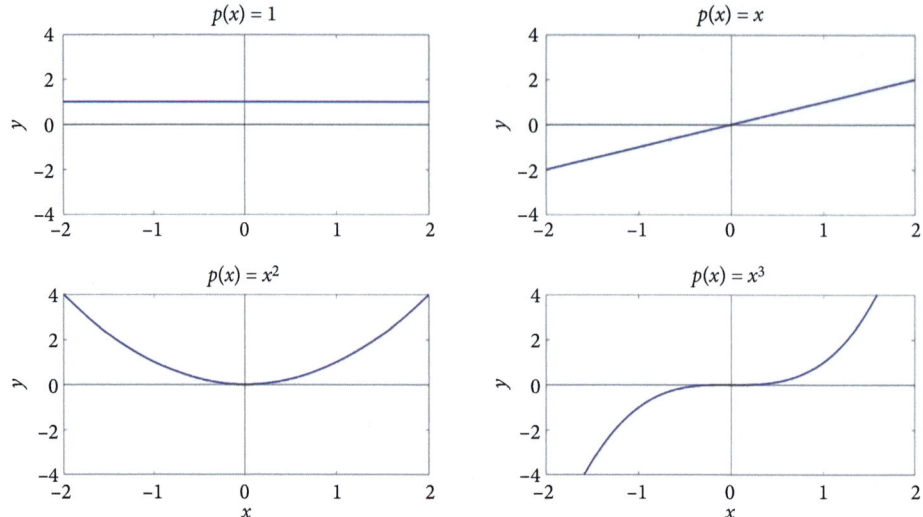

Figure A.7 Graphs of the first four monomials. This graph can be reproduced using the MATLAB script presented in Code A.1.

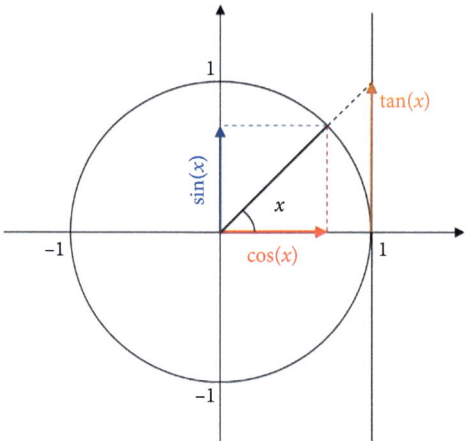

Figure A.8 The trigonometric circle.

Other trigonometric functions can be defined using the basic trigonometric functions, such as, for example, the **secant** and **cosecant**:

$$\sec(x) = \frac{1}{\cos(x)} \text{ and } \csc(x) = \frac{1}{\sin(x)} \tag{A.15}$$

Trigonometric functions are examples of **periodic functions**. Periodic functions have the property that they repeat their values at regular intervals or periods T. Therefore, given a periodic function f of period $T > 0, f(x + T) = f(x)$ for all x. The $\sin(x)$ and $\cos(x)$ function graphs, with respect to an independent variable x, are depicted in Figure A.9, while the $\tan(x)$ and $\cot(x)$ functions are presented in Figure A.10. Codes A.2 and A.3 can be used to plot them in MATLAB.

Code A.2 *MATLAB code to plot the* sin *(x) and* cos *(x) function graphs. This code produces Figure A.9.*

```matlab
clear all; clc; close all;

set(0,'defaultaxesfontsize',16);

x=-15:0.01:15;
y=-2:0.01:2;

y1=sin(x);
y2=cos(x);

figure(1)
plot(x,y1,'b-',x,y2,'r--','linewidth',1.5)
hold on
plot(x,zeros(size(x)),'k-','linewidth',0.2)
plot(zeros(size(y)),y,'k-','linewidth',0.2)
hold off
legend('sin(x)','cos(x)','fontsize',16)
ylim([-2 2])
xlabel('x','FontSize',18,'FontWeight','bold')
ylabel('y','FontSize',18,'FontWeight','bold')
```

From Figures A.9 and A.10, it is easily verified that the sine and cosine functions are periodic with period $T = 2\pi \approx 6.283$ and the tangent and cotangent functions are periodic with period $T = \pi \approx 3.142$. The vertical lines in the graphs of the tangent and cotangent functions are called **asymptotes** (Figure A.10), because the function never touches them. When the function approaches these lines, its values become very large and ultimately tend to infinity.

Finally, we mention that under specific constraints, trigonometric functions have inverse functions that will return the initial argument; that is, the angle that produces specific values of the trigonometric functions. These inverse functions are denoted as

$$\arcsin(x), \arccos(x), \arctan(x), \text{ and } \operatorname{arccot}(x) \tag{A.16}$$

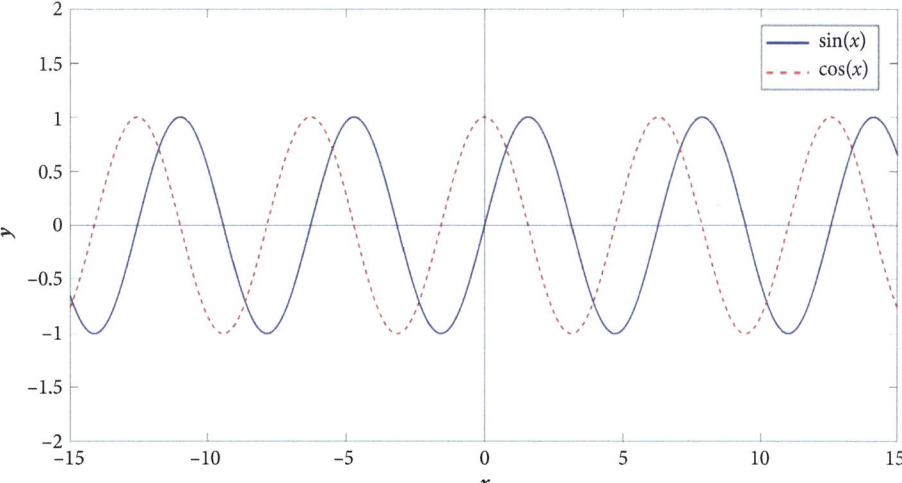

Figure A.9 The graphs for the sin (x) and cos (x) functions. This graph can be reproduced using the MATLAB script presented in Code A.2.

Code A.3 *MATLAB code to plot the* **tan** (x) *and cot* (x) *function graphs. This code produces Figure A.10.*

```
clear all; clc; close all;

set(0,'defaultaxesfontsize',16);

x=-15:0.01:15;
y=-5:0.1:5;

y1=tan(x);
y2=cot(x);

figure(1)
subplot(2,1,1)
plot(x,y1,'b-','linewidth',1.5)
hold on
plot(x,zeros(size(x)),'k-','linewidth',0.2)
plot(zeros(size(y)),y,'k-','linewidth',0.2)
hold off
ylim([-5 5])
xlabel('x','FontSize',18,'FontWeight','bold')
ylabel('y','FontSize',18,'FontWeight','bold')
title('tan(x)','FontSize',18,'FontWeight','bold')

subplot(2,1,2)
plot(x,y2,'r-','linewidth',1.5)
hold on
plot(x,zeros(size(x)),'k-','linewidth',0.2)
plot(zeros(size(y)),y,'k-','linewidth',0.2)
hold off
ylim([-5 5])
xlabel('x','FontSize',18,'FontWeight','bold')
ylabel('y','FontSize',18,'FontWeight','bold')
title('cot(x)','FontSize',18,'FontWeight','bold')
```

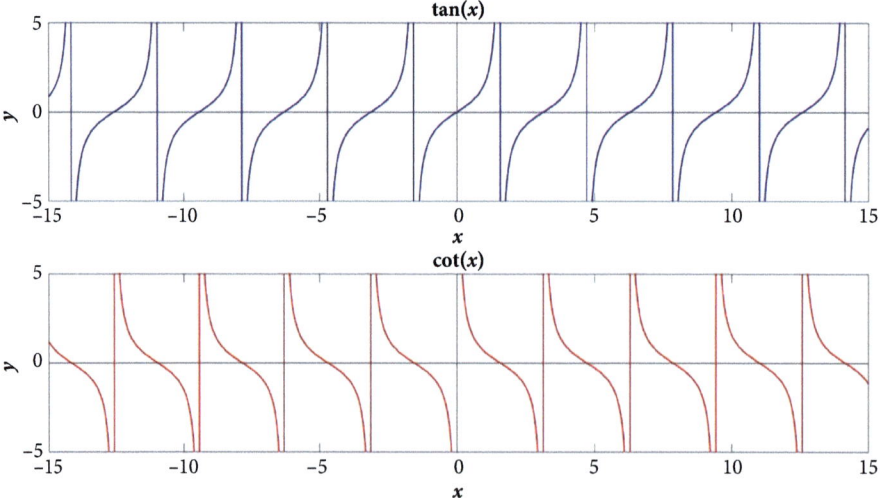

Figure A.10 The tan (x) and cot (x) function graphs. The vertical lines in these graphs are called asymptotes. This graph can be reproduced using the MATLAB script presented in Code A.3.

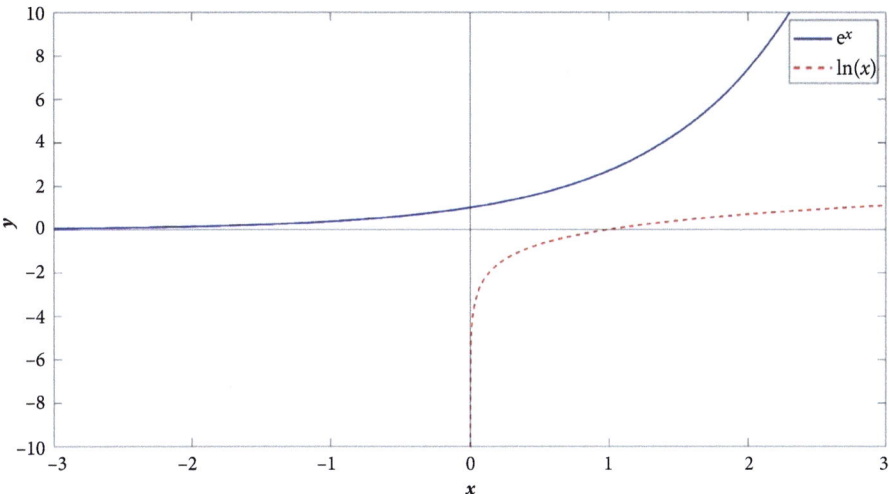

Figure A.11 The exponential (solid line) and natural logarithm (dashed line) function graphs. This graph can be reproduced using the MATLAB script presented in Code A.4.

Exponential function and natural logarithm

Two important functions in engineering mathematics are the **exponential function** e^x and the **natural logarithm** $\ln(x)$. The number $e \approx 2.718$ is an irrational number. The graphs of the exponential function and the natural logarithm are plotted in Figure A.11 using Code A.4. The exponential function and the natural logarithm are inverse functions of each other. That means,

$$\ln(e^x) = e^{\ln(x)} = x \tag{A.17}$$

Code A.4 *MATLAB code to plot the exponential, e^x, and natural logarithm, $\ln(x)$, function graphs. This code produces Figure A.11.*

```
clear all; clc; close all;

set(0,'defaultaxesfontsize',16);

xa=-3:0.01:3;
xb=0:0.00001:3;
y=-10:0.1:20;
y1=exp(xa);
y2=log(xb);

figure(1)
plot(xa,y1,'b-','linewidth',1.5)
hold on
plot(xb,y2,'r--','linewidth',1.5)
plot(xa,zeros(size(xa)),'k-','linewidth',0.2)
plot(zeros(size(y)),y,'k-','linewidth',0.2)
hold off
legend('e^x','ln(x)','FontSize',16)
ylim([-10 10])
xlabel('x','FontSize',18,'FontWeight','bold')
ylabel('y','FontSize',18,'FontWeight','bold')
```

It can be easily verified that, for real numbers, the codomain of e^x includes only positive numbers. At the same time, $\ln(x)$ is defined only for positive arguments (Figure A.11). Note that the domain of one function coincides with the codomain of the other, as is always the case for inverse functions. The negative y axis in the $\ln(x)$ function (lower part of Figure A.11) is an asymptote for the natural logarithm because the values there tend to become minus infinity. The negative x axis is an asymptote for the exponential function (upper part of Figure A.11), although the values there do not tend to become very large but rather very small, gradually approaching zero. The $\ln(x)$ function has one root at $x = 1$ (Figure A.11). This is expected since $e^{\ln(1)} = e^0 = 1$.

The basic properties of the exponential and natural logarithm functions are:

$$e^{a+b} = e^a \cdot e^b \text{ and } e^{-a} = \frac{1}{e^a} \tag{A.18}$$

and

$$\ln(a \cdot b) = \ln(a) + \ln(b) \text{ and } \ln\left(a^{-1}\right) = -\ln(a) \tag{A.19}$$

Hyperbolic functions

Using the exponential function, we can define the **hyperbolic functions**. The basic hyperbolic functions are:

- Hyperbolic sine : $\sinh(x) = \dfrac{e^x - e^{-x}}{2}$ $\qquad$ (A.20a)

- Hyperbolic cosine : $\cosh(x) = \dfrac{e^x + e^{-x}}{2}$ $\qquad$ (A.20b)

- Hyperbolic tangent : $\tanh(x) = \dfrac{e^x - e^{-x}}{e^x + e^{-x}}$ $\qquad$ (A.20c)

- Hyperbolic cotangent : $\coth(x) = \dfrac{e^x + e^{-x}}{e^x - e^{-x}}$ $\qquad$ (A.20d)

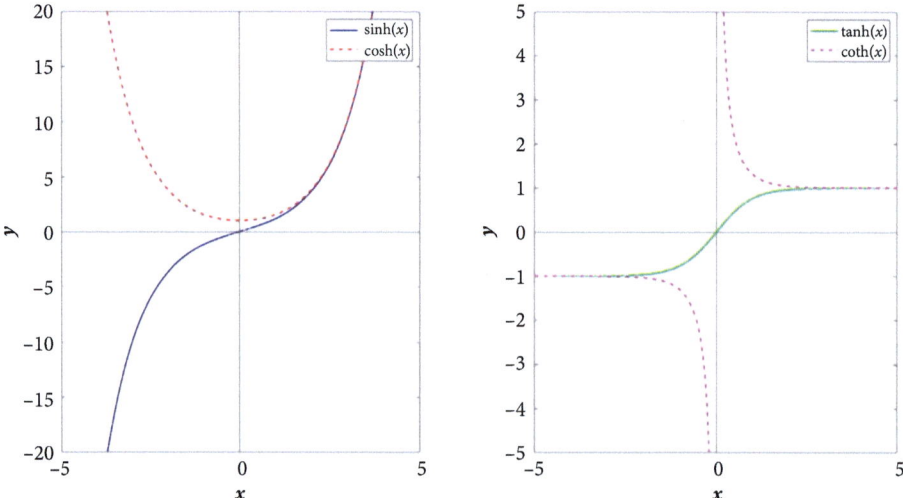

Figure A.12 Graphs of the four basic hyperbolic functions. This graph can be reproduced using the MATLAB script presented in Code A.5.

Plots of these hyperbolic functions are shown in Figure A.12 as produced using the MATLAB script in Code A.5. The hyperbolic functions have the following important property:

$$\cosh^2(x) - \sinh^2(x) = 1 \qquad\qquad (A.21)$$

This property resembles closely the one for trigonometric functions: $\cos^2(x) + \sin^2(x) = 1$.

Code A.5 *MATLAB code to plot the graphs of the four basic hyperbolic functions,* $\sinh(x)$, $\cosh(x)$, $\tanh(x)$, *and* $\coth(x)$. *This code generates Figure A12.*

```
clear all; clc; close all;

set(0,'defaultaxesfontsize',16);

x=-100:0.01:100;
y=-100:0.1:100;

y1=sinh(x);
y2=cosh(x);
y3=tanh(x);
y4=coth(x);
%
figure(1)
subplot(1,2,1)
plot(x,y1,'b-','linewidth',1.5); hold on;
plot(x,y2,'k--','linewidth',1.5)
hold on
plot(x,zeros(size(x)),'k-','linewidth',0.2)
plot(zeros(size(y)),y,'k-','linewidth',0.2)
hold off
xlim([-5 5]); ylim([-20 20]);
xlabel('x','FontSize',18,'FontWeight','bold');
ylabel('y','FontSize',18,'FontWeight','bold');
legend('sinh(x)','cosh(x)','FontSize',18,'FontWeight','bold')
set(gca,'linewidth',1.0);
%%%%%%%%%%%%%%%%%%%%%%%%%%%%%%%%%%%%%%%%%%%%%%%%%%%%%%
subplot(1,2,2)
plot(x,y3,'b-','linewidth',1.5);hold on;
plot(x,y4,'k--','linewidth',1.5);
hold on
plot(x,zeros(size(x)),'k-','linewidth',0.2)
plot(zeros(size(y)),y,'k-','linewidth',0.2)
hold off
xlim([-5 5]); ylim([-5 5]);
xlabel('x','FontSize',18,'FontWeight','bold')
ylabel('y','FontSize',18,'FontWeight','bold')
legend('tanh(x)','coth(x)','FontSize',18,'FontWeight','bold')
set(gca,'linewidth',1.0);
%%%%%%%%%%%%%%%%%%%%%%%%%%%%%%%%%%%%%%%%%%%%%%%%%%%%%%
```

A.2.3 Roots of functions

An important concept is that of the roots of a function. The **roots** of a function $f(x)$ are the independent variable x values that satisfy $f(x) = 0$. Note that a function might have no roots, one root, or multiple

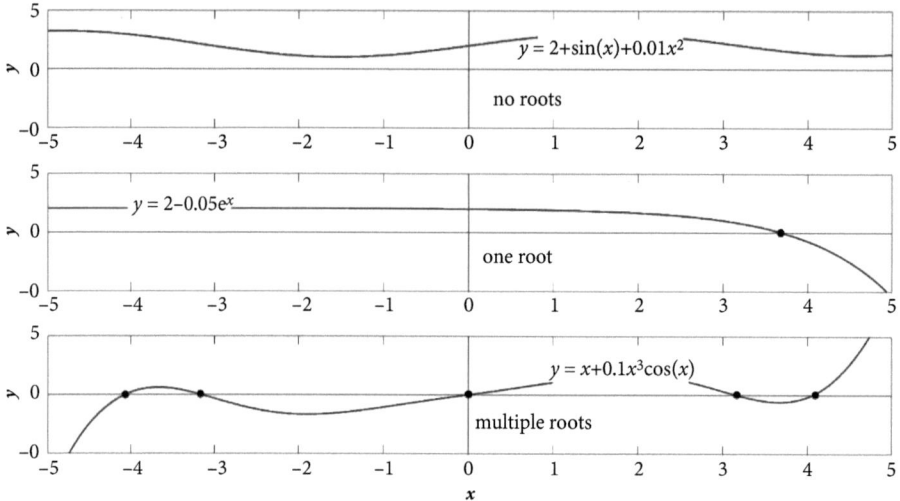

Figure A.13 Graphs of functions with no roots, one root, and multiple roots. The dots on the horizontal axes show the roots of the functions.

roots. These three cases are depicted in Figure A.13, demonstrating that a root is the x co-ordinate of the point where the function crosses the horizontal axis (x axis).

The roots of a polynomial function are the values of the independent variable x for which $p(x) = 0$. If all the roots $x_1, x_2, \ldots x_{n-1}, x_n$ of an n-degree polynomial are known, the polynomial can be written in the following form:

$$p(x) = c_n (x - x_1)(x - x_2) \cdots (x - x_{n-1})(x - x_n) \tag{A.22}$$

For a polynomial of first degree, $p(x) = c_0 + c_1 x$, there is a single root which can be computed as the solution of the equation $c_0 + c_1 x = 0$, which is $x_1 = -\frac{c_0}{c_1}$.

A polynomial of second degree (or quadratic polynomial), $p(x) = c_0 + c_1 x + c_2 x^2$, which can also be written in the form $p(x) = c + bx + ax^2$, always has two roots. In some cases these roots can coincide or be imaginary numbers. The two roots (x_1 and x_2) can be calculated as follows:

$$x_1 = \frac{-b + \sqrt{b^2 - 4ac}}{2a} \text{ and } x_2 = \frac{-b - \sqrt{b^2 - 4ac}}{2a} \tag{A.23}$$

The quantity appearing under the square root $(b^2 - 4ac)$ is called **discriminant** of the quadratic polynomial and is denoted with a capital delta: $\Delta = b^2 - 4ac$. There are three possible cases:

1. $\Delta > 0$: there are **two distinct real roots** as shown in Eq (A.23).
2. $\Delta = 0$: the two roots coincide, and we say that there is **one double root**, $x_{1,2} = \frac{-b}{2a}$.
3. $\Delta < 0$: there are **two imaginary roots** of the following form:

$$x_1 = \frac{-b + i\sqrt{|b^2 - 4ac|}}{2a} \text{ and } x_2 = \frac{-b - i\sqrt{|b^2 - 4ac|}}{2a} \tag{A.24}$$

For the latter case, note that the two roots are complex conjugate numbers. In general, if a polynomial has a complex root, then the conjugate complex number is a root as well. That is, complex roots always appear in pairs.

Box A.1: Example

Find the roots of the polynomial $p(x) = x^3 - 1$.

Solution

The polynomial can be rearranged as $x^3 - 1^3 = (x - 1) \cdot (x^2 + x + 1) = 0$. There is a real root $x_1 = 1$ plus the two roots of the quadratic polynomial $x^2 + x + 1$.

The discriminant of the quadratic polynomial $x^2 + x + 1$ is $\Delta = 1^2 - 4 \cdot 1 \cdot 1 = 1 - 4 = -3 < 0$. Therefore, there are two complex roots for the quadratic term:

$$x_{2,3} = \frac{-1 \pm i\sqrt{|\Delta|}}{2} = \frac{-1 \pm i\sqrt{|-3|}}{2} = \frac{-1 \pm i\sqrt{3}}{2}$$

In summary, the three roots of the function are

$$x_1 = 1, \quad x_2 = \frac{-1 + i\sqrt{3}}{2}, \quad x_3 = \frac{-1 - i\sqrt{3}}{2}.$$

The MATLAB script in Code A.6 can be used to calculate the roots of this function using the computer.

Code A.6 *MATLAB code to calculate the roots of the function* $p(x) = x^3 - 1$.

```
% find the roots of the qubic polynomial p(x)=x^3-1

clear all
clc

% define polynomial
p=[1 0 0 -1];
% find the roots
r=roots(p);
% display the roots in command window
disp('the roots are')
r
```

Note that if a polynomial is of degree n, then the polynomial has exactly n roots, although some of them might be complex. Note that since complex roots always appear in pairs, a polynomial of odd degree has at least one real root.

The roots of trigonometric functions are the values where the sine, cosine, tangent, and cotangent functions, with respect to the independent variable x, are equal to zero. From the graphs of the basic trigonometric functions and their periodicity (Figures A.9 and A.10), it is clear that these functions have an infinite number of roots. By simple inspection, the roots of these functions are as follows:

$$\sin(x) \text{ roots}: 0, \pm\pi, \pm 2\pi, \pm 3\pi, \pm 4\pi,$$

$$\cos(x) \text{ roots}: \pm\pi/2, \pm 3\pi/2, \pm 5\pi/2, \dots$$

$$\tan(x) \text{ roots}: 0, \pm\pi, \pm 2\pi, \pm 3\pi, \pm 4\pi, \dots$$

$$\cot(x) \text{ roots}: \pm\pi/2, \pm 3\pi/2, \pm 5\pi/2, \dots$$

To calculate the roots of a trigonometric function, such as $f(x) = \sin(x)$, the MATLAB script presented in Code A.7 can be employed. Note that this MATLAB code and other scripts in this book can be downloaded at: http://www.oup.com/AdvancedMathematicalModelling.

Code A.7 *MATLAB code to calculate the roots of the function $f(x) = \sin(x)$.*

```
% find the roots of f(x)= sin(x) in the interval [-10,10]
clear all; clc; close all;

% find the roots of f(x)= sin(x) in the interval [-5,5]
clear all; clc; close all;

% find roots of nonlinear functions in intervals where the function sign
% changes.
x1=fzero(@sin,[-5,-2])
x2=fzero(@sin,[-2,2])
x3=fzero(@sin,[2,5])
```

A.2.4 Derivatives and differentiation

Consider a function $f(x)$ of a single independent variable with the graph shown in Figure A.14. For a fixed value of the independent variable $x = x_0$, it is evident that as point x moves along the horizontal axis towards x_0, the value of the function at point $x, f(x)$, approaches the value $f(x_0)$. As point x moves towards x_0 along the horizontal axis, the straight line that passes through both x and x_0 (the dashed line in Figure A.14) tends to become the tangent to the graph of the function $y = f(x)$ (the solid line in Figure A.14). The slope of the grey line is the tangent of angle θ which can be written in terms of $x, x_0, f(x),$ and $f(x_0)$ as follows:

$$\tan(\theta) = \frac{f(x) - f(x_0)}{x - x_0} \tag{A.25}$$

The **derivative** of function $f(x)$ at the point x_0, if it exists (the function must be continuous), is denoted as $\left.\frac{df(x)}{dx}\right|_{x=x_0}$ or $\left.\frac{df}{dx}\right|_{x=x_0}$ or $f'(x_0)$ and it is the following limit:

$$\left.\frac{df}{dx}\right|_{x=x_0} = \lim_{x \to x_0} \frac{f(x) - f(x_0)}{x - x_0} \tag{A.26}$$

Note that we need to consider the limit value because if we directly substitute $x = x_0$ in the fraction, the fraction will produce $0/0$, which is an indeterminate form. The derivative of function $f(x)$ at the point x_0 is thus the slope of the function at the point x_0.

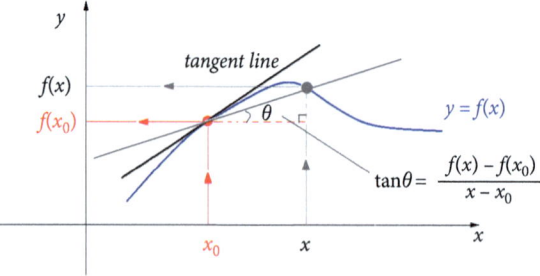

Figure A.14 Tangent line of a function at the point x_0.

Table A.1 Derivatives of some basic mathematical functions.

Function, $f(x)$	Derivative, df/dx
x^n	nx^{n-1}
$\sin(x)$	$\cos(x)$
$\cos(x)$	$-\sin(x)$
$\tan(x)$	$1 + \tan^2(x) = \sec^2(x) = \dfrac{1}{\cos^2(x)}$
$\cot(x)$	$-\csc^2(x) = -\dfrac{1}{\sin^2(x)}$
e^x	e^x
a^x	$a^x \ln(a)$
$\ln(x)$	$\dfrac{1}{x}$
$\sqrt{x}$	$\dfrac{1}{2\sqrt{x}}$
$\arcsin(x)$	$\dfrac{1}{\sqrt{1-x^2}}$
$\arccos(x)$	$-\dfrac{1}{\sqrt{1-x^2}}$
$\arctan(x)$	$\dfrac{1}{1+x^2}$
$\text{arccot}(x)$	$-\dfrac{1}{1+x^2}$
$\sinh(x)$	$\cosh(x)$
$\cosh(x)$	$\sinh(x)$

The process of finding the derivative of a function is called **differentiation**. The derivative of $f(x)$, denoted as $\frac{df}{dx}$ or $f'(x)$, will be a new function describing the slope of $f(x)$ at each point x. The derivatives of some basic functions are provided in Table A.1.

The following rules can be used for the calculation of derivatives. Let $f(x), g(x)$ be two differentiable functions. Then we have:

$$\frac{d(cf)}{dx} = c\frac{df}{dx} \text{ (derivative of a function times a constant } c)$$

$$\frac{d(f \pm g)}{dx} = \frac{df}{dx} + \frac{dg}{dx} \text{ (derivative of the sum or difference of two functions)}$$

$$\frac{d(f \cdot g)}{dx} = \frac{df}{dx} \cdot g + \frac{dg}{dx} \cdot f \text{ (derivative of the product of two functions)}$$

$$\frac{d(f/g)}{dx} = \frac{\frac{df}{dx} \cdot g - \frac{dg}{dx} \cdot f}{g^2} \text{ (derivative of the quotient of two functions)}$$

$$\frac{d[f(g(x))]}{dx} = \frac{df}{dg} \cdot \frac{dg}{dx} \text{ (the \textbf{chain rule} for function composition)}$$

Using the rules of differentiation and the derivatives of the basic functions given in Table A.1, we can calculate the derivatives of several other functions. An example is given in Box A.2 for finding the derivative of a function; this can also be solved using MATLAB by applying Code A.8.

Box A.2: Example

Find the derivative of the function

$$f(x) = -5x^3 + 2x^2 + 12x - \sin(x)$$

Solution

Using the basic rules of differentiation, it is

$$\frac{df}{dx} = -5\frac{dx^3}{dx} + 2\frac{dx^2}{dx} + 12\frac{dx}{dx} - \frac{d\sin(x)}{dx} = -15x^2 + 4x + 12 - \cos(x)$$

Code A.8 *MATLAB code to find the derivative of the function* $f(x) = -5x^3 + 2x^2 + 12x - \sin(x)$.

```
% find the derivative of function f(x) = -5x³ + 2x² + 12x - sin (x).
clear all; clc; close all;
% define symbolic x
syms x
% define function
f=-5*x^3+2*x^2+12*x-sin(x);
% calculate first derivative
diff(f,x)
```

Strictly increasing and strictly decreasing functions

A differentiable function for which the derivative is always positive, that is, $\frac{df}{dx} > 0$, at all points always has positive slope and thus is **strictly increasing**. A differentiable function for which the derivative is always negative, that is, $\frac{df}{dx} < 0$, at all points always has negative slope and thus is **strictly decreasing**. Figure A.15a shows a strictly increasing function, $f(x) = e^x$, while Figure A.15b shows a strictly decreasing function, $f(x) = e^{-x}$.

Higher-order derivatives

Since the derivative of a function is another function, it can also be differentiated. For a function $f(x)$ with derivative $\frac{df}{dx}$, we define the derivative of $\frac{df}{dx}$ as $\frac{d\left(\frac{df}{dx}\right)}{dx}$, called the second derivative of $f(x)$ and denoted as $\frac{d^2f}{dx^2}$. Similarly, we can find higher-order derivatives, e.g., $\frac{d^3f}{dx^3}, \frac{d^4f}{dx^4}, ..., \frac{d^nf}{dx^n}$. Higher-order derivatives are also denoted as $f^{(2)}(x), f^{(3)}(x), ..., f^{(n)}(x)$.

Local maxima, minima, and points of inflection

When studying the behaviour of differentiable functions on open sets, the first derivative can be used to determine the minimum and maximum values of the function, also known as extreme values. The following steps can be followed to find the extreme values of a function.

Step 1 Calculate the first derivative of the function, $f'(x) = \frac{df}{dx}$.

Step 2 Find the roots of the first derivative, that is, find the values of x such that $f'(x) = 0$. These roots are called **stationary points** of $f(x)$ and they correspond to either **local maxima**, **local minima**, or **points of inflection**.

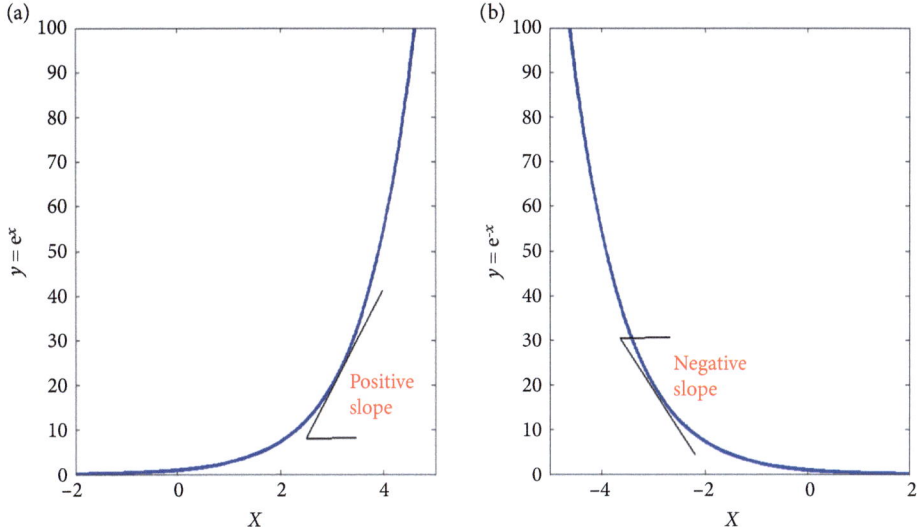

Figure A.15 Plots showing (a) a strictly increasing function (positive slope), and (b) a strictly decreasing function (negative slope).

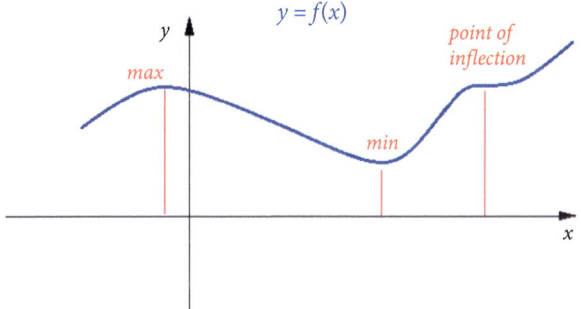

Figure A.16 Local maximum, minimum, and point of inflection.

Step 3 Calculate the second derivative of function $f(x)$, $\frac{d^2f}{dx^2}$. At the stationary points, check the sign of the second derivative.

- If at a stationary point and $\frac{d^2f}{dx^2} < 0$, then this point is a local maximum.
- If at a stationary point and $\frac{d^2f}{dx^2} > 0$, then this point is a local minimum.
- If at a stationary point and $\frac{d^2f}{dx^2} = 0$, then this point is a point of inflection.

The nature of local maxima, minima, and inflection points is shown in Figure A.16.

Taylor and Maclaurin series

The Taylor or Maclaurin series expansions allow the approximation of sufficiently smooth functions in the neighbourhood of a point by polynomials. This is useful since polynomials are well-studied functions and easy to compute. The quality of the approximation increases when more terms are retained in the series expansion, i.e., higher-degree polynomials are employed.

According to Taylor's theorem, the value of every sufficiently smooth function at a point $x_0 + \Delta x$ near x_0 can be calculated as a Taylor series in the following form:

$$f(x_0 + \Delta x) = f(x_0) + \left.\frac{df}{dx}\right|_{x_0} \cdot \Delta x + \frac{1}{2}\left.\frac{d^2f}{dx^2}\right|_{x_0} \cdot \Delta x^2 + \cdots + \frac{1}{n!}\left.\frac{d^nf}{dx^n}\right|_{x_0} \cdot \Delta x^n + \cdots$$

$$= \sum_{n=0}^{\infty} \frac{1}{n!}\left.\frac{d^nf}{dx^n}\right|_{x_0} \cdot \Delta x^n \qquad (A.27)$$

where $n!$ is the factorial of the natural number n, i.e., $n! = 1 \cdot 2 \cdot 3 \cdots n$ and $0! = 1$. It is evident from Eq (A.27) that as $\Delta x \to 0$, i.e., when the point $x_0 + \Delta x$ is very close to the point x_0, the terms with factors Δx^n become very small as n increases. In that sense, their contribution to the above sum is very small and, in a first approximation, can be neglected. Thus, in order to analyse the behaviour of a function at a point x, it suffices to know the behaviour of the function and its derivatives at a nearby point x_0 such that $x - x_0 = \Delta x$. Substituting the value of Δx in the Taylor series formula, we have:

$$f(x) \cong f(x_0) + \left.\frac{df}{dx}\right|_{x_0} \cdot (x - x_0) + \frac{1}{2}\left.\frac{d^2f}{dx^2}\right|_{x_0} \cdot (x - x_0)^2 + \cdots + \frac{1}{n!}\left.\frac{d^nf}{dx^n}\right|_{x_0} \cdot (x - x_0)^n$$

$$= \sum_{k=0}^{n} \frac{1}{k!}\left.\frac{d^kf}{dx^k}\right|_{x_0} \cdot (x - x_0)^k \qquad (A.28)$$

In the case where $x_0 = 0$, the above series expansion takes the form

$$f(x) = f(x_0) + \left.\frac{df}{dx}\right|_{x_0} \cdot x + \frac{1}{2}\left.\frac{d^2f}{dx^2}\right|_{x_0} \cdot x^2 + \frac{1}{6}\left.\frac{d^3f}{dx^3}\right|_{x_0} \cdot x^3 + \cdots + \frac{1}{n!}\left.\frac{d^nf}{dx^n}\right|_{x_0} \cdot x^n + \cdots \qquad (A.29)$$

Equation A.29 is called a Maclaurin series. The Maclaurin series for some common functions are:

$$e^x = \sum_{n=0}^{\infty} \frac{x^n}{n!} = 1 + x + \frac{x^2}{2!} + \frac{x^3}{3!} + \frac{x^4}{4!} + \cdots$$

$$\sin(x) = \sum_{n=0}^{\infty} \frac{(-1)^n x^{2n+1}}{(2n+1)!} = x - \frac{x^3}{6} + \frac{x^5}{120} + \cdots$$

$$\cos(x) = \sum_{n=0}^{\infty} \frac{(-1)^n x^{2n}}{(2n)!} = 1 - \frac{x^2}{2} + \frac{x^4}{24} + \cdots$$

$$\sinh(x) = \sum_{n=0}^{\infty} \frac{x^{2n+1}}{(2n+1)!} = x + \frac{x^3}{6} + \frac{x^5}{120} + \cdots$$

$$\cosh(x) = \sum_{n=0}^{\infty} \frac{x^{2n}}{(2n)!} = 1 + \frac{x^2}{2} + \frac{x^4}{24} + \cdots$$

The Taylor or Maclaurin series expansion is not always applicable. In certain cases, it is only possible if the distance of the point x from the point x_0 is less than a critical value. As an example, the following Maclaurin expansions are valid only in certain domains:

$$\frac{1}{1-x} = \sum_{n=0}^{\infty} x^n = 1 + x + x^2 + x^3 + x^4 + \cdots, \text{valid for} -1 < x < 1 \text{ (geometric series)}$$

$$\ln(1+x) = \sum_{n=1}^{\infty} (-1)^{n+1} \frac{x^n}{n} = x - \frac{x^2}{2} + \frac{x^3}{3} - \frac{x^4}{4} + \cdots, \text{valid for} -1 < x < 1$$

Using these formulas, it is also possible to evaluate the series expansion of more complex functions. A specific example is presented in Box A.3.

Box A.3: Example

Find a series expansion for the function $f(x) = (1 + x)\,e^{2x}$ around the point $x_0 = 0$.

Solution

The expansion for the exponential can be evaluated by replacing x by $2x$ in the respective exponential Maclaurin series, as follows:

$$e^{2x} = \sum_{n=0}^{\infty} \frac{2^n x^n}{n!} = 1 + 2x + 4\frac{x^2}{2!} + 8\frac{x^3}{3!} + \cdots$$

It is then rearranged as $(1 + x)\,e^{2x} = (1 + x)\sum_{n=0}^{\infty}\frac{2^n x^n}{n!} = 1 + 3x + 4x^2 + \frac{4}{3}x^3 + \cdots$

A.2.5 Integrals and integration

Consider a function $f(x)$ of a single independent variable with the graph shown in Figure A.17. **Integration** of the function $f(x)$ is a procedure associated with the calculation of the area between the graph of the function and the x axis (the grey shaded area in Figure A.17). For the calculation of this area defined by the lateral boundaries $x = a$ and $x = b$, where $b > a$, we assume that the area is divided in a set of rectangles as shown in Figure A.18. Each of these rectangles has width Δx. The area is then approximated as the sum of the areas of these rectangles (see Figure A.18).

If the height of each rectangle is taken as the value of the function at the first point of the associated interval (Figure A.18a), we obtain the **lower Riemann sum** approximation expressed as

$$Area \cong area1 + area2 + area3 = f(a) \cdot \Delta x + f(a + \Delta x) \cdot \Delta x + f(a + 2 \cdot \Delta x) \cdot \Delta x$$

$$= \sum_{n=0}^{2} f(a + n \cdot \Delta x) \cdot \Delta x$$

If the height of each rectangle is taken as the value of the function at the end point of the associated interval (Figure A.18b), we obtain the **upper Riemann sum** approximation as follows:

$$Area \cong Area1 + Area2 + Area3 = f(a + \Delta x) \cdot \Delta x + f(a + 2 \cdot \Delta x) \cdot \Delta x + f(b) \cdot \Delta x$$

$$= \sum_{n=1}^{3} f(a + n \cdot \Delta x) \cdot \Delta x$$

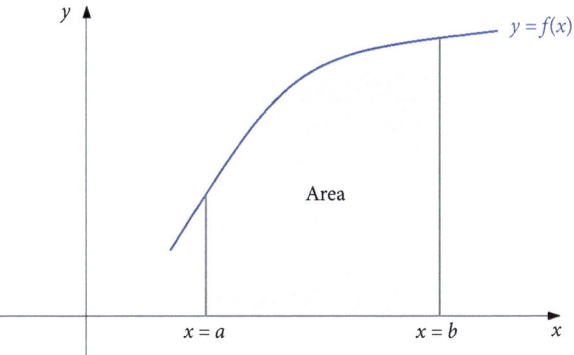

Figure A.17 The integral of the function $f(x)$ is the area between the function $f(x)$ and the x axis confined between the lateral boundaries $x = a$ and $x = b$.

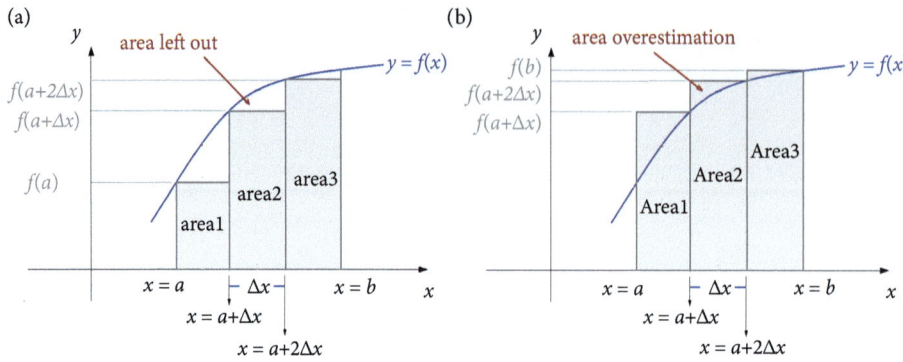

Figure A.18 Approximation of the area below a function using the upper and lower sums.

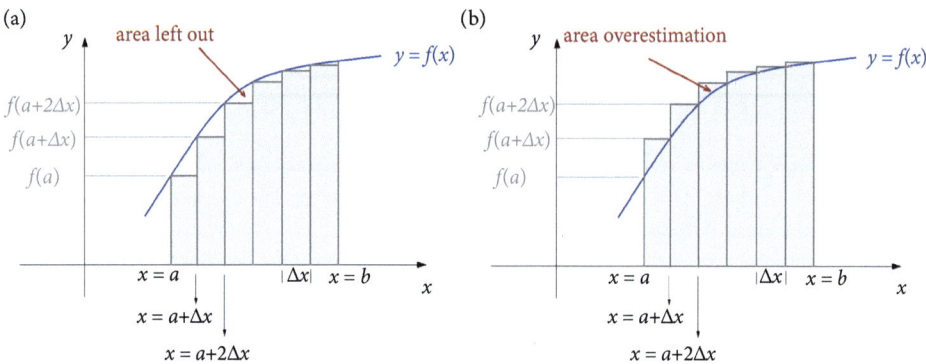

Figure A.19 Refined approximation of the area below a function using upper and lower sums. This is achieved by reducing the width of each rectangle.

It is evident from Figure A.18 that when the lower sum is used, then the actual area of the given function is underestimated. If the upper sum is used, then the actual area is overestimated. If we now consider a smaller width for each rectangle, the approximation of the area using the lower and upper sums is expected to become more accurate. This situation is depicted in Figure A.19. In this case, the larger number of rectangles used makes the approximation of the area more accurate. Thus, the areas left out (when the lower sum is used) or overestimated (when the upper sum is used) become smaller.

As the width of each rectangle approaches zero, that is as $\Delta x \to 0$, the two sums (lower and upper) approach the same value, which is the actual area between $f(x)$ and the x axis. The limit value of the two sums, that is, the actual area defined by the lateral boundaries $x = a$ and $x = b$, is called the **integral of** $f(x)$ **from** $x = a$ **to** $x = b$ and it is denoted as

$$\int_a^b f(x)\,dx \tag{A.30}$$

Note that if the values of $f(x)$ are negative, then the products $f(a + n \cdot \Delta x) \cdot \Delta x$ will be negative and thus the integral will have a negative value. This means that the integral is equivalent to the signed area between $f(x)$ and the x axis.

When the boundaries of integration $x = a$ and $x = b$ have been defined, the integral $\int_a^b f(x)\,dx$ has a specific value, that is, a number. In this case the integral is called a **definite integral**. If the boundaries of integration are not specified, then again the integral can be defined and, in this case, it is a function of x. This function, denoted as $F(x)$, is written as

$$F(x) = \int f(x)\, dx + c \tag{A.31}$$

where c is a constant. Through this last equation, the process of integration can be defined. If we are given a function $f(x)$ and we integrate it, we get a new function $F(x)$. This function $F(x)$ is called the **indefinite integral** of $f(x)$. If for a function $f(x)$ we have the indefinite integral function $F(x)$, then the definite integral between any two boundaries $x = a$ and $x = b$ is given by the following expression:

$$\int_a^b f(x)\, dx = [F(x)]_a^b = F(b) - F(a) \tag{A.32}$$

The problem of integration of a function consists of finding the indefinite integral. For this purpose, the process of differentiation can be used. For the indefinite integral function, we have

$$\frac{dF(x)}{dx} = f(x) \tag{A.33}$$

This relation indicates that integration and differentiation are the reverse of each other. This means, given a function $f(x)$, if we calculate the indefinite integral we get $F(x)$. If we now differentiate $F(x)$ we retrieve the function $f(x)$. Table A.2 presents the integrals for some standard functions.

Rules of integration

Let $f(x), g(x)$ be two integrable functions. The following rules can be used for the calculation of integrals:

$$\int_a^b c \cdot f(x)\, dx = c \cdot \int_a^b f(x)\, dx \quad \text{(integral of a function times a constant } c\text{)}$$

$$\int_a^b [f(x) \pm g(x)]\, dx = \int_a^b f(x)\, dx \pm \int_a^b g(x)\, dx \quad \text{(integral of the sum or difference of two functions)}$$

$$\int_a^b f(x)\, dx = \int_a^d f(x)\, dx + \int_d^b f(x)\, dx \quad \text{(where } d \text{ is any point along the } x \text{ axis)}$$

$$\int_b^a f(x)\, dx = - \int_a^b f(x)\, dx \quad \text{(inversion of the integration limits)}$$

$$\int_a^a f(x)\, dx = 0 \quad \text{(an application of the preceding integration rule if we use } b = a\text{)}$$

$$\frac{d\left(\int_a^x f(z)\, dz\right)}{dx} = \frac{d[F(x) - F(a)]}{dx} = \frac{dF(x)}{dx} - \frac{dF(a)}{dx} = f(x) - 0 = f(x)$$

$$\int_a^b \frac{df}{dx} \cdot g\, dx + \int_a^b \frac{dg}{dx} \cdot f\, dx = f(b) \cdot g(b) - f(a) \cdot g(a) = [f(x) \cdot g(c)]_a^b \quad \text{(integration by parts formula)}$$

Integration can be used in order to find the mean value of a function $f(x)$ in an interval (a, b). The mean value of the function in (a, b) is given by

$$\bar{f} = \frac{1}{b-a} \int_a^b f(x)\, dx \tag{A.34}$$

Rational functions are functions in the form of fractions where the numerator and denominator are polynomials. Given a polynomial $P_m(x)$ of degree m and a polynomial $Q_n(x)$ of degree n, a rational

Table A.2 Integrals of basic functions. In this table, c denotes a constant.

Function, $f(x)$	Integral, $F(x) = \int f(x)\, dx + c$
x^n, when $n \neq -1$	$\dfrac{1}{n+1}x^{n+1} + c$
$\sin(x)$	$-\cos(x) + c$
$\cos(x)$	$\sin(x) + c$
$\tan(x)$	$-\ln|\cos(x)| + c$
$\cot(x)$	$\ln|\sin(x)| + c$
e^x	$e^x + c$
$a^x,\ a \neq 1$	$\dfrac{a^x}{\ln(a)} + c$
$\dfrac{1}{x}$	$\ln|x| + c$
$\dfrac{1}{ax+b},\ a \neq 0$	$\dfrac{1}{a}\ln|ax+b| + c$
$\dfrac{1}{(ax+b)^n},\ a \neq 0$ and $n \neq 1$	$\dfrac{1}{a}\dfrac{(ax+b)^{1-n}}{(1-n)} + c$
$\dfrac{1}{x^2+R^2},\ R \neq 0$	$\dfrac{\arctan\left(\frac{x}{R}\right)}{R} + c$
$\dfrac{x}{x^2+R^2},\ R \neq 0$	$\dfrac{1}{2}\ln|x^2+R^2| + c$
$\sinh(x)$	$\cosh(x) + c$
$\cosh(x)$	$\sinh(x) + c$
$\tanh(x)$	$\ln(\cosh(x)) + c$
$\coth(x)$	$\ln|\sinh(x)| + c$

function has the following form:

$$f(x) = \frac{P_m(x)}{Q_n(x)} \tag{A.35}$$

In such cases, integration can be done using **partial fraction decomposition** and using the basic integrals provided in Table A.2. In the partial fraction decomposition of a rational function $f(x) = \frac{P_m(x)}{Q_n(x)}$, we find the roots $r_1, r_2, \ldots, r_n$ of the denominator and factorize $Q_n(x)$ as

$$Q_n(x) = c_n(x - r_1)(x - r_2)(x - r_3)\cdots(x - r_n)$$

- If one of the roots, e.g. r_j, has multiplicity k, then the factorization is of the following form:

$$Q_n(x) = c_n(x - r_1)(x - r_2)\cdots(x - r_j)^k\cdots(x - r_{n-k})$$

We work similarly if there are more multiple roots.

- If there are complex roots, the polynomial factorization will include quadratic polynomials as well.

Partial fraction decomposition is a decomposed form of the rational function in the following forms:

- If all roots are distinct:

$$\frac{P_n(x)}{c_n(x - r_1)(x - r_2)(x - r_3)\cdots(x - r_n)} = \frac{1}{c_n}\left(\frac{D_1}{x - r_1} + \frac{D_2}{x - r_2} + \frac{D_2}{x - r_3} + \cdots + \frac{D_n}{x - r_n}\right) \tag{A.36}$$

- For a root with multiplicity n:

$$\frac{P_m(x)}{c_n(x - r)^n} = \frac{1}{c_n}\left(\frac{D_1}{x - r} + \frac{D_2}{(x - r)^2} + \frac{D_2}{(x - r)^3} + \cdots + \frac{D_n}{(x - r)^2}\right) \tag{A.37}$$

- For the case of a complex root:

$$\frac{P_m(x)}{c_n(x - r_1)(x - r_2)(x - r_3)\cdots(x - r_{n-2})(ax^2 + bx + c)} = \frac{1}{c_n}\left(\frac{D_1}{x - r_1}\right.$$

$$\left. +\frac{D_2}{x - r_2} + \cdots + \frac{D_{n-2}}{x - r_{n-2}} + \frac{Ax + B}{ax^2 + bx + c}\right) \tag{A.38}$$

To calculate the constants that appear in the decomposed form we equate the decomposed form to the initial form of the rational function, multiply both sides of the resulting equation by $Q_n(x)$, and match the coefficients of the resulting polynomials.

To practise integration, an example is presented in Box A.4 which can also be solved using the MATLAB script in Code A.9.

Box A.4: Example

Calculate the integral of the following function in the domain $[-2, 8]$:

$$f(x) = 3x^4 + 5x^2 - 3$$

Solution

Using the basic rules of integration, the definite integral can be calculated as

$$\int_{-2}^{8} f(x)\,dx = \left[\frac{3}{5}x^5 + \frac{5}{3}x^3 - 3x\right]_{-2}^{8} = \frac{61550}{3}$$

Code A.9 *MATLAB code to calculate the integral of $f(x) = 3x^4 + 5x^2 - 3$ in the domain $[-2, 8]$.*

```
% find the integral of function f(x) = 3x^4+5x^2-3
clear all; clc; close all;
% define symbolic x
syms x
% define function
f=3*x^4+5*x^2-3;
% calculate the integral
int(f,x,-2,8)
```

A.2.6 Further reading

The interested reader can find more information on relevant topics in Stroud and Booth (2007), Yang (2018), and Singh (2011).

A.3 Vectors and matrices

A.3.1 Introduction

A **matrix** is a rectangular array of numbers, symbols, or expressions arranged in rows and columns. The array is placed within square brackets. A matrix composed of m rows and n columns (where m, n are natural numbers) is designated with dimensions $m \times n$, which is read 'm by n'. Matrices are typically denoted using capital letters (e.g. A, B, C), while their elements are denoted by lowercase letters followed by a double index (e.g., a_{ij}, b_{ij}, c_{ij}). The first index (i) indicates the row, while the second index (j) indicates the column. As an example, consider the following matrix:

$$A = \begin{bmatrix} 1 & 5 & 2 \\ 0 & -3 & 8.5 \end{bmatrix}$$

Matrix A is a 2×3 matrix, with elements $a_{11} = 1$, $a_{12} = 5$, $a_{13} = 2$, $a_{21} = 0$, $a_{22} = -3$, $a_{23} = 8.5$.
The general notation for an $m \times n$ matrix is of the following form:

$$A = A_{m \times n} = \left(a_{ij} \right)_{i=1, j=1}^{i=m, j=n} = \begin{bmatrix} a_{11} & & \downarrow jcolumn & a_{1n} \\ \vdots & irow \rightarrow a_{ij} & & \vdots \\ a_{m1} & & \cdots & a_{mn} \end{bmatrix} \tag{A.39}$$

- A matrix with dimensions $1 \times n$, $A = [a_{11} \cdots a_{1n}]$, is called a **row vector**.
- A matrix with dimensions $m \times 1$, $A = \begin{bmatrix} a_{11} \\ \vdots \\ a_{m1} \end{bmatrix}$, is called a **column vector**.
- A matrix that has the same number of rows and columns is called a **square matrix**:

$$A = A_{n \times n} = \left(a_{ij} \right)_{i=1, j=1}^{i=n, j=n} = \begin{bmatrix} a_{11} & \cdots & a_{1n} \\ \vdots & \ddots & \vdots \\ a_{n1} & \cdots & a_{nn} \end{bmatrix}.$$

- A matrix with dimensions $m \times n$ that has all its elements equal to zero is called a **zero matrix** and denoted as $O_{m \times n}$.

A.3.2 Operations with matrices

In the following, several operations with matrices are presented.

Matrix transposition

Given an $m \times n$ matrix $A = A_{m \times n} = \left(a_{ij} \right)_{i=1, j=1}^{i=m, j=n} = \begin{bmatrix} a_{11} & \cdots & a_{1n} \\ \vdots & \ddots & \vdots \\ a_{m1} & \cdots & a_{mn} \end{bmatrix}$, the matrix with dimensions $n \times m$ that is produced by changing the rows to columns and the columns to rows in A is called the transpose of A, denoted as A^T. It therefore takes the following form:

$$A^\mathrm{T} = \begin{bmatrix} a_{11} & \cdots & a_{1m} \\ \vdots & \ddots & \vdots \\ a_{1n} & \cdots & a_{mn} \end{bmatrix}$$

As an example, consider the matrix $A = \begin{bmatrix} 1 & 5 & 2 \\ 0 & -3 & 8.5 \end{bmatrix}$. Its transpose is $A^\mathrm{T} = \begin{bmatrix} 1 & 0 \\ 5 & -3 \\ 2 & 8.5 \end{bmatrix}$.

Notice that the transpose of a transpose is the original matrix. That is, $(A^\mathrm{T})^\mathrm{T} = A$.

Matrix addition

Addition of two matrices is only feasible when they possess identical dimensions. The resultant matrix, of matching dimensions, comprises entries derived from summing corresponding elements in the added matrices. This means:

$$A_{m \times n} + B_{m \times n} = \begin{bmatrix} a_{11} & \cdots & a_{1n} \\ \vdots & \ddots & \vdots \\ a_{m1} & \cdots & a_{mn} \end{bmatrix} + \begin{bmatrix} b_{11} & \cdots & b_{1n} \\ \vdots & \ddots & \vdots \\ b_{m1} & \cdots & b_{mn} \end{bmatrix} = \begin{bmatrix} a_{11} + b_{11} & \cdots & a_{1n} + b_{1n} \\ \vdots & \ddots & \vdots \\ a_{m1} + b_{m1} & \cdots & a_{mn} + b_{mn} \end{bmatrix}.$$

As an example, for the matrices $A = \begin{bmatrix} 1 & 5 & 2 \\ 0 & -3 & 8.5 \end{bmatrix}$ and $B = \begin{bmatrix} 1 & 2 & 4 \\ 4 & 5 & -4 \end{bmatrix}$, we have

$$A + B = \begin{bmatrix} 1 & 5 & 2 \\ 0 & -3 & 8.5 \end{bmatrix} + \begin{bmatrix} 1 & 2 & 4 \\ 4 & 5 & -4 \end{bmatrix} = \begin{bmatrix} 1 + 1 & 5 + 2 & 2 + 4 \\ 0 + 4 & -3 + 5 & 8.5 - 4 \end{bmatrix} = \begin{bmatrix} 2 & 7 & 6 \\ 4 & 2 & 4.5 \end{bmatrix}.$$

The transpose of a sum of two matrices is the sum of the transposes. In other words, $(A + B)^\mathrm{T} = A^\mathrm{T} + B^\mathrm{T}$.

Matrix multiplication by a number

A matrix $A_{m \times n}$ can be multiplied by any number c. The result is a new matrix with the same dimensions that has as entries the products of the respective elements of $A_{m \times n}$ with the number c:

$$c \cdot A_{m \times n} = c \cdot \begin{bmatrix} a_{11} & \cdots & a_{1n} \\ \vdots & \ddots & \vdots \\ a_{m1} & \cdots & a_{mn} \end{bmatrix} = \begin{bmatrix} c \cdot a_{11} & \cdots & c \cdot a_{1n} \\ \vdots & \ddots & \vdots \\ c \cdot a_{m1} & \cdots & c \cdot a_{mn} \end{bmatrix}.$$

As an example, consider the matrix $A = \begin{bmatrix} 1 & 5 & 2 \\ 0 & -3 & 8.5 \end{bmatrix}$ multiplied by 2. We have:

$$2A = 2 \begin{bmatrix} 1 & 5 & 2 \\ 0 & -3 & 8.5 \end{bmatrix} = \begin{bmatrix} 2 & 10 & 4 \\ 0 & -5 & 17 \end{bmatrix}.$$

Matrix multiplication

Two matrices can be multiplied only if the number of columns of the first one is the same as the number of rows of the second. For example, a matrix $A_{m \times n}$ can be multiplied with a matrix $B_{n \times q}$. The result will be a new matrix C with dimensions $m \times q$, as $C_{m \times q} = A_{m \times n} \cdot B_{n \times q}$. The entries of matrix $C_{m \times q}$ are calculated as follows:

$$c_{ij} = a_{i1} \cdot b_{1j} + a_{i2} \cdot b_{2j} + a_{i3} \cdot b_{3j} + \cdots + a_{in} \cdot b_{nj} \tag{A.40}$$

In other words, the elements of the new matrix in the ith row and jth column are the inner product of the corresponding row of A and column of B.

For the transpose of a product of two matrices A and B we have $(A \cdot B)^\mathrm{T} = B^\mathrm{T} \cdot A^\mathrm{T}$.

A.3.3 Square matrices

A matrix that has the same number of rows and columns is called a **square matrix**,

$$A = A_{n \times n} = \left(a_{ij}\right)_{i=1,\,j=1}^{i=n,\,j=n} = \begin{bmatrix} a_{11} & \cdots & a_{1n} \\ \vdots & \ddots & \vdots \\ a_{n1} & \cdots & a_{nn} \end{bmatrix}, \text{ e.g., } A = \begin{bmatrix} 1 & 2 \\ 3 & 4 \end{bmatrix}. \text{ Square matrices have}$$

important applications. The elements of a square matrix that have the same indices are called the elements of the **principal diagonal**. These elements are shown in bold in the following matrix:

$$A = \begin{bmatrix} a_{11} & & \cdots & & a_{1n} \\ & a_{22} & & & \\ \vdots & & a_{33} & & \vdots \\ & & & \ddots & \\ a_{n1} & & \cdots & & a_{nn} \end{bmatrix}$$

The sum of all the elements in the principal diagonal is called the **trace** of the matrix, denoted tr. It is therefore

$$tr(A) = a_{11} + a_{22} + a_{33} + a_{44} + \cdots + a_{nn} \tag{A.41}$$

- A square matrix that has nonzero entries only in the principal diagonal is called a **diagonal matrix**,
  e.g., $\begin{bmatrix} 2 & 0 & 0 \\ 0 & -1 & 0 \\ 0 & 0 & 3.4 \end{bmatrix}$.

- A diagonal matrix that has all the elements in the principal diagonal equal to 1 is called the **identity matrix**. It is denoted as I or $I_{n \times n}$. Examples are $I_{2 \times 2} = \begin{bmatrix} 1 & 0 \\ 0 & 1 \end{bmatrix}$ and $I_{3 \times 3} = \begin{bmatrix} 1 & 0 & 0 \\ 0 & 1 & 0 \\ 0 & 0 & 1 \end{bmatrix}$.

  The identity matrix has the property $A \cdot I = I \cdot A = A$ for all matrices A, when the multiplication is admissible.

- If, for a square matrix A, there exists another matrix D such that

$$A \cdot D = D \cdot A = I \tag{A.42}$$

then we say that matrix A is invertible and its inverse is the matrix D. At the same time, the inverse of D will be A. Note that not all square matrices have inverses. The inverse of a matrix A, if it exists, is denoted as A^{-1} (the exponent minus one is just a notation for the inverse). Therefore, we have

$$A \cdot A^{-1} = A^{-1} \cdot A = I \tag{A.43}$$

- A square matrix that has zero entries below the principal diagonal is called an **upper triangular matrix**, e.g., $\begin{bmatrix} 2 & 1 & 5 \\ 0 & -1 & 0 \\ 0 & 0 & 3.4 \end{bmatrix}$.

- A square matrix that has zero entries above the principal diagonal is called a **lower triangular matrix**, e.g., $\begin{bmatrix} 2 & 0 & 0 \\ 0 & -1 & 0 \\ 1 & 0.5 & 3.4 \end{bmatrix}$.

- A square matrix that is equal to its transpose is called a **symmetric matrix**. That is, A is symmetric if $A = A^{\mathrm{T}}$. The elements of a symmetric matrix have the property that $a_{ij} = a_{ji}$, Therefore, all the elements symmetrically placed with respect to the principal diagonal are the same. As an example, consider the matrix $A = \begin{bmatrix} 1 & 3 & 8 \\ 3 & -2 & -5 \\ 8 & -5 & 0 \end{bmatrix}$ and observe that $A^{\mathrm{T}} = \begin{bmatrix} 1 & 3 & 8 \\ 3 & -2 & -5 \\ 8 & -5 & 0 \end{bmatrix} = A$.

- A square matrix that is equal to minus its transpose is called a **skew-symmetric matrix** or **anti-symmetric matrix**. A is skew-symmetric if $A = -A^{\mathrm{T}}$. The elements of a skew-symmetric matrix have the property that $a_{ij} = -a_{ji}$, so all elements symmetrically placed with respect to the principal diagonal are opposite numbers. As an example, consider the matrix $A = \begin{bmatrix} 0 & 1 & -8 \\ -1 & 0 & 5 \\ 8 & -5 & 0 \end{bmatrix}$ and

observe that $A^T = \begin{bmatrix} 0 & -1 & 8 \\ 1 & 0 & -5 \\ -8 & 5 & 0 \end{bmatrix} = -A$.

The symmetric part of a matrix A is the matrix $\frac{1}{2}(A + A^T)$. The skew-symmetric part of A is the matrix $\frac{1}{2}(A - A^T)$.

Note that all the elements on the principal diagonal of a skew-symmetric matrix are always equal to zero. This can be easily verified from the element properties of a skew-symmetric matrix. Since $a_{ij} = -a_{ji}$ for all elements, this should be so for the elements in the principal diagonal, where the row and column indices are the same. Thus, $a_{ii} = -a_{ii}$ and then $2a_{ii} = 0$, which implies that $a_{ii} = 0$.

Determinants

Linked to square matrices is a quantity called the **determinant**. The determinant of a square matrix A is a number, denoted $\det(A)$ or $|A|$. The following rules apply for the calculation of the determinant:

- If the dimensions of a matrix A are 1×1, the determinant is the number in the matrix.
- If the dimensions of a matrix A are 2×2, that is, $A = \begin{bmatrix} a_{11} & a_{12} \\ a_{21} & a_{22} \end{bmatrix}$, then the determinant of A is calculated as follows:

$$|A| = \begin{vmatrix} a_{11} & a_{12} \\ a_{21} & a_{22} \end{vmatrix} = a_{11} \cdot a_{22} - a_{12} \cdot a_{21} \qquad (A.44)$$

- If the dimensions of a matrix A are 3×3, that is, $A = \begin{bmatrix} a_{11} & a_{12} & a_{13} \\ a_{21} & a_{22} & a_{23} \\ a_{31} & a_{32} & a_{33} \end{bmatrix}$, then the determinant of A is calculated as

$$|A| = \begin{vmatrix} a_{11} & a_{12} & a_{13} \\ a_{21} & a_{22} & a_{23} \\ a_{31} & a_{32} & a_{33} \end{vmatrix} = a_{11} \cdot a_{22} \cdot a_{33} + a_{23} \cdot a_{12} \cdot a_{31} + a_{32} \cdot a_{21} \cdot a_{13} - a_{13} \cdot a_{22} \cdot a_{31}$$

$$- a_{12} \cdot a_{21} \cdot a_{33} - a_{23} \cdot a_{32} \cdot a_{11} \qquad (A.45)$$

Determinants can be applied in the solution of simultaneous equations using **Cramer's method**. This method works using specific steps, and is demonstrated in Box A.5.

Box A.5: Example (Solution of three simultaneous equations using Cramer's method)

Find the solution of following system of equations using Cramer's method:

$$\begin{cases} 2x + 5y - z = 0 \\ 2x + y + 3z = 1 \\ -x + = -2 \end{cases}$$

Box A.5 *Continued*

Solution

Write the system of equations in its matrix form as follows:

$$\begin{cases} 2x + 5y - z = 0 \\ 2x + y + 3z = 1 \\ -x + z = -2 \end{cases} \rightarrow \begin{bmatrix} 2 & 5 & -1 \\ 2 & 1 & 3 \\ -1 & 0 & 1 \end{bmatrix} \cdot \begin{bmatrix} x \\ y \\ z \end{bmatrix} = \begin{bmatrix} 0 \\ 1 \\ -2 \end{bmatrix}$$

Then calculate the determinants D, D_x, D_y, and D_z, where D is the determinant of the system coefficients matrix:

$$D = \begin{vmatrix} 2 & 5 & -1 \\ 2 & 1 & 3 \\ -1 & 0 & 1 \end{vmatrix}$$

$$= 2 \cdot 1 \cdot 1 + 5 \cdot 3 \cdot (-1) + (-1) \cdot 2 \cdot 0 - (-1) \cdot 1 \cdot (-1) - 0 \cdot 3 \cdot 2 - 1 \cdot 2 \cdot 5$$

$$= -24$$

D_x is the same as D but the **first column** is replaced in the determinant by the right-hand side vector of the system:

$$D_x = \begin{vmatrix} 0 & 5 & -1 \\ 1 & 1 & 3 \\ -2 & 0 & 1 \end{vmatrix}$$

$$= 0 \cdot 1 \cdot 1 + 5 \cdot 3 \cdot (-2) + (-1) \cdot 1 \cdot 0 - (-2) \cdot 1 \cdot (-1) - 0 \cdot 3 \cdot 0 - 1 \cdot 1 \cdot 5$$

$$= -37$$

D_y is the same as D but the **second column** is replaced in the determinant by the right-hand side vector of the system:

$$D_y = \begin{vmatrix} 2 & 0 & -1 \\ 2 & 1 & 3 \\ -1 & -2 & 1 \end{vmatrix}$$

$$= 2 \cdot 1 \cdot 1 + 0 \cdot 3 \cdot (-1) - 1 \cdot 2 \cdot (-2) + 1 \cdot 1 \cdot (-1) - (-2) \cdot 3 \cdot 2 - 1 \cdot 2 \cdot 0$$

$$= 17$$

D_z is the same as D but the **third column** is replaced in the determinant by the right-hand side vector of the system:

$$D_z = \begin{vmatrix} 2 & 5 & 0 \\ 2 & 1 & 1 \\ -1 & 0 & -2 \end{vmatrix}$$

$$= 2 \cdot 1 \cdot (-2) + 5 \cdot 1 \cdot (-1) + 0 \cdot 2 \cdot 0 - (-1) \cdot 1 \cdot 0 - 0 \cdot 1 \cdot 2 - (-2) \cdot 2 \cdot 5$$

$$= 11$$

Now we can solve the system of equations as:

$$x = D_x/D = -37/-24 = 1.542$$
$$y = D_y/D = 17/-24 = -0.708$$
$$z = D_z/D = 11/-24 = -0.458$$

This example can also be solved using the MATLAB script in Code A.10.

Code A.10 *MATLAB code to solve a system of three simultaneous equations involving* $2x + 5y - z = 0$, $2x + y + 3z = 1$, *and* $-x + z = -2$.

```
% this Matlab code solve a system of three simultaneous equations
% involving 2x+5y-z=0, 2x+y+3z=1,
% and -x+z=-2.
clear all; clc; close all;
% define system coefficient matrix
A=[2 5 -1;2 1 3;-1 0 1];
% define right-hand-side vector
b=[0;1;-2];
% solve system
A\b
```

A.3.4 Eigenvalues and eigenvectors

The eigenvalue problem for a square matrix A of dimensions $n \times n$ is to find numbers, denoted by λ, the eigenvalues, and column vectors (that is, matrices of dimensions $n \times 1$), denoted by x, the eigenvectors, such that the following equation holds true:

$$A \cdot x = \lambda \cdot x \tag{A.46}$$

Equation (A.46) can be equivalently written as $(A - \lambda \cdot I) \cdot x = 0$, where I is the $n \times n$ identity matrix. The eigenvalues of matrix A are calculated by solving the following equation:

$$|A - \lambda \cdot I| = 0 \tag{A.47}$$

To calculate the eigenvalues of A, we need to compute the determinant of the matrix, $A - \lambda \cdot I$, and set it equal to zero. This will result in what is called the **characteristic polynomial**, as demonstrated in Box A.6.

Box A.6: Example (Eigenvalues and eigenvectors)

Find the eigenvalues and eigenvectors of the following matrix:

$$A = \begin{bmatrix} 2 & 1 \\ 1 & 2 \end{bmatrix}$$

Solution

To find the eigenvalues of A, we need to construct the matrix

$$A - \lambda \cdot I = \begin{bmatrix} 2 & 1 \\ 1 & 2 \end{bmatrix} - \lambda \cdot \begin{bmatrix} 1 & 0 \\ 0 & 1 \end{bmatrix} = \begin{bmatrix} 2-\lambda & 1 \\ 1 & 2-\lambda \end{bmatrix}.$$

The determinant of this matrix is now set equal to zero:

$$|A - \lambda \cdot I| = \begin{vmatrix} 2-\lambda & 1 \\ 1 & 2-\lambda \end{vmatrix} = (2-\lambda)^2 - 1 = \lambda^2 - 4\lambda + 3$$

Thus, $|A - \lambda \cdot I| = 0$ implies $\lambda^2 - 4\lambda + 3 = 0$. This expression is a quadratic polynomial for λ. This is the characteristic polynomial for the specific matrix $A = \begin{bmatrix} 2 & 1 \\ 1 & 2 \end{bmatrix}$. The eigenvalues are the roots of this polynomial. For the quadratic polynomial there are the following two roots:

Box A.6 *Continued*

$$\lambda_1 = \frac{4 + \sqrt{16 - 12}}{2} = \frac{6}{2} = 3 \text{ and } \lambda_2 = \frac{4 - \sqrt{16 - 12}}{2} = \frac{2}{2} = 1.$$

Therefore, the eigenvalues of the matrix $A = \begin{bmatrix} 2 & 1 \\ 1 & 2 \end{bmatrix}$ are $\lambda_1 = 3$ and $\lambda_2 = 1$.

For the computation of the eigenvector corresponding to each eigenvalue, we substitute the eigenvalue found in the original system and solve the simultaneous equations produced as follows:

Computation of the first eigenvector

It is $A \cdot x = \lambda_1 \cdot x$, which for a vector $x = \begin{bmatrix} x \\ y \end{bmatrix}$

can be written as $\begin{bmatrix} 2 & 1 \\ 1 & 2 \end{bmatrix} \begin{bmatrix} x \\ y \end{bmatrix} - 3 \cdot \begin{bmatrix} x \\ y \end{bmatrix} = \begin{bmatrix} 0 \\ 0 \end{bmatrix}.$

This matrix equation, upon performing the matrix-vector multiplications, is equivalent to the following simultaneous system:

$$\left\{ \begin{array}{l} 2x + y - 3x = 0 \\ x + 2y - 3y = 0 \end{array} \right\}$$

Solving this system, we obtain

$$\left\{ \begin{array}{l} x - y = 0 \\ x - y = 0 \end{array} \right\} \rightarrow \left\{ \begin{array}{l} y = x \\ y - x = 0 \end{array} \right\} \rightarrow \left\{ \begin{array}{l} y = x \\ y - y = 0 \end{array} \right\} \rightarrow \left\{ \begin{array}{l} y = x \\ 0 = 0 \end{array} \right\}$$

The first equation implies that any vector having the y coordinate equal to the x coordinate is a solution. That means any vector $\begin{bmatrix} x \\ y \end{bmatrix}$, where $y = x$ is a solution. Thus, the solution vectors will

have the form $\begin{bmatrix} x \\ x \end{bmatrix} = x \cdot \begin{bmatrix} 1 \\ 1 \end{bmatrix}$. We say that any vector of the form $x \cdot \begin{bmatrix} 1 \\ 1 \end{bmatrix}$ is an eigenvector corresponding to eigenvalue $\lambda_1 = 3$.

Computation of the second eigenvector

It is $A \cdot x = \lambda_2 \cdot x$, which for a vector $x = \begin{bmatrix} x \\ y \end{bmatrix}$ can be written as $\begin{bmatrix} 2 & 1 \\ 1 & 2 \end{bmatrix} \begin{bmatrix} x \\ y \end{bmatrix} - 1 \cdot \begin{bmatrix} x \\ y \end{bmatrix} = \begin{bmatrix} 0 \\ 0 \end{bmatrix}.$

This matrix equation, upon performing the matrix–vector multiplications, is equivalent to the following simultaneous system:

$$\left\{ \begin{array}{l} 2x + y - x = 0 \\ x + 2y - y = 0 \end{array} \right\}$$

Solving this system, we have

$$\left\{ \begin{array}{l} x + y = 0 \\ x + y = 0 \end{array} \right\} \rightarrow \left\{ \begin{array}{l} y = -x \\ x + y = 0 \end{array} \right\} \rightarrow \left\{ \begin{array}{l} y = -x \\ y - y = 0 \end{array} \right\} \rightarrow \left\{ \begin{array}{l} y = -x \\ 0 = 0 \end{array} \right\}$$

The first equation implies that any vector having the y coordinate equal to minus the x coordinate is a solution. That means any vector $\begin{bmatrix} x \\ y \end{bmatrix}$, where $y = -x$ is a solution. Thus, the solution vectors

will have the form $\begin{bmatrix} x \\ -x \end{bmatrix} = x \cdot \begin{bmatrix} 1 \\ -1 \end{bmatrix}$. We say that any vector of the form $x \cdot \begin{bmatrix} 1 \\ -1 \end{bmatrix}$ is an eigenvector corresponding to eigenvalue $\lambda_2 = 1$.

The MATLAB code presented in Code A.11 can be used to solve this question.

Code A.11 *MATLAB code to calculate the eigenvalues and eigenvectors of the 2×2 matrix* $A = \begin{bmatrix} 2 & 1 \\ 1 & 2 \end{bmatrix}$.

```
% this Matlab code calculates the eigenvectors and
% eigenvalues of a 2×2 matrix.
clear all; clc; close all;
% define matrix
A=[2 1;1 2];
% find eigenvalues and eigenvectors
[V,D]=eig(A)
```

Notice that, in the previous example, the sum of the eigenvalues is $\lambda_1 + \lambda_2 = 3 + 1 = 4$. At the same time, the trace of $A = \begin{bmatrix} 2 & 1 \\ 1 & 2 \end{bmatrix}$ is tr$(A) = 2 + 2 = 4$. Thus, the sum of the eigenvalues is equal to the trace of the matrix. This is a general result and is valid for all matrices. In other words, in all matrices, **the sum of the eigenvalues is equal to the trace of the matrix**. Furthermore, notice that the product of the eigenvalues is $\lambda_1 \cdot \lambda_2 = 3 \cdot 1 = 3$. At the same time, the determinant of A is $|A| = 2 \cdot 2 - 1 \cdot 1 = 4 - 1 = 3$. Thus, the product of the eigenvalues is equal to the determinant of the matrix. This again is a general result and valid for all matrices. In all matrices, **the product of the eigenvalues is equal to the determinant of the matrix**.

In the example we found a characteristic polynomial of second degree and thus two eigenvalues and two eigenvectors for a 2×2 matrix. In general, a square matrix A of dimensions $n \times n$ has a characteristic polynomial of degree n and thus n eigenvalues and n eigenvectors.

A.3.5 Inverse matrix

We have already studied the solution of systems of simultaneous equations. The method of elimination and Cramer's method using determinants have been discussed. In this section, the solution of simultaneous equations using inverse matrices is presented.

First, consider a linear system of simultaneous equations with n equations and n unknowns of the following form:

$$
\begin{aligned}
a_{11}x_1 + a_{12}x_2 \quad &\cdots \quad + a_{1n}x_n = b_1 \\
\vdots \quad &\ddots \quad \vdots \\
a_{n1}x_1 + a_{n2}x_2 \quad &\cdots \quad + a_{nn}x_n = b_n
\end{aligned}
\tag{A.48}
$$

This system can be written equivalently in the matrix form $A \cdot x = b$, where

$$
A = \begin{bmatrix} a_{11} & \cdots & a_{1n} \\ \vdots & \ddots & \vdots \\ a_{n1} & \cdots & a_{nn} \end{bmatrix}, \quad x = \begin{bmatrix} x_1 \\ \vdots \\ x_n \end{bmatrix}, \quad b = \begin{bmatrix} b_1 \\ \vdots \\ b_n \end{bmatrix}
$$

Let us recall the concept of the inverse matrix A^{-1}. If, for a square matrix A, there exists another matrix D such that $A \cdot D = D \cdot A = I$, then matrix A is invertible and its inverse is the matrix D. At the same time, the inverse of matrix D will be A. Note that not all square matrices are invertible. The inverse of a matrix A, if it exists, is denoted as A^{-1}; here, the exponent minus one is just a notation for the inverse. We therefore have $A \cdot A^{-1} = A^{-1} \cdot A = I$.

From the above, we can infer that, given a system of simultaneous equations set in matrix form $A \cdot x = b$, if the inverse of matrix A exists then by multiplying from the left by the inverse we get the following:

$$
A^{-1} \cdot A \cdot x = A^{-1} \cdot b
\tag{A.49}
$$

which means $I \cdot x = A^{-1} \cdot b$ and thus the solution x is $x = A^{-1} \cdot b$. This means the solution of a system of simultaneous equations can be found by calculating the inverse of the coefficient matrix A and multiplying vector b from the left by this inverse matrix. It remains to find a way to calculate the inverse. This can be done using the characteristic polynomial. The steps to be followed are:

Step 1: Check if the inverse exists. This can be done by any of the following criteria:
- **(a) A square matrix is invertible if and only if its determinant is different from zero.** In other words, given an $n \times n$ matrix A, in order for A^{-1} to exist we must have $|A| \neq 0$.
- **(b)** Equivalent to criterion (a) is to ensure that the **$n \times n$ matrix A has all its eigenvalues different from zero.** Criterion (b) is equivalent to (a) since the product of the eigenvalues is equal to the determinant of the matrix.

Step 2: After ensuring that matrix A is invertible, we calculate the characteristic polynomial of A in terms of λ, as $|A - \lambda \cdot I| = 0$.

Step 3: Assuming that the characteristic polynomial of A is of the form

$$p(\lambda) = \lambda^n + c_{n-1}\lambda^{n-1} + \cdots + c_2\lambda^2 + c_1\lambda + c_0 \tag{A.50}$$

then, according to the Cayley–Hamilton theorem, we have

$$p(A) = A^n + c_{n-1}A^{n-1} + \cdots + c_2A^2 + c_1A + c_0I = O \tag{A.51}$$

In other words, every square matrix satisfies its characteristic polynomial.

Step 4: We multiply the expression $A^n + c_{n-1}A^{n-1} + \cdots + c_2A^2 + c_1A + c_0I = O$ by the inverse A^{-1}, which leads to the following equation:

$$A^{n-1} + c_{n-1}A^{n-2} + \cdots + c_2A + c_1I + c_0A^{-1} = O \tag{A.52}$$

Consequently, the inverse can be calculated using only matrix multiplications as follows:

$$A^{-1} = -\frac{1}{c_0}\left(A^{n-1} + c_{n-1}A^{n-2} + \cdots + c_2A + c_1I\right) \tag{A.53}$$

An example calculation to find the inverse of a matrix is shown in Box A.7.

Box A.7: Example (Finding the inverse of a matrix)

Find the inverse of the following matrix, which usually appears in the analysis of structures in a term representing the structural member mass:

$$A = \begin{bmatrix} 2 & 1 \\ 1 & 2 \end{bmatrix}$$

Solution

Step 1 $|A| = \begin{vmatrix} 2 & 1 \\ 1 & 2 \end{vmatrix} = 4 - 1 = 3 \neq 0$, hence the inverse matrix A^{-1} exists.

Step 2 $|A - \lambda \cdot I| = \begin{vmatrix} 2-\lambda & 1 \\ 1 & 2-\lambda \end{vmatrix} = (2-\lambda)^2 - 1 = \lambda^2 - 4\lambda + 3$. Thus, $|A - \lambda \cdot I| = 0$ implies $\lambda^2 - 4\lambda + 3 = 0$. This expression is a quadratic polynomial for λ. This is the characteristic polynomial for the specific matrix $A = \begin{bmatrix} 2 & 1 \\ 1 & 2 \end{bmatrix}$.

Step 3 Since the matrix satisfies the characteristic polynomial, we have $A^2 - 4A + 3I = O$.

Step 4 The previous equation is multiplied by the inverse of A, which is A^{-1}, to obtain $A^{-1}A^2 - 4A^{-1}A + 3 \cdot A^{-1} = O$. This equation is rearranged to $A - 4I + 3A^{-1} = O$.

Finally, we have:

$$A^{-1} = -\frac{1}{3}(A - 4I) = -\frac{1}{3}\left(\begin{bmatrix} 2 & 1 \\ 1 & 2 \end{bmatrix} - 4\begin{bmatrix} 1 & 0 \\ 0 & 1 \end{bmatrix}\right) = -\frac{1}{3}\begin{bmatrix} -2 & 1 \\ 1 & -2 \end{bmatrix} = \begin{bmatrix} \frac{2}{3} & -\frac{1}{3} \\ -\frac{1}{3} & \frac{2}{3} \end{bmatrix}$$

We can verify this last computation by checking that $A^{-1}A = I$. Indeed, a simple calculation shows:

$$\begin{bmatrix} 2 & 1 \\ 1 & 2 \end{bmatrix} \cdot \begin{bmatrix} \frac{2}{3} & -\frac{1}{3} \\ -\frac{1}{3} & \frac{2}{3} \end{bmatrix} = \begin{bmatrix} \frac{4}{3} - \frac{1}{3} & 0 \\ 0 & -\frac{1}{3} + \frac{4}{3} \end{bmatrix} = \begin{bmatrix} 1 & 0 \\ 0 & 1 \end{bmatrix}$$

This example can be solved using the MATLAB script in Code A.12.

Code A.12 *MATLAB code to calculate the inverse of the 2×2 matrix* $A = \begin{bmatrix} 2 & 1 \\ 1 & 2 \end{bmatrix}$.

```
% this Matlab code calculate the inverse of a 2×2 matrix.
clear all; clc; close all;
% define matrix
A=[2 1;1 2];
% find the inverse
inv(A)
```

Note that the result used in step 3 in Box A.7 is the **Cayley–Hamilton theorem**,

$$p(A) = A^n + c_{n-1}A^{n-1} + \cdots + c_2A^2 + c_1A + c_0I = O$$

and can be intuitively shown in the following manner. When λ is an eigenvalue, the characteristic polynomial is equal to zero:

$$\lambda^n + c_{n-1}\lambda^{n-1} + \cdots + c_2\lambda^2 + c_1\lambda + c_0 = 0$$

By multiplying both sides by any eigenvector x of matrix A, it becomes

$$\lambda^n x + c_{n-1}\lambda^{n-1}x + \cdots + c_2\lambda^2 x + c_1\lambda x + c_0 x = 0$$

Since for eigenvalues and eigenvectors we have $Ax = \lambda x$, we can write

$$\lambda^{n-1}Ax + c_{n-1}\lambda^{n-2}Ax + \cdots + c_2\lambda Ax + c_1Ax + c_0Ix = 0$$

And, by repeating the same process, we obtain

$$\lambda^{n-2}A^2 x + c_{n-1}\lambda^{n-3}A^2 x + \cdots + c_2A^2 x + c_1Ax + c_0Ix = O$$

After successive substitutions we finally get the following equation:

$$\left(A^n + c_{n-1}A^{n-1} + \cdots + c_2A^2 + c_1A + c_0I\right)x = O$$

which implies $p(A) = O$.

A.3.6 Further reading

The interested reader can find more information on relevant topics in Stroud and Booth (2007) and Horn and Johnson (1985).

A.4 Ordinary differential equations

A.4.1 Linear differential equations with constant coefficients

Differential equations are equations where the unknown quantity is a function. Among the terms appearing in a differential equation are derivatives of the unknown function. In the case where this function has one independent variable, the differential equation is termed an **ordinary differential equation** (ODE). ODEs are extremely useful for several applications in mechanics, physics, and engineering.

The highest-order derivative of the unknown function that appears in a differential equation defines the order of the equation. As an example, consider the following differential equations with unknown function $f(x)$:

- $\frac{df}{dx} = 2f + 8f^2$

- $\frac{d^2f}{dx^2} + 3x\frac{df}{dx} + 4f = 4$

The first one is a first-order equation since the highest-order derivative of $f(x)$ is $\frac{df}{dx}$. The second equation is a second-order differential equation since the highest-order derivative of $f(x)$ that appears in it is $\frac{d^2f}{dx^2}$. An important class of ordinary differential equations are the so-called **linear ordinary differential equations**. Linear ODEs of order n, where n is a natural number, have the following general form:

$$a_n(x)\frac{d^nf}{dx^n} + a_{n-1}(x)\frac{d^{n-1}f}{dx^{n-1}} + \cdots + a_2(x)\frac{d^2f}{dx^2} + a_1(x)\frac{df}{dx} + a_0(x)f = b(x) \tag{A.54}$$

where $a_n(x)$, $a_{n-1}(x)$, ..., $a_2(x)$, $a_1(x)$, $a_0(x)$, and $b(x)$ are functions of the independent variable x. In the case where $b(x) = 0$, the equation is called **homogeneous**. If $b(x) \neq 0$, the equation is called **nonhomogeneous**.

A linear ODE of the form

$$a_n\frac{d^nf}{dx^n} + a_{n-1}\frac{d^{n-1}f}{dx^{n-1}} + \cdots + a_2\frac{d^2f}{dx^2} + a_1\frac{df}{dx} + a_0f = b(x) \tag{A.55}$$

where $a_n, a_{n-1}, \ldots, a_2, a_1$, and a_0 are constants is a linear ODE of order n with constant coefficients.

The solution of a homogeneous linear ODE with constant coefficients can be found using the following procedure:

Step 1 It is assumed that the solutions are of the form $f(x) = Ce^{\lambda x}$. For this type of solution, the derivatives of $f(x)$ are $\frac{df}{dx} = C\lambda e^{\lambda x}$, $\frac{d^2f}{dx^2} = C\lambda^2 e^{\lambda x}$, ..., $\frac{d^nf}{dx^n} = C\lambda^n e^{\lambda x}$.

Step 2 Since the equation is homogeneous, $b(x) = 0$, and therefore we have

$$a_n\frac{d^nf}{dx^n} + a_{n-1}\frac{d^{n-1}f}{dx^{n-1}} + \cdots + a_2\frac{d^2f}{dx^2} + a_1\frac{df}{dx} + a_0f = 0$$

By substituting the derivatives calculated in step 1, the equation takes the following form:

$$a_nC\lambda^n e^{\lambda x} + a_{n-1}C\lambda^{n-1}e^{\lambda x} + \cdots + a_2C\lambda^2 e^{\lambda x} + a_1C\lambda e^{\lambda x} + a_0Ce^{\lambda x} = 0$$

Step 3 If $C = 0$, then the solution would be identically zero, which is of no interest. We thus assume that $C \neq 0$. Furthermore, $e^{\lambda x} \neq 0$. Then, we can divide the last equation in step 2 by $Ce^{\lambda x}$ and obtain $a_n\lambda^n + a_{n-1}\lambda^{n-1} + \cdots + a_2\lambda^2 + a_1\lambda + a_0 = 0$. This equation is a polynomial in the unknown number λ, called the **characteristic polynomial** of the differential equation. The polynomial will have n roots, where the roots might be real, purely imaginary, or complex.

- If all the roots are distinct, and denoted as $\lambda_1, \lambda_2, ..., \lambda_{n-1}, \lambda_n$, then the solution of the homogeneous equation is

$$f(x) = C_1 e^{\lambda_1 x} + C_2 e^{\lambda_2 x} + \cdots + C_{n-1} e^{\lambda_{n-1} x} + C_n e^{\lambda_n x} \tag{A.56a}$$

Note that the solution is the sum of n different terms. When the solution is expressed in terms of the general constants $C_1, C_2, ..., C_n$, it is called the **general solution** of the differential equation.

- If some of the roots are multiple roots, then the form of the general solution is altered. Assume that root λ_j appears m times, where $m < n$. Then, the solution, which must be the sum of n terms, does not include the term $C e^{\lambda_j x}$ m times. Instead, it includes (among the total of n terms) m terms of the following form:

$$C_j e^{\lambda_j x}, C_{j+1} x e^{\lambda_j x}, C_{j+2} x^2 e^{\lambda_j x}, ..., C_{j+m-2} x^{m-2} e^{\lambda_j x}, C_{j+m-1} x^{m-1} e^{\lambda_j x} \tag{A.56b}$$

The solution of a nonhomogeneous linear ODE can be found as the sum of the solution of the homogeneous form and a **partial solution** of the full form. A partial solution is any function that satisfies the nonhomogeneous form.

There is no general methodology to obtain a partial solution of a nonhomogeneous ODE. The procedures applied are case dependent, and depend on the particular form of the nonhomogeneous term $b(x)$.

Example solutions are given in Box A.8 for a number of nonhomogeneous ODEs.

Box A.8: Example

Find the general solution of the following nonhomogeneous linear ODEs with constant coefficients.

(a) $2\dfrac{df}{dx} + 3f = 3$

(b) $\dfrac{d^2 f}{dx^2} - 4f = 7x$

(c) $\dfrac{d^2 f}{dx^2} - 4f = 5x^2$

Solution

(a) This is a nonhomogeneous linear first-order equation. The solution will be the solution of the homogeneous form, plus a partial solution. The solution of the homogeneous form is $Ce^{-\frac{3}{2}x}$. To find a partial solution, we observe that if the partial solution was a constant, i.e., $f_{partial}(x) = a$, the first derivative would be zero. Substituting into the nonhomogeneous equation, we obtain $2\frac{da}{dx} + 3a = 3$, which leads to $3a = 3$, and we have $a = 1$. Therefore, $f_{partial}(x) = a = 1$ and the general solution of the differential equation becomes

$$f(x) = Ce^{-\frac{3}{2}x} + 1$$

(b) This is a nonhomogeneous linear second-order equation. The solution will be the solution of the homogeneous form, plus a partial solution. The solution of the homogeneous form is $f(x) = C_1 e^{2x} + C_2 e^{-2x}$. To find a partial solution of the nonhomogeneous form, we observe that if the partial solution was a polynomial of the form ax, then, the second derivative would be zero. By substituting the function $f_{partial}(x) = ax$ into the nonhomogeneous differential equation, we obtain $\frac{d^2 ax}{dx^2} - 4ax = 7x$ and thus $-4ax = 7x$. This last equation is valid for every value of the independent variable x if and only if $-4a = 7$, which gives: $a = -7/4$. Finally, the

Box A.8 *Continued*

---

partial solution becomes: $f_{partial}(x) = -\frac{7}{4}x$ and the general solution of the nonhomogeneous equation takes the form

$$f(x) = C_1 e^{2x} + C_2 e^{-2x} - \frac{7}{4}x$$

(c) This is a nonhomogeneous linear second-order equation. The solution will be the solution of the homogeneous form, plus a partial solution. The solution of the homogeneous form is $f(x) = C_1 e^{2x} + C_2 e^{-2x}$. The procedure to find the partial solution is more challenging in this case, as the degree of the polynomial on the right-hand side of the equation is higher. We assume a partial solution of the form $f_{partial}(x) = ax^2 + bx + c$. Substituting this solution into the nonhomogeneous differential equation, it takes the following form:

$$\frac{d^2(ax^2 + bx + c)}{dx^2} - 4(ax^2 + bx + c) = 5x^2$$

which, after differentiating the quadratic polynomial, gives

$$2a - 4(ax^2 + bx + c) = 5x^2$$

or, by gathering together all the same powers of x,

$$-4ax^2 - 4bx + (2a - 4c) = 5x^2$$

For the right-hand side of this equation to be equal to the left-hand side for all values of the independent variable x, it must satisfy $-4a = 5$, $-4b = 0$, and $2a - 4c = 0$. These equations imply that

$$a = -\frac{5}{4}$$

$$b = 0$$

$$c = \frac{a}{2} = -\frac{5}{8}$$

Thus, the general solution of the differential equation becomes

$$f(x) = C_1 e^{2x} + C_2 e^{-2x} - \frac{5}{4}x^2 - \frac{5}{8}$$

The MATLAB script in Code A.13 can be used to solve the nonhomogeneous linear ODE in part (b).

Code A.13 *A MATLAB code to solve the non-homogeneous linear ordinary differential equation $\frac{d^2f}{dx^2} - 4f = 7x$.*

```
% this Matlab code solves a non-homogeneous, linear
% second order ordinary differential equation.
clear all; clc; close all;
% define symbolic x and f(x)
syms x f(x)
% solve the ordinary differential equation
dsolve(diff(f,x,2)-4*f==7*x)
```

Further reading

The interested reader can find more information on relevant topics in Stroud and Booth (2007), Yang (2018), Singh (2011), and Boyce and DiPrima (1992).

A.4.2 Some applications in engineering

Hydrostatic pressure

Assume a bulk of fluid (e.g., water) in static equilibrium inside a reservoir (Figure A.20). The fluid density is denoted as ρ. The upper surface of the fluid, located at $z = 0$, is at atmospheric pressure, denoted as p_{atm}. The depth of the reservoir is measured along the z axis and ranges between 0 at the water surface and h at the bottom of the reservoir (see Figure A.20).

The pressure of the fluid at depth is desired as the depth increases. The answer can be found if we analyse the static equilibrium of a thin slice of the fluid located at any point along the z axis. A thin slice with thickness dz and area A is considered, located between points z and dz (Figure A.21). The forces acting along the z direction on this portion of the fluid are the weight and those due to the water pressure (p) on the upper and lower surfaces. Denoting the acceleration of gravity as g, the weight of the fluid slice is given by

$$W = mg = \rho Vg = \rho gAdz \tag{A.57}$$

The force due to pressure is calculated as $F = pA$. Figure A.21 shows the free body diagram for the fluid slice.

Using the conservation of momentum (Newton's law), the total force acting on the fluid must be equal to zero. This leads to the following equation:

$$p(z)A - p(z + dz)A + \rho gAdz = 0 \tag{A.58}$$

Using Taylor's expansion and retaining only the constant and the linear term in dz, since the differential length is very small, Eq (A.58) becomes

$$p(z)A - p(z)A - \frac{dp}{dz}dzA + \rho gAdz = 0 \tag{A.59}$$

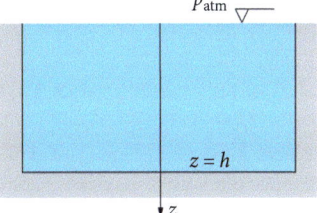

Figure A.20 Reservoir with a fluid of depth h in static equilibrium.

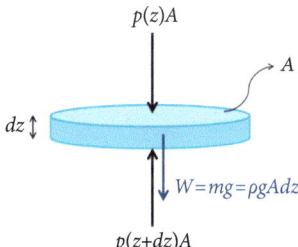

Figure A.21 Free body diagram for the fluid in static equilibrium. A thin layer of fluid with thickness dz and surface area A is considered.

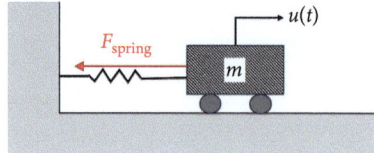

Figure A.22 Single degree of freedom mechanical system with no dissipation with a mass (m) and a spring with stiffness coefficient k.

or, equivalently,

$$\frac{dp}{dz} = \rho g \tag{A.60}$$

This is a nonhomogeneous first-order differential equation. Integrating Eq (A.60) with respect to z gives the following equation for the pressure $p\,(z)$:

$$p\,(z) = \rho g z + c \tag{A.61}$$

where c is the integration constant. Since $p\,(0) = p_{atm}$, we must have $c = p_{atm}$.

Single degree of freedom mechanical system

Consider the dynamic response of a single degree of freedom mechanical system consisting of a spring and a mass (Figure A.22). The spring stiffness is denoted as k and has units of N/m. The mass is denoted m with units of kg. The mass displacement is a function of time and is denoted as $u\,(t)$. Using Newton's law, the balance of forces on the mass is written as

$$\text{mass} \cdot \text{acceleration} = F_{spring} \tag{A.62}$$

Since the spring force is a restoring force, it is always opposed to the displacement $u\,(t)$, hence we always have $F_{spring} = -k \cdot u\,(t)$. The acceleration is the second derivative of the displacement with respect to time (i.e., $\frac{d^2 u}{dt^2}$). Finally, Newton's law becomes

$$M\frac{d^2 u}{dt^2} + ku\,(t) = F\,(t) \tag{A.63}$$

The natural frequency of the single degree of freedom system is calculated as $\sqrt{k/M}$.

A.4.3 Equations of separable variables

Another important category of differential equations for which a general solution methodology exists are equations with separable variables. These are equations of the form

$$\frac{df}{dx} = h\,(f) \cdot g(x) \tag{A.64}$$

where h is a function of f, and g is a function of x. Note that separable variable equations are of second order and can be nonlinear. Examples of differential equations with separable variables are:

$$\frac{df}{dx} = f^2 \cdot x^2$$

$$\frac{df}{dx} = f^{-2} \cdot x^2$$

$$e^f \frac{df}{dx} = \frac{1}{x}$$

To solve this type of equation, we split the variables as follows:

$$\frac{df}{h(f)} = g(x)\,dx \tag{A.65}$$

and integrate the left-hand side with respect to f and the right-hand side with respect to x, as follows:

$$\int \frac{df}{h(f)} = \int g(x)\,dx + C \tag{A.66}$$

where C is a constant. Finally, we solve the resulting form for f. The example in Box A.9 presents an application to the isothermal, heterocatalytic curing of thermosetting resins (Kamal and Sourour 1973; Vergnaud and Bouzon 1992).

Box A.9: Example (Degree of cure for isothermal, heterocatalytic curing of thermosetting resins)

Curing of thermosetting resins is typically an exothermal reaction. The degree of cure is a function of time, with values in the range of $(0, 1)$. It is defined as the ratio of the heat generated by the exothermal reaction up to time t to the total amount of heat generated when the curing reaction is complete. For a fully complete reaction, we need $t \to \infty$.

Under isothermal conditions (constant temperature T_0 K), the heterocatalytic curing reaction of thermosetting resins can be modelled by an equation of the form

$$\frac{da}{dt} = Ae^{-Q/RT_0}(1-a)^n \tag{A.67}$$

where A is a constant, e^{-Q/RT_0} is the Arrhenius term, including the activation energy Q, temperature T_0, and the universal gas constant $R \cong 8.314$ J/K/mol.

Solution

The cure equation is of separable variables and can be written as

$$(1-a)^{-n}\,da = Ae^{-Q/RT_0}\,dt$$

Integrating with respect to time, we can distinguish between two cases:

(a) $n = 1$. By integrating the equation and adding the integration constant, we have

$$-\ln(1-a) = Ate^{-\frac{Q}{RT_0}} + C$$

so

$$a(t) = 1 - e^{-Ate^{-\frac{Q}{RT_0}}} + C \tag{A.68}$$

(b) $n \neq 1$. By integrating the equation and adding the integration constant, we have

$$\frac{1}{1-n}(1-a)^{-n+1} = Ate^{-\frac{Q}{RT_0}} + C$$

so

$$a(t) = 1 - (1-n)^{\frac{1}{1-n}}\left(Ate^{-\frac{Q}{RT_0}} + C\right)^{\frac{1}{1-n}} \tag{A.69}$$

The integration constant C can be calculated if the degree of cure at time instant $t = 0$ is known, i.e., $a(0) = a_0$. Apart from their significance for isothermal curing, the analytical solutions above can be used to design very accurate numerical integration formulae for nonisothermal curing (Vergnaud and Bouzon 1992; Papathanasiou and Tsamasphyros 2013).

A.4.4 Further reading

The interested reader can find more information on relevant topics in Stroud and Booth (2007), Yang (2018), Singh (2011), Polyanin and Zaitsev (2003), and Logan (2013).

A.5 Functions of several variables

A.5.1 Introduction

We introduce functions of several variables using the following two examples.

Example We need to measure the temperature inside a building. It is expected that the temperature will not be uniform throughout the building, but will change from storey to storey and room to room (Figure A.23). Even small temperature variations are expected from point to point, e.g., if we consider a location near an opening, a window, or a door. Furthermore, the temperature will change depending on the hour of the day and the month within the year, since lower temperatures are expected during winter. This example suggests that the temperature inside the building depends on the specific location examined and the specific time instant of measurement. Hence, temperature in this example is a function of the location of measurement and time. We can denote temperature as T and thus we have a function of several independent variables such as location x, y, z and time t:

$$\text{Temperature} = T(x, y, z, t)$$

Example We need to measure the flow velocity of water in a river (Figure A.24). The velocity values change according to the specific points where measurements are taken. Furthermore, they change with respect to time. However, the velocity at any point of the river is a vector and has three components, namely the velocity along direction x, the velocity along direction y, and the velocity along direction z. Each of these three velocity components is a function of the spatial coordinates and time.

We denote the velocity component in direction x as u, the velocity component in direction y as v, and the velocity component in direction z as w. Consequently, we have the following expressions for the various velocity components:

$$\text{Velocity along } x = u(x, y, z, t)$$
$$\text{Velocity along } y = v(x, y, z, t)$$
$$\text{Velocity along } z = w(x, y, z, t)$$

The velocity at each point is thus given by the following vector:

$$\text{Velocity} = u(x, y, z, t)\,\boldsymbol{i} + v(x, y, z, t)\,\boldsymbol{j} + w(x, y, z, t)\,\boldsymbol{k}$$

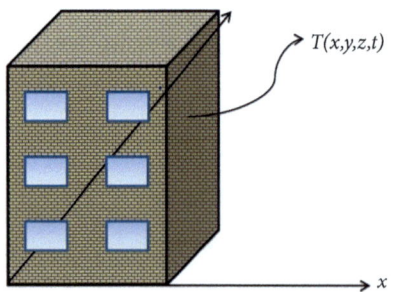

Figure A.23 Temperature distribution inside a building.

Figure A.24 Velocity of water flow inside a river. This photo shows water flowing in the River Avon in Bath (UK).

In this case, we have a function of several independent variables (x, y, z, t), with several components (u, v, w) that are functions of these independent variables. Such a function is called a **vector field**.

Our study thus focuses on the following two cases:

- Functions that have several variables in their domain and map them to a single value (**scalar function**), e.g., the temperature inside a three-dimensional solid:

$$\text{Temperature} = T(x, y, z, t)$$

- Functions that have several variables in their domain and map them to more than one value (**vector field**), e.g., the velocity of a three-dimensional flow:

$$\text{Velocity} = u(x, y, z, t)\,\boldsymbol{i} + v(x, y, z, t)\,\boldsymbol{j} + w(x, y, z, t)\,\boldsymbol{k}$$

where, $\boldsymbol{i}, \boldsymbol{j}, \boldsymbol{k}$ are the unit vectors along the x, y, z axes, respectively. When a scalar or vector field depends on time (variable t), it describes **transient phenomena**, that is, phenomena that evolve with respect to time. When a scalar or vector field does not depend on time it expresses **steady-state** conditions.

A.5.2 Partial derivatives

The concept of the derivative can be generalized to cover the case of functions that have several variables. In this case, under specific assumptions, we can define a derivative with respect to each of the independent variables. This quantity is called a **partial derivative**. Consider the following scalar function of three variables:

$$f(x, y, z) = 2x + 3y - z + xy + yz + zx$$

For this function, we can define the derivative with respect to x, the derivative with respect to y, and the derivative with respect to z. To calculate each one of these derivatives, we differentiate the function with respect to the specific variable, treating all other independent variables as constants. The partial derivatives with respect to each independent variable are denoted as $\frac{\partial f}{\partial x}$, $\frac{\partial f}{\partial y}$, and $\frac{\partial f}{\partial z}$.

Note that all the standard rules of differentiation apply to the partial derivatives as well.

For the function above, the partial derivatives are

$$\frac{\partial f}{\partial x} = 2 + y + z, \frac{\partial f}{\partial y} = 3 + x + x, \frac{\partial f}{\partial z} = -1 + y + x$$

A.5.3 Gradient and divergence

A useful quantity associated with scalar fields is the **gradient**. The gradient of a scalar function of several variables is a vector field. The components of this vector are the partial derivatives with respect to the respective variable. This means:

- For a problem in two dimensions, the gradient of a function of two independent variables $f(x, y)$ is

$$\text{grad} f = \frac{\partial f}{\partial x} i + \frac{\partial f}{\partial y} j \qquad (A.70a)$$

- For a problem in three dimensions, the gradient of a function of three independent variables $f(x, y, z)$ is

$$\text{grad} f = \frac{\partial f}{\partial x} i + \frac{\partial f}{\partial y} j + \frac{\partial f}{\partial z} k \qquad (A.70b)$$

The gradient vector of a scalar function at a specific location points to the direction of the greatest rate of increase of this function. The measure (magnitude or length) of the gradient is associated with the slope of the function at this location. The gradient of a scalar function at a specific point is a vector.

An important quantity associated with vector fields is the **divergence**. The divergence of a vector field is a scalar function of several variables. This scalar function is defined as the sum of the partial derivatives of the vector field components, where each partial derivative is with respect to the associated independent variable, as follows:

- For a problem in two dimensions characterized by the vector field $q(x, y, z) = u(x, y, z, t) i + v(x, y, z, t) j$, the divergence is given by the following expression:

$$\text{div} q = \frac{\partial u}{\partial x} + \frac{\partial v}{\partial y} \qquad (A.71a)$$

- For a problem in three dimensions characterized by the vector field $q(x, y, z) = u(x, y, z, t) i + v(x, y, z, t) j + w(x, y, z, t) k$:

$$\text{div} q = \frac{\partial u}{\partial x} + \frac{\partial v}{\partial y} + \frac{\partial w}{\partial z} \qquad (A.71b)$$

The divergence of a vector field at a specific location is a number, associated with the source intensity of the vector field at this specific location.

Given a scalar function $f(x, y, z)$, its gradient is the vector $\frac{\partial f}{\partial x} i + \frac{\partial f}{\partial y} j + \frac{\partial f}{\partial z} k$. This vector is denoted as $\text{grad} f$, or sometimes as ∇f, where ∇ is the symbol of the gradient operator $\frac{\partial}{\partial x} i + \frac{\partial}{\partial y} j + \frac{\partial}{\partial z} k$ acting on f. Since $\text{grad} f$ is a vector, we can consider its divergence, which is the following:

$$\text{div} (\text{grad} f) = \nabla^2 f = \frac{\partial^2 f}{\partial x^2} + \frac{\partial^2 f}{\partial y^2} + \frac{\partial^2 f}{\partial z^2} \qquad (A.72)$$

Table A.3 Examples of partial differential equations in civil engineering.

Equation	Application
Laplace $\nabla^2 u = 0$	• Potential fluid flow • Pure torsion of prismatic structural members
Poisson $\lambda \nabla^2 u = -f$	• Steady-state conduction of heat in a solid with thermal conductivity λ and distributed heat sources f • Steady-state Fickean mass diffusion with diffusivity constant λ and distributed mass sources f
Helmholtz $\nabla^2 u + k^2 u = 0$ where $k = \frac{\omega}{c}$ is the wave number, with ω the eigenfrequency and c the speed of sound	• Acoustic eigenfrequencies • Frequency domain analysis of wave propagation
Heat $\frac{\partial u}{\partial t} - a \nabla^2 u = f$	• Conduction of heat in a solid with thermal diffusivity a and distributed heat sources f • Mass diffusion in a solid with mass diffusivity a and distributed mass sources f
Wave $\frac{\partial^2 u}{\partial t^2} - c^2 \nabla^2 u = f$ where c is the speed of sound	• Vibration of thin membranes with distributed load, where $c = \sqrt{T/\rho h}$, T is the membrane tension, h is the membrane thickness, and ρ the material density • Pressure waves in solids, where $c = \sqrt{\left(K + \frac{4\mu}{3}\right)/\rho}$, K is the bulk modulus, μ is the shear modulus, and ρ the density • Small-amplitude waves in shallow water, where $c = \sqrt{gH}$, g is the acceleration of gravity, and H the depth
Biharmonic $D \nabla^2 (\nabla^2 u) = f$	• Bending of thin plates with flexural rigidity D and distributed load f • For $D = 1, f = 0$, stress state in plane problems of elasticity

which is the sum of the second partial derivatives with respect to each variable. This quantity is called the Laplacian of the function f and is an important concept in mechanics and physics. The function f in these subject areas is typically called a potential.

A.5.4 Partial differential equations in civil engineering

The Laplacian appears in many mathematical models that describe phenomena of importance in physics, mechanics, and engineering. Of particular interest in civil engineering are models of solid and structural mechanics, heat and mass diffusion, fluid flow and water waves, and acoustics. Specific examples are included in Table A.3.

In several applications of practical importance that feature rectangular domains, Laplace's equation can be solved using the method of separation of variables, or Fourier's method. To demonstrate the method, a two-dimensional example is considered in Box A.10.

Box A.10: Example (Laplace's equation)

Solve Laplace's equation $\nabla^2 u\,(x,y) = 0$ in the rectangular domain $[0,a] \times [0,b]$ such that

$$u\,(0,y) = 0, \quad u\,(a,y) = U, \quad u\,(x,0) = 0, \quad \frac{\partial u}{\partial y}\,(x,b) = 0$$

Solution

The equation is

$$\frac{\partial^2 u}{\partial x^2} + \frac{\partial^2 u}{\partial y^2} = 0$$

Assume that the solution can be written as the product of a function that depends only on x and a function that depends only on y, i.e., $u\,(x,y) = X(x) \cdot Y(y)$. Substituting into the equation, we have

$$Y\frac{\partial^2 X}{\partial x^2} + X\frac{\partial^2 Y}{\partial y^2} = 0$$

Dividing by $X \cdot Y$ and changing the partial derivatives to ordinary derivatives produces $\frac{1}{X}\frac{d^2 X}{dx^2} = -\frac{1}{Y}\frac{d^2 Y}{dy^2}$. We notice now that one side of the equation depends only on x, while the other side depends only on y. Since these are independent variables, this is only possible if each term is constant. In other words,

$$\frac{d^2 X}{dx^2} = -kX \text{ and } \frac{d^2 Y}{dy^2} = kY$$

These ODEs can be easily solved. We have the following cases:

- $k = 0$: $X(x) = C_1 x + C_0$, $Y(y) = D_1 y + D_0$
- $k = r^2 > 0$: $X(x) = C_1\cos(rx) + C_2\sin(rx)$, $Y(y) = D_1 e^{ry} + D_2 e^{-ry}$
- $k = -r^2 < 0$: $X(x) = C_1 e^{rx} + C_2 e^{-rx}$, $Y(y) = D_1\cos(ry) + D_2\sin(ry)$

Assume first that $k = 0$. Since $u\,(x,0) = 0$ we must have $D_0 = 0$. Since $\frac{\partial u}{\partial y}\,(x,b) = 0$ we therefore have $Y(y) = 0$, and the only possible solution is $u\,(x,y) = X(x) \cdot Y(y) = 0$. This is not feasible.

Similar considerations apply in the case where $k = r^2 > 0$ and $D_1 = D_2 = 0$. However, the final case (i.e. $k = -r^2$) is different and can lead to meaningful answers for the equations.

If $k = -r^2 < 0$, $Y(y) = D_1\cos(ry) + D_2\sin(ry)$, then $u\,(x,0) = 0$ implies that $D_1 = 0$. Then, we have

$$\frac{\partial Y}{\partial y}\,(x,b) = rD_2 \cos(rb) = 0$$

For this last equation to be true, it does not have to that $D_2 = 0$, provided that $\cos(rb) = 0$ and therefore

$$r_n = \frac{(2n + 1)\,\pi}{2b}, \quad n = 0, 1, 2, \ldots$$

Since there are infinite valid values r_n and the problem is linear, superposition implies that

$$u\,(x,y) = \sum_{n=0}^{\infty} \left(C_{1n}e^{r_n x} + C_{2n}e^{-r_n x}\right) D_{2n} \sin(r_n y) = \sum_{n=0}^{\infty} \left(E_{1n}e^{r_n x} + E_{2n}e^{-r_n x}\right) \sin(r_n y).$$

Since $u(0, y) = 0$ we must have $\sum_{n=0}^{\infty} (E_{1n} + E_{2n}) \sin(r_n y) = 0$ and therefore $E_{1n} = -E_{2n} = E_n/2$. The solution now becomes

$$u(x, y) = \sum_{n=0}^{\infty} E_n \sinh(r_n x) \sin(r_n y)$$

This is the solution in the form of a Fourier series. Finally, since $u(a, y) = U$, we have $\sum_{n=0}^{\infty} E_n \sinh(r_n a) \sin(r_n y) = U$. Multiplying both sides of this last equation by $\sin(r_m y)$ and integrating from $y = 0$ to $y = b$, we observe that

$$\int_0^b \sin(r_m y) \cdot \sin(r_n y)\, dy = \begin{cases} 0, & \text{if } m \neq n \\ \frac{b}{2}, & \text{if } m = n \end{cases}$$

The E_n constants are then calculated as $E_n \sinh(r_n a)\, b = 2U \int_0^b \sin(r_n y)\, dy$, and therefore

$$E_n = \frac{2U}{b \cdot \sinh(r_n a)} \int_0^b \sin(r_n y)\, dy = \frac{4 \cdot U}{(2n + 1) \cdot \pi \cdot \sinh(r_n a)}$$

Therefore, the solution of the Laplace equation becomes

$$u(x, y) = \sum_{n=0}^{\infty} \left[\frac{4 \cdot U}{(2n + 1) \cdot \pi \cdot \sinh(r_n a)} \right] \sinh(r_n x) \sin(r_n y)$$

This solution is shown in Figure A.25 by assuming $a = 1$, $b = 1$, and $U = 5$, using the MATLAB script in Code A.14.

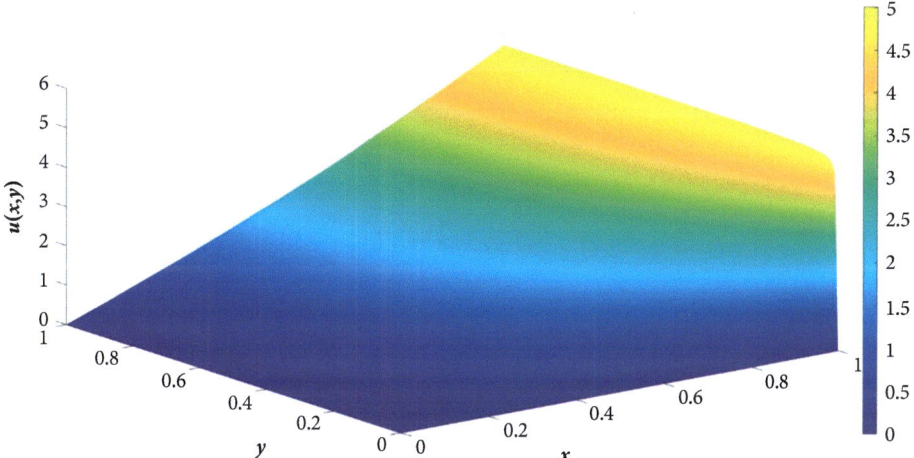

Figure A.25 Plot of the results of Laplace's equation solution in the domain $[0, 1] \times [0, 1]$ with boundary conditions $u(0, y) = 0$, $u(1, y) = 5$, $u(x, 0) = 0$, $\frac{\partial u}{\partial y}(x, 1) = 0$.

Code A.14 *MATLAB code to plot the results of the solution to Laplace's equation in the domain* $[0, 1] \times [0, 1]$ *with boundary conditions* $u(0, y) = 0$, $u(1, y) = 5$, $u(x, 0) = 0$, *and* $\frac{\partial u}{\partial y}(x, 1) = 0$.

```matlab
% this Matlab code plots the results of Laplace's
% equation solution in domain [0,1] × [0,1] for Figure A.25
clear all; clc; close all;

set(0,'defaultaxesfontsize',16);

U=5;
a=1;
b=1;
N=200; % Number of terms in Fourier series


[X,Y]=meshgrid(0:a/100:a,0:b/100:b);

u=4*U/(pi*sinh(pi*a/(2*b)))*sinh(pi/(2*b)*X).*sin(pi/(2*b)*Y);

for n=1:N

    r=(2*n+1)*pi/(2*b);

    u=u+4*U/((2*n+1)*pi*sinh(r*a))*sinh(r*X).*sin(r*Y);

end

figure(1)
surf(X,Y,u)
colorbar
shading interp
xlabel('x','FontSize',18,'FontWeight','bold')
ylabel('y','FontSize',18,'FontWeight','bold')
zlabel('u(x,y)','FontSize',18,'FontWeight','bold')
```

A.5.3 Further reading

The interested reader can find more information on relevant topics in Stroud and Booth (2007), Yang (2018), Singh (2011), Logan (2013), Polyanin (2002), and Asmar (2005).

A.6 Sets of discrete and continuous data

In the following, we define the concepts of **data** and **variables**.

- **Data** are sets of recorded observations or values.
- **Variables** are quantities that can have a number of values, like the number of people inside a bus. This number is different every day of the week.

Data can be discrete or continuous. **Discrete data** are data that can be counted or have a fixed set of values. Examples of discrete data are:

- number of people in a room
- number of phone calls during a specific time interval
- number of views of a specific webpage
- number of cars passing a specific traffic light during one month.

Table A.4 Tally diagram and frequency or relative frequency values for a discrete dataset.

Variable (number of people in the bus, x)	Tally marks (number of days within the 30-day period that x people were in the bus)	Frequency (number of days within the 30-day period that x people were in the bus, f)	Relative frequency (percentage of relative frequency)
4	//	2	$2/30 = 0.0667 = 6.67\%$
5	//// ///	8	$8/30 = 0.2667 = 26.67\%$
6	//// ////	10	$10/30 = 0.3333 = 33.33\%$
7	//// ////	9	$9/30 = 0.3000 = 30.00\%$
8	/	1	$1/30 = 0.0333 = 3.33\%$
	Sum:	30	1 (100%)

Continuous data are data that take values on a continuous scale. Examples of continuous data are:

- heights of people
- magnitude of loading actions on a structure
- amplitude of water waves
- average height of rainfall in the UK during 2016.

A.6.1 Sets of discrete data

Discrete data can be conveniently arranged in the form of tables using the **Tally diagram**. Consider the following example.

Example For statistical purposes regarding bus routes, the number of people on a late evening bus route to the airport is measured for 30 days. The values measured are presented in Table A.4. Note that the total sum of the numbers in the frequency column is the total number of measurements that were taken. For example, the number of people inside the bus was monitored for 30 days, one route per day. This is a total of 30 measurements. The **relative frequency** of a specific event (e.g., four people in the bus) is the ratio of the times that this event occurred to the total number of observations. In the above case, on two evenings there were four people in the bus, so the relative frequency of this event (four people in a bus) is $2/30 = 0.0667 = 6.67\%$. The sum of all the numbers in the relative frequency column is always 1, or equivalently 100%.

Another convenient way to represent discrete data is the frequency histogram. The frequency histogram corresponding to the above example is presented in Figure A.26.

For every set of discrete data, we can define certain measures of central tendency. **Measures of central tendency** are indicators of the most probable value that will occur during some observations. There are three basic measures of central tendency, the **mean**, the **mode**, and the **median**, defined as:

- The **mean** is the average of a set of observations, defined as

$$\bar{x} = \frac{\text{sum of all observed values}}{\text{number of observations}} = \frac{\sum fx}{\sum f} \tag{A.73}$$

- The **mode** is the value of the variable that occurs most often.
- The **median** is the value of the middle term when all the observations are arranged in ascending or descending order.

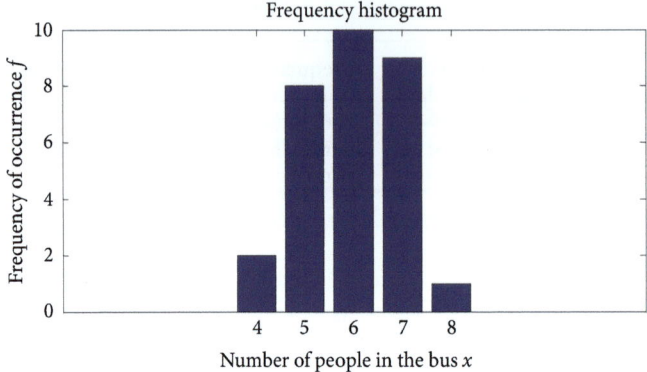

Figure A.26 Histogram of the data in Table A.4.

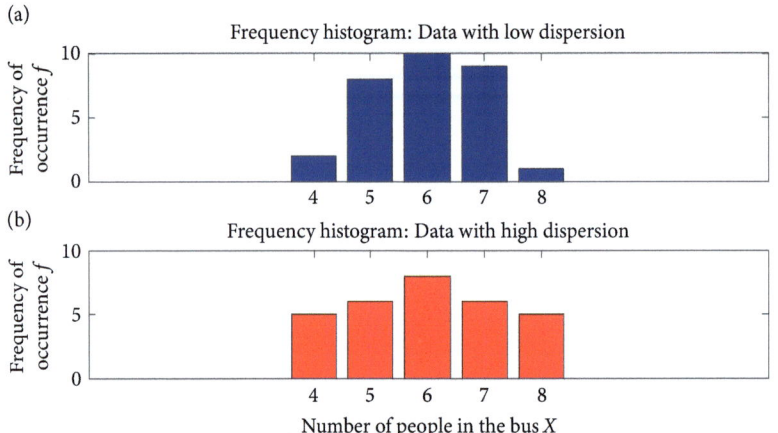

Figure A.27 Histograms of data with (a) low dispersion and (b) high dispersion.

Apart from measures of central tendency, for a set of data we can also calculate **measures of dispersion**. These indicate how scattered the data are with respect to the most probable value. Figure A.27a shows data with low dispersion, and Figure A.27b shows data with large dispersion, which would correspond to another set of measurements (e.g., 30 days of another month).

The basic measures of dispersion are the **spread** and the **standard deviation**, defined as:

- **Spread** is the difference between the highest and the lowest value in a set of observations.
- **Standard deviation** is denoted by σ and is defined as

$$\sigma = \sqrt{\frac{\sum f \cdot (x - \bar{x})^2}{\sum f}} \tag{A.74}$$

where the quantity $\frac{\sum f \cdot (x-\bar{x})^2}{\sum f}$ is called the **variance**.

An example of calculating measures of central tendency and dispersion for discrete data is shown in Box A.11.

Box A.11: Example

Calculate the measures of central tendency and dispersion for the data in Table A.4 indicating number of people on the bus route to the airport.

Solution

The **mean** is

$$\bar{x} = \frac{2 \cdot 4 + 8 \cdot 5 + 10 \cdot 6 + 9 \cdot 7 + 1 \cdot 8}{2 + 8 + 10 + 9 + 1} = \frac{179}{30} = 5.97$$

The **mode** is the value occurring most often. Based on the data in Table A.4 the mode is 6, which occurs 10 times.

The **median** is found by arranging the numbers in ascending order:

$$4, 4, 5, 5, 5, 5, 5, 5, 5, 5, 6, 6, 6, 6, 6, 6, 6, 6, 6, 6, 7, 7, 7, 7, 7, 7, 7, 7, 7, 8$$

$$\text{median} = \frac{6 + 6}{2} = 6$$

The **spread** is the difference between the highest and the lowest values in a set of observations. Based on the data in Table A.4, we have: 8 – 4 = 4.

The **standard deviation** is calculated as

$$\sigma = \sqrt{\frac{2 \cdot (4 - 5.97)^2 + 8 \cdot (5 - 5.97)^2 + 10 \cdot (6 - 5.97)^2 + 9 \cdot (7 - 5.97)^2 + 1 \cdot (8 - 5.97)^2}{2 + 8 + 10 + 9 + 1}} = 0.97$$

For calculating the median, if there is only one number in the middle, then the median is this number. The mean is always larger than the minimum value observed and less than the maximum value observed.

A.6.2 Sets of continuous data

Continuous data are data that take values on a continuous scale. Continuous data can be represented using histograms if we define specific data classes. Consider an example where the maximum height of waves in a harbour at a specific sea state are measured. The results are presented in Table A.5.

Note that the total sum of the numbers in the frequency column is the total number of values that occurred. The **relative frequency** of a specific event (e.g., maximum value of wave height between 0 m and 0.5 m) is the ratio of the times that this event occurred to the total number of events. In the above case, the average height of waves was in the interval 0–0.5 m twice. Therefore, the relative frequency of this event is 2/38 = 0.0526 = 5.26%. The sum of all the numbers in the relative frequency column is always 1, or equivalently 100%.

Another convenient way to represent discrete data is the frequency histogram. In the case of continuous data, the frequency histogram is plotted using bars with a width that corresponds to the width of the data class intervals, placed symmetrically with respect to the data class average value x_m. The frequency histogram corresponding to the above example is shown in Figure A.28.

If the frequency value at the centre of each class interval is connected to the respective value at the centre of the following class interval using a line segment, then a **frequency polygon** is constructed. The frequency polygon corresponding to the histogram in Figure A.28 is plotted in Figure A.29. It is expected that the more data classes used, the smoother the frequency polygon becomes. This means that if more data classes are used, the width of each rectangle in Figure A.29 will be smaller. Consequently, more line segments, each one having smaller width, will appear. In the limit of a very large number of classes, the polygon will tend to a smooth curve.

Table A.5 Tally diagram and frequency or relative frequency values for a continuous dataset.

Variable (maximum height of waves, x)	Average value in the interval (x_m)	Frequency (f)	Relative frequency (percentage of relative frequency)
0–0.5 m	(0.5+0)/2 = 0.25 m	2	2/38 = 0.0526 = 5.26%
0.5–1 m	(1+0.5)/2 = 0.75 m	5	5/38 = 0.1316 = 13.16%
1–1.5 m	(1+1.5)/2 = 1.25 m	12	12/38 = 0.3158 = 31.58%
1.5–2 m	(1.5+2)/2 = 1.75 m	9	9/38 = 0.2368 = 26.38%
2–2.5 m	(2.5+2)/2 = 2.25 m	6	6/38 = 0.1579 = 15.79%
2.5–3 m	(3+2.5)/2 = 2.75 m	3	3/38 = 0.0789 = 7.89%
3–3.5 m	(3.5+3)/2 = 3.25 m	1	1/38 = 0.0263 = 2.63%
Sum:		**38**	**1 (100%)**

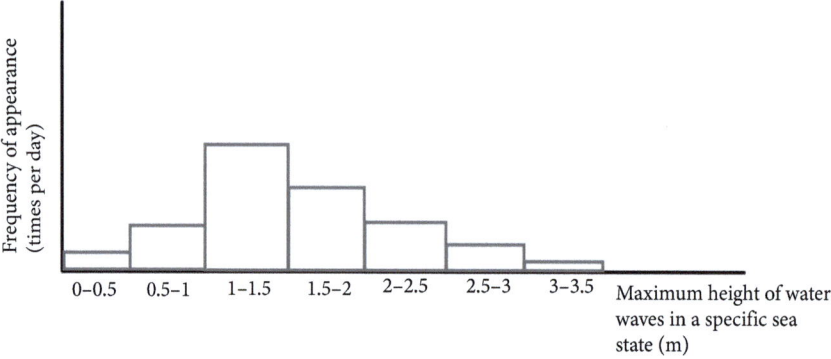

Figure A.28 Frequency histogram for continuous data analysis.

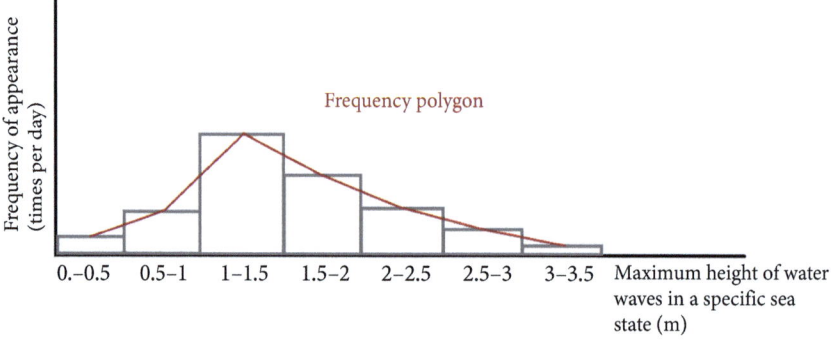

Figure A.29 Frequency histogram and frequency polygon for continuous data analysis.

For sets of continuous data (similarly to the case of discrete data), we can define certain measures of central tendency, i.e., indicators of the most probable value that will occur during some observations. There are three basic measures of central tendency, the **mean**, the **mode**, and the **median**.

- The **mean** is the average of a set of observations, defined as

$$\bar{x} = \frac{\text{sum of all observed values}}{\text{number of observations}} = \frac{\sum f x_m}{\sum f} \tag{A.75}$$

where Σ denotes the sum and $\sum f = n$ is the total number of observations.

- The **mode** is the value of the variable that occurs most often. The class that has the highest frequency is the **modal class** (in the example above it is class 3, which is 1–1.5 m). Having identified the modal class, the mode can be calculated using the following formula:

$$\text{mode} = L + \left(\frac{a}{a + b}\right)W \qquad (A.76)$$

where L is the lower boundary of the modal class, W is the modal class interval width, a is the height difference between the modal class and the previous class, and b is the height difference between the modal class and the next class.

- The **median** is the value of the variable that divides the histogram into two equal areas.

Apart from measures of central tendency, for a set of data we can also calculate **measures of dispersion**. The measures of dispersion indicate how scattered the data are with respect to the most probable value.

The most commonly used measure of dispersion is the **standard deviation** (σ), defined as

$$\sigma = \sqrt{\frac{\sum f \cdot (x_m - \bar{x})^2}{\sum f}} \qquad (A.77)$$

The quantity $\frac{\sum f \cdot (x_m - \bar{x})^2}{\sum f}$ is called the variance.

An example of calculating measures of central tendency and dispersion for continuous data is shown in Box A.12.

Box A.12: Example

Calculate the measures of central tendency and dispersion for the data indicating wave height in the harbour (see Table A.5).

Solution

The **mean** is

$$\bar{x} = \frac{0.25 \cdot 2 + 0.75 \cdot 5 + 1.25 \cdot 12 + 1.75 \cdot 9 + 2.25 \cdot 6 + 2.75 \cdot 3 + 3.25 \cdot 1}{2 + 5 + 12 + 9 + 6 + 3 + 1} = \frac{60}{38} = 1.58\,m$$

The **mode** is the value occurring most often. In this case it is

$$\text{mode} = 1 + \left(\frac{7}{7 + 3}\right) \cdot 0.5 = 1 + \frac{7}{10} \cdot 0.5 = 1 + 0.35 = 1.35\,m$$

The **median** can be found by adding the areas of the rectangles and seeing which value divides the total area in two:

Area 1 = 0.5 × 2 = 1,   Area 2 = 0.5 × 5 = 2.5,   Area 3 = 0.5 × 12 = 6, Area 4 = 0.5 × 9 = 4.5,

Area 5 = 0.5 × 6 = 3,   Area 6 = 0.5*3 = 1.5,   Area 7 = 0.5 × 1 = 0.5

Note that Area 1 + Area 2 + Area 3 = 1 + 2.5 + 6 = 9.5 and Area 4 + Area 5 + Area 6 + Area 7 = 4.5 + 3 + 1.5 + 0.5 = 9.5

This means the median is the value at the interface of Area 3 and Area 4. Consequently, the median is 1.5 m.

Box A.12 *Continued*

The **standard deviation** is

$$
\sigma = \sqrt{\frac{\begin{array}{c} 2 \cdot (0.25 - 1.58)^2 + 5 \cdot (0.75 - 1.58)^2 + 12 \cdot (1.25 - 1.58)^2 + \\ 9 \cdot (1.75 - 1.58)^2 + 6 \cdot (2.25 - 1.58)^2 \\ + 3 \cdot (2.75 - 1.58)^2 + 1 \cdot (3.25 - 1.58)^2 \end{array}}{2 + 5 + 12 + 9 + 6 + 3 + 1}} = 0.69 \ m.
$$

In this last example, the median happened to coincide with the interface value of two classes. In general, the median might be inside a certain class. In this case, the following procedure is followed:

We assume that the median has the value m. The area of the specific class is then the sum of two different rectangles, as follows:

$$
M_1 = (m - \text{lower bound of the class}) \times \text{frequency of the class}
$$

and

$$
M_2 = (\text{upper bound of the class} - m) \times \text{frequency of the class}
$$

- We add M_1 to the areas of the classes located to the left of the class of the median to obtain the total area left of the median, A_{left}. We add to the areas of the classes located to the right with respect to the class of the median, to get the total area on the right of the median (A_{right}).
- Since the median divides the total area into to equal parts, we have the equation $A_{left} = A_{right}$.
- Finally, we solve the above equation for m.

A.6.3 The normal distribution

In several cases of continuous datasets that include many narrow-banded classes, the frequency polygon tends to become a smooth, bell-like curve as shown in Figure A.30. This curve is known as the normal distribution curve and is symmetrical about its centre line. The centre line for the normal distribution curve coincides with the mean value of the observations $(\bar{x})$.

It can be shown that specific percentages of the total number of observations are confined in intervals around the mean value that are related to the standard deviation (σ). As illustrated in Figure A.30, 99.7% of all observations are included in the interval $(\bar{x} - 3\sigma, \bar{x} + 3\sigma)$.

The normal distribution curve corresponds to the following function:

$$
f(x) = \frac{1}{\sigma\sqrt{2\pi}} e^{-\frac{(x - \bar{x})^2}{2\sigma^2}} \tag{A.78}
$$

Setting $\frac{x-\bar{x}}{\sigma=1} = z$ and $\sigma = 1$, the standard normal distribution is produced as follows:

$$
f(z) = \frac{1}{\sqrt{2\pi}} e^{-\frac{z^2}{2}} \tag{A.79}
$$

The standard normal distribution has mean value equal to zero and probability density equal to $f(z)$. Consequently, the probability of an event in a sample space defined by the interval $[a, b]$ is given by

$$
P(a \leqslant Z \leqslant b) = \int_a^b f(z) \, dz \tag{A.80}
$$

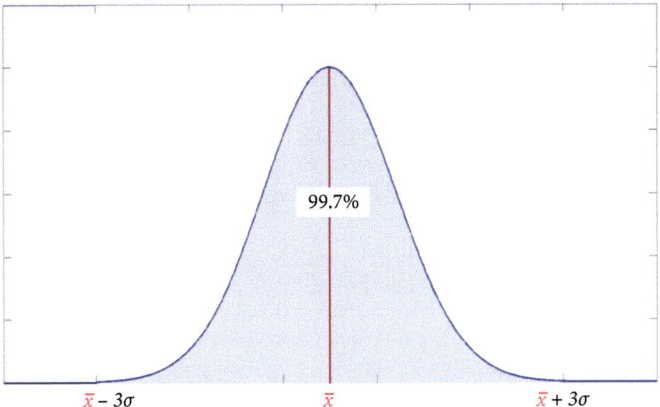

99.7%

$\bar{x} - 3\sigma$ $\qquad\qquad\qquad$ $\bar{x}$ $\qquad\qquad\qquad$ $\bar{x} + 3\sigma$

Figure A.30 Normal distribution showing the percentage of observed values in specific intervals. Here, $\bar{x}$ is the mean value and σ is the standard deviation.

Probability is a measure of the likelihood of a specific event occurring within trials or experiments conducted under prescribed conditions.

The integral in the formula above for the standard normal distribution can be found tabulated for different values of the integration limits, or can be calculated using MATLAB with the following commands (see also Appendix B):

```
>> syms z
>> int(f(z),z,a,b)
```

The MATLAB script provided in Code A.15 calculates the probability using the standard normal distribution for the following expression:

$$P(2 \leq Z \leq 3) = \int_{2}^{3} \frac{1}{\sqrt{2\pi}} e^{-\frac{z^2}{2}} dz$$

The result is approximately 0.0214 or 2.14%.

Code A.15 *MATLAB code for the probability calculation using the standard normal distribution for*

$P(2 \leq Z \leq 3) = \int_{2}^{3} \frac{1}{\sqrt{2\pi}} e^{-\frac{z^2}{2}} dz.$

```
clear; clc;
close all;
%
syms z
P= int(1/sqrt(2*pi)*exp(-z^2/2),z,2,3);
%
% Convert fraction to decimal
decimal_approximation = double(P);
disp(decimal_approximation);
```

A.6.4 Further reading

The interested reader can find more information on relevant topics in Stroud and Booth (2007), Spiegel and Stephens (2014), and Ross (2021).

A.7 Interpolation of functions and curve fitting

A.7.1 Piecewise linear Lagrange interpolation

In the case where a mathematical model (e.g., differential equations) is used for the simulation of a physical process, the solution pursued is a function. On several occasions, the solution function can have a complicated form. In such cases it is desirable to approximate these functions using simpler ones that are easier to handle, e.g., polynomials. This approximation process is called **interpolation.** When polynomial functions are used for interpolation, the process is termed **polynomial interpolation**.

The type of polynomial interpolation most often employed is **Lagrange interpolation**. Lagrange interpolation is based on equating the function values at specific locations (called the **nodes**) with the value of the approximation polynomial. This procedure leads to the formulation of equations that can be used for the determination of appropriate polynomial coefficients. The case of **piecewise linear Lagrange interpolation** is studied in detail here.

Consider the function $u(x)$ with the graph shown in Figure A.31, where its values at specific locations, called 'nodes', are marked. Using only a few nodes, the piecewise linear interpolation (the sequence of straight lines) is not a good approximation. However, increasing the number of nodal interpolation points increases the quality of the approximation. In the following, the form of the linear interpolation between two neighbouring nodes will be investigated.

Consider the function $u(x)$ between the nodal coordinates x_A and x_B, where $x_A < x_B$ (Figure A.32). The nodal values of the function are $u(x_A) = u_A$ and $u(x_B) = u_B$. The general form of a linear function approximating $u(x)$ between the nodal coordinates x_A and x_B is as follows:

$$\bar{u}(x) = ax + b \tag{A.81}$$

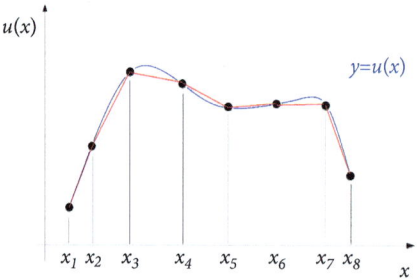

Figure A.31 Piecewise linear interpolation of a function $u(x)$.

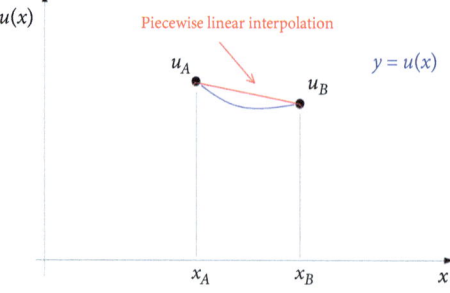

Figure A.32 Linear interpolation between two successive nodes.

If we impose the condition that the two functions are equal at the nodes, two equations are derived:

$$u_A = ax_A + b \text{ and } u_B = ax_B + b$$

Using these two equations, we can express the approximation function in terms of the nodal coordinates x_A and x_B and the nodal values u_A and u_B as $\bar{u}(x) = \frac{u_B - u_A}{x_B - x_A} x + u_A - \frac{u_B - u_A}{x_B - x_A} x_A$.

Upon rearranging the terms in this last equation, we obtain the following:

$$\bar{u}(x) = \left(1 - \frac{x - x_A}{x_B - x_A}\right) u_A + \left(\frac{x - x_A}{x_B - x_A}\right) u_B \tag{A.82}$$

which is the interpolation function written in terms of the nodal values u_A and u_B and the following linear Lagrange polynomials:

$$N_A(x) = \left(1 - \frac{x - x_A}{x_B - x_A}\right) \text{ and } N_B(x) = \frac{x - x_A}{x_B - x_A}$$

Note that the Lagrange polynomials are equal to 1 at the associated node and are 0 at the other node:

$$N_A(x_A) = 1 \text{ and } N_A(x_B) = 0$$

$$N_B(x_A) = 0 \text{ and } N_B(x_B) = 1$$

Furthermore, the sum of the Lagrange polynomials produced is always equal to 1:

$$N_A(x) + N_B(x) = 1 \tag{A.83}$$

This is called the **partition of unity** property. Finally, note that polynomials of higher degree (e.g., quadratic, cubic, etc.) can be used for the Lagrange interpolation of functions.

An example linear interpolation is given in Box A.13.

Box A.13: Example

Find the linear interpolation function for $u(x) = e^x$ in the interval $x = 0$ to $x = 0.5$. Find the relative error of the approximation at the points $x = 0.1$ and $x = 0.25$.

Solution

We set $x_A = 0$ and $x_B = 0.5$ We then have $u_A = u(x_A) = u(0) = e^0 = 1$ and $u_B = u(x_B) = u(0.5) = e^{0.5} \cong 1.6487$.

The linear interpolation in the interval from $x = 0$ to $x = 0.5$ is given by

$$\bar{u}(x) = \left(1 - \frac{x - 0}{0.5 - 0}\right) 1 + \left(\frac{x - 0}{0.5 - 0}\right) 1.6487$$

which can be rearranged to

$$\bar{u}(x) = \left(1 - \frac{x}{0.5}\right) + \left(\frac{x}{0.5}\right) 1.6487 = 1 + 0.6487 \frac{x}{0.5}$$

A plot of the exponential function and the linear interpolation in the interval from $x = 0$ to $x = 0.5$ is shown in Figure A.33.

We now calculate the relative error of the approximation using the interpolation function. At, $x = 0.1$ we have

$$\frac{\bar{u}(0.1) - u(0.1)}{u(0.1)} = \frac{1 + 0.6487 \frac{0.1}{0.5} - e^{0.1}}{e^{0.1}} = 0.0222, \text{ or } 2.22\%$$

Box A.13 *Continued*

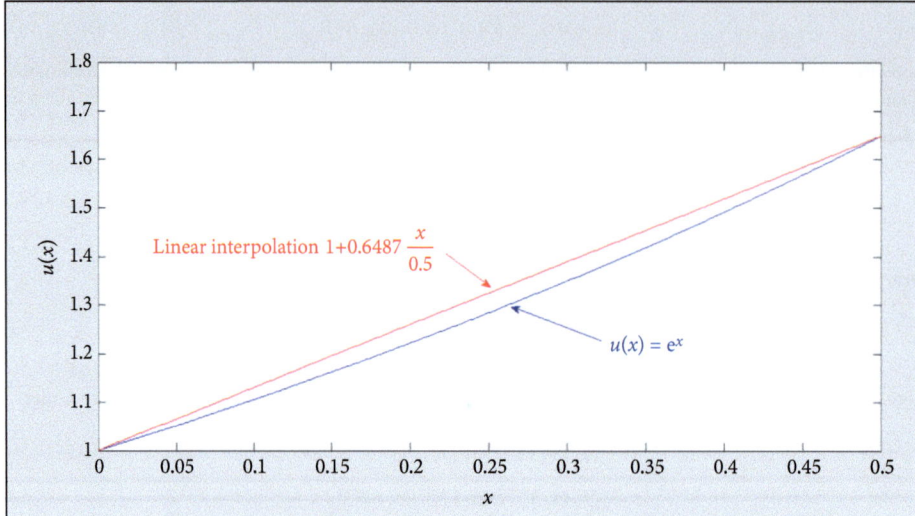

Figure A.33 Plots of the exponential function (lower line) and its interpolation (upper line).

At, $x = 0.25$ the error is

$$\frac{\bar{u}\,(0.25) - u\,(0.25)}{u\,(0.25)} = \frac{1 + 0.6487\frac{0.25}{0.5} - e^{0.25}}{e^{0.25}} = 0.0314, \text{ or } 3.14\%$$

We observe that the relative error is small. Hence, the linear interpolation provides a good approximation of the exponential function in the selected interval.

A.7.2 Curve fitting and the least squares method

It is common in practice to obtain and record measurements of various phenomena. Engineers use these data to understand the properties of systems. However, it is not always easy to determine the relationship that describes the behaviour of the system under investigation based on the data collected.

Regression analysis can be used to identify the relation among variables. Regression analysis includes many techniques. One of the most used is the curve fitting technique. Through curve fitting, we attempt to find a function that can model the measurements as accurately as possible.

The aim is to identify a function that will model all the data collected with the smallest possible overall error. Several functions can be used for curve fitting. Typically, polynomial, exponential, and power functions are employed.

Method of least squares

Here, we introduce a simple example to demonstrate how curve fitting works with the method of least squares. Assume that a dataset containing five measurements is given. The measurements are given as pairs of x and y values. Assume that the pairs are: (1,1), (2,2), (4,2), (5,3), and (6,4). Although an increasing trend can be identified in the small dataset of the example, the relationship between the x and y values is not apparent. The goal is to try and find a line that can describe as accurately as possible the given dataset. The question is, how can we obtain a good fit that is at the same time simple enough?

The simplest possible curve that we could use is a line with the general equation $y = ax + b$. The goal now is to calculate the constants a and b in an optimum manner. The **method of least squares** can be used to achieve this goal. The method of least squares provides the coefficients of the linear function

$y = ax + b$ that best approximates all the points in a set in an overall sense. For that, first we need to denote the error related to each of the points in the set.

The method of least squares states that, given a set of n measured data for the pair of quantities (x, y), that is, the residuals (r_i) from using a linear equation to approximate the measured data set are

$$r_1 = y_1 - (ax_1 + b)$$
$$r_2 = y_2 - (ax_2 + b)$$
$$\vdots$$
$$r_n = y_n - (ax_n + b)$$

The sum of the squares of all the residuals is then calculated:

$$I = [y_1 - (ax_1 + b)]^2 + [y_2 - (ax_2 + b)]^2 + [y_3 - (ax_3 + b)]^2$$
$$+ \cdots + [y_n - (ax_n + b)]^2 = \sum_{i=1}^{n} [y_n - (ax_n + b)]^2. \tag{A.72}$$

This sum, which is related to the total error of the approximation, is a minimum when its partial derivatives with respect to a and with respect to b are zero, i.e.,

$$\frac{\partial I}{\partial a} = 0 \text{ and } \frac{\partial I}{\partial b} = 0 \tag{A.73}$$

The partial derivatives are

$$\frac{\partial I}{\partial a} = \frac{\partial [y_1 - (ax_1 + b)]^2}{\partial a} + \frac{\partial [y_2 - (ax_2 + b)]^2}{\partial a} + \frac{\partial [y_3 - (ax_3 + b)]^2}{\partial a} + \cdots + \frac{\partial [y_n - (ax_n + b)]^2}{\partial a}$$
$$= -2x_1 [y_1 - (ax_1 + b)] - 2x_2 [y_2 - (ax_2 + b)] - 2x_3 [y_3 - (ax_3 + b)] - \cdots - 2x_n [y_n - (ax_n + b)]$$

and

$$\frac{\partial I}{\partial b} = \frac{\partial [y_1 - (ax_1 + b)]^2}{\partial b} + \frac{\partial [y_2 - (ax_2 + b)]^2}{\partial b} + \frac{\partial [y_3 - (ax_3 + b)]^2}{\partial b} + \cdots + \frac{\partial [y_n - (ax_n + b)]^2}{\partial b}$$
$$= -2 [y_1 - (ax_1 + b)] - 2 [y_2 - (ax_2 + b)] - 2 [y_3 - (ax_3 + b)] - \cdots - 2 [y_n - (ax_n + b)]$$

Hence, we have

$$-2x_1 [y_1 - (ax_1 + b)] - 2x_2 [y_2 - (ax_2 + b)] - 2x_3 [y_3 - (ax_3 + b)] - \cdots - 2x_n [y_n - (ax_n + b)] = 0$$

and

$$-2 [y_1 - (ax_1 + b)] - 2 [y_2 - (ax_2 + b)] - 2 [y_3 - (ax_3 + b)] - \cdots - 2 [y_n - (ax_n + b)] = 0$$

This is a set with two equations for the two unknowns a, b. Upon defining the mean values,

$$\bar{x} = \frac{1}{n} \left(\sum_{i=1}^{n} x_i \right) \text{ and } \bar{y} = \frac{1}{n} \left(\sum_{i=1}^{n} y_i \right) \tag{A.74}$$

the two constants a, b are calculated as follows:

$$a = \frac{\sum_{i=1}^{n} (x_i - \bar{x}) \cdot (y_i - \bar{y})}{\sum_{i=1}^{n} (x_i - \bar{x})^2} \tag{A.75}$$

$$b = \bar{y} - a \cdot \bar{x} \tag{A.76}$$

For the case of the dataset with five pairs of (1,1), (2,2), (4,2), (5,3), and (6,4), the MAT-LAB script provided in Code A.16 calculates the coefficients a and b for a fitting line of the form $y = ax + b$.

Code A.16 *MATLAB code to fit a line for a dataset of five pairs of (1,1), (2,2), (4,2), (5,3), and (6,4) using the least squares approach. The code calculates the coefficients a and b for the fitting line of the form $y = ax + b$.*

```matlab
% Least Squares Fit to a data set

clear all;clc;close all;

set(0,'defaultaxesfontsize',16);

% x data coordinates
x=[1 2 4 5 6];
% y data coordinates
y=[1 2 2 3 4];
% mean values
xm=mean(x); ym=mean(y);
% least squate solution
a=sum((x-xm).*(y-ym))/sum((x-xm).^2); b=ym-a*xm;

figure(1)
plot(x,y,'Ob',x,a*x+b,'k-','linewidth',1.5)
hold on
plot([x(1) x(1)],[y(1) a*x(1)+b],'k--','linewidth',1)
plot([x(2) x(2)],[y(2) a*x(2)+b],'k--','linewidth',1)
plot([x(3) x(3)],[y(3) a*x(3)+b],'k--','linewidth',1)
plot([x(4) x(4)],[y(4) a*x(4)+b],'k--','linewidth',1)
plot([x(5) x(5)],[y(5) a*x(5)+b],'k--','linewidth',1)
hold off
xlabel('x','FontSize',18,'FontWeight','bold')
ylabel('y','FontSize',18,'FontWeight','bold')
```

When executed, Code A.16 will produce Figure A.34, where the fitted line is plotted as a solid line, the dataset points as circles, and the distances between the points and the fitted line with dashed lines.

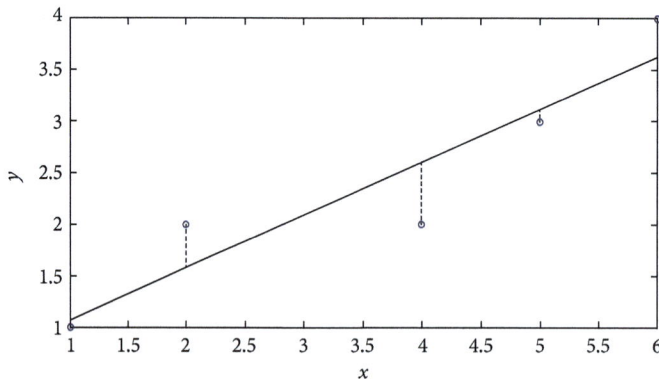

Figure A.34 Least squares fitting (solid line) of the form $y = ax + b$ for the dataset (1,1), (2,2), (4,2), (5,3), and (6,4). The distances are denoted with dashed lines.

In the case where the data correspond not to a linear relation but to a power law of the form

$$y = ax^b \tag{A.77}$$

there is an easy way to apply the least squares method like we did in the previous example. In this case, we use the natural logarithm to get $\ln(y) = \ln\left(ax^b\right)$. This implies:

$$\ln(y) = \ln(a) + b\ln(x) \tag{A.78}$$

By setting

$$\ln(y) = Y$$

$$\ln(a) = A$$

$$\ln(x) = X$$

we have the equation $Y = A + bX$, where we can apply the least squares method as before.

A.7.3 Further reading

The interested reader can find more information on relevant topics in Stroud and Booth (2007).

References

Asmar, N.H. (2005). *Partial Differential Equations with Fourier Series and Boundary Value Problems*, 2nd edn. Pearson Prentice Hall.

Boyce, W.E., and DiPrima, R.C. (1992). *Elementary Differential Equations*, 5th edn. John Wiley.

Horn, P., and Johnson, C. (1985). *Matrix Analysis*. Cambridge University Press.

Kamal, M.R., and Sourour, S. (1973). Kinetics and Thermal Characterization of Thermoset Cure. *Polymer Engineering and Science*, 13(1):59–64.

Logan, J.D. (2013). *Applied Mathematics*, 4th edn. John Wiley & Sons, Inc

Papathanasiou, T.K., and Tsamasphyros, G.J. (2013). On Vergnaud's time integration method for autocatalytic cure rate equations. *Applied Mathematics and Computation*, 220:748–755.

Polyanin, A.D. (2002). *Handbook of Linear Partial Differential Equations for Engineers and Scientists*. Chapman & Hall/CRC.

Polyanin, A.D., and Zaitsev, V.F. (2003). *Handbook of Exact Solutions for Ordinary Differential Equations*, 2nd edn. Chapman & Hall/CRC.

Ross, S.M. (2021). *Introduction to Probability and Statistics for Engineers and Scientists*, 6th edn. Academic Press.

Singh, K. (2011). *Engineering Mathematics Through Applications*, 2nd edn. Palgrave Macmillan.

Spiegel, M.R., and Stephens, L.J. (2014). *Schaum's Outline of Statistics*, 5th edn. McGraw-Hill Education.

Stoll, R.R. (1963). *Set Theory and Logic*. W.H. Freeman & Co.

Stroud, K.A, and Booth, D.J. (2007). *Engineering Mathematics*, 6th edn. Palgrave Macmillan.

Vergnaud, J.M., and Bouzon, J. (1992). *Cure of Thermosetting Resins: Modelling and Experiments*. Springer.

Yang, X.S. (2018). *Mathematics for Civil Engineers: An Introduction*, Dunedin Academic Press.

An Introduction to MATLAB

B.1 Introduction to MATLAB

MATLAB (MATrix LABoratory) is a powerful interactive tool for numerical computations. It allows users to program their own algorithms by creating scripts that can execute powerful operations using only a few commands. MATLAB's strengths include its rather user-friendly graphical interface, its numerous powerful toolboxes for performing various types of mathematical analyses, and its simple programming language which has made it an attractive tool for engineers (e.g., Heidarzadeh et al. 2017; Papathanassiou et al. 2011).

In this chapter it is assumed that MATLAB is already installed on the reader's computer; otherwise, the reader is advised to contact their IT department to get MATLAB installed.

B.1.1 Getting started

MATLAB can be launched by finding the software icon from the Start Menu of the PC or by finding the MATLAB shortcut (Figure B.1) and double-clicking on it.

After the software loads, the MATLAB interface, called the MATLAB desktop, appears as shown in Figure B.2. The MATLAB interface consists of smaller windows that provide access to certain actions. In general, the MATLAB desktop may initially appear different depending on the default settings or

Figure B.1 MATLAB shortcut (Mathworks®); the program can be initiated by double-clicking on it.

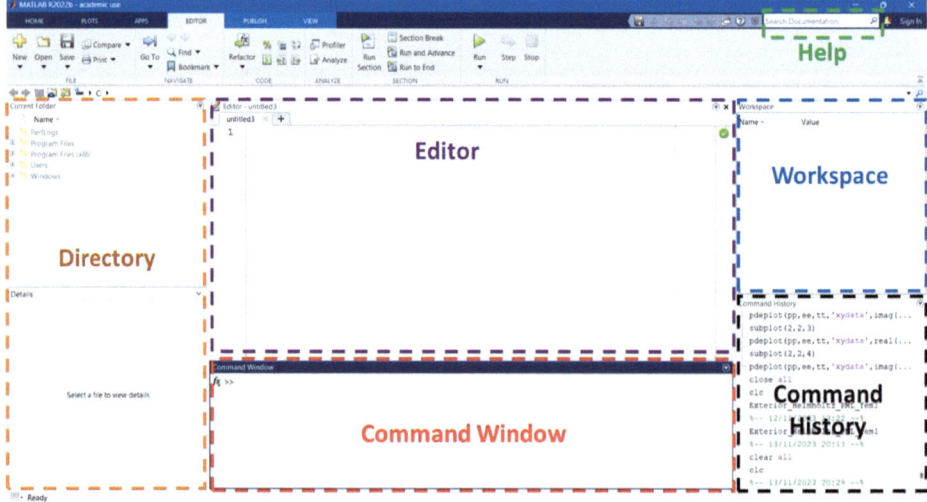

Figure B.2 An example of a typical MATLAB desktop showing its various components.

the specific edition. It is, however, modifiable to suit everyone's personal preferences. In Figure B.2, some important windows are highlighted: **Directory** (left), **Command Window** (middle/bottom), **Command History** (right/bottom), **Workspace** (right/top), **Editor** (middle/top), and **Help** (right/-top). A brief description of each window is given here, but more information will be provided in the later sections.

The Directory window specifies the current folder the user is working in, as well as the files that it contains. If the user requires files located in a different folder, they can navigate from the current directory to the desired location.

The Command Window allows the user to type commands that are executed immediately after they are inserted. The symbol '≫' is called a 'prompt' and commands are inserted after it. Results are also printed here.

All recently inserted commands are displayed in Command History. This window allows the user to review the latest commands that were executed, and easily repeat any previous command. This feature in MATLAB can significantly save a user's time.

The Workspace contains a list of all the variables that are currently stored in the computer memory. Various information regarding them can also be found here, such as data structure, and size of data. The data stored here can be viewed in the Command Window by simply typing the name of a specific data item, as shown in the Workspace.

The Editor allows the creation of MATLAB scripts. Each command inserted here is not executed immediately as in the Command Window, but instead they are executed together, in the sequence they were written, when the user executes the script file. This allows the execution of commands without retyping them in the Command Window.

The Help window provides information on the use and the syntax of various commands. The help function can be conveniently accessed by typing relevant keywords in the Help window related to the specific assistance needed (MathWorks Inc. 2022).

B.1.2 Basic calculations

In this section, some basic calculations will be performed using MATLAB's Command Window to provide an understanding of how to execute simple commands. Table B.1 provides a list of some symbols that can be used to perform basic calculations.

Suppose we aim to calculate the following expression: $2 + 3 \times 2 - 1$. To perform this calculation in the Command Window, type the following:

```
>> 2+3*2-1
```

Table B.1 A list of symbols for basic calculations and mathematical constants.

Operation	Symbol
Addition	+
Subtraction	−
Multiplication	*
Division	/
Power	∧
Imaginary Unit, $i = \sqrt{-1}$	1i
π	pi

After entering the command, MATLAB prints the result on the command line. The answer should be as follows:

```
ans =
    7
```

Note that in the Workspace, a variable named 'ans' was created and contains the value 7. Suppose that we now want to evaluate the expression $5^3 + 2^{-2}/0.25$. After entering this expression in the Command Window, the new result appears as follows:

```
>> 5^3+2^(-2)/0.25
ans =
    126
```

Notice how the previous value of the variable 'ans' has been overwritten with the value of the new calculation. Furthermore, from the above results, it is easy to see that calculations are performed following the usual priority (brackets, powers, division/multiplication, and lastly addition/subtraction). Again, the output of the expression appeared in the Command Window, just as it did previously. When a lot of calculations are involved, printing each calculation executed leads to increased computational time as well as incomprehensible outputs. Adding the semicolon symbol (;) at the end of the command suppresses the output of the command.

For instance, we execute the following command:

```
>> 2*3+9/3;
```

Although no result was printed on the screen, the value of 'ans' in the memory of MATLAB will be changed to 9, which is the result of the preceding command.

In many cases it is advantageous to store the results of a calculation in a variable for use in subsequent calculations. For instance, consider the earlier calculation. We can define variables $a = 2*3$ and $b = 9/3$. Subsequently, the previous calculation can then be rewritten as $a + b$. To assign a value to the variable the '=' sign is used, like writing an equation.

We insert the following commands in the Command Window:

```
>> a=2*3;
>> b=9/3;
```

Two new variables, named a and b, are now created and stored in the workspace. However, we must be careful with the syntax! On the left-hand side of the equation, only the variable name can be specified. Although the expressions $9/3 = b$ and $3 \times b = 9$ have the same mathematical meaning as the one we used, they are invalid in MATLAB.

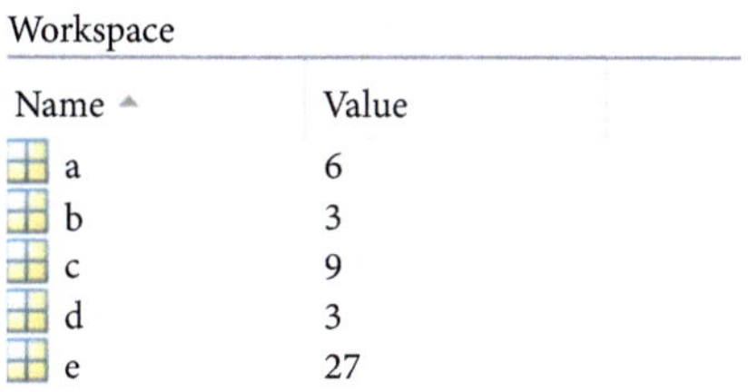

Workspace

Name ▲	Value
a	6
b	3
c	9
d	3
e	27

Figure B.3 An example MATLAB workspace.

Now that *a* and *b* are defined, we can calculate their addition by typing:

```
>> a+b
ans =
     9
```

If we want to store the result in a variable and use it later, we can define a third variable *c*, and write:

```
>> c=a+b;
```

Now we can do further calculations as follows:

```
>> d=a-b;
>> e=c*d;
```

What is happening now in the workspace? All the variables created during the previous calculations are still there. The formatting might vary, but the workspace should resemble that shown in Figure B.3.

Every time we want to start a new series of calculations, we should ensure that our results are not mixed with or affected by those from previous work (e.g., when starting to solve a new exercise). To delete all the variables that are currently stored in the workspace, we use the following command:

```
>> clear all
```

Similarly, to clear the Command Window from previous commands, we type:

```
>> clc
```

It is recommended to always type those two commands (i.e., clear all; clc) before starting to work.

B.1.3 Basic functions

Table B.2 summarizes some common functions in MATLAB. Each function is invoked in MATLAB by typing its name followed by parentheses, i.e., '( )', such as $\exp(x)$. The parentheses define the input arguments on which the function operates. These functions are relatively simple and require only one input. In subsequent sections, more complex functions that require multiple arguments are discussed.

Table B.2 A list of some basic mathematical functions in MATLAB.

Mathematical function	In MATLAB
Exponential: e^x	exp (x)
Sine (radians): $\sin(x)$	sin (x)
Cosine (radians): $\cos(x)$	cos (x)
Tangent (radians): $\tan(x)$	tan (x)
Cotangent (radians): $\cot(x)$	cot (x)
Sine (degrees): $\sin(x)$	sind (x)
Cosine (degrees): $\cos(x)$	cosd (x)
Tangent (degrees): $\tan(x)$	tand (x)
Cotangent (degrees): $\cot(x)$	cotd (x)
Natural logarithm: $\ln(x)$	log (x)
Logarithm base 10: $\log_{10}(x)$	log10 (x)
Absolute value: $|x|$ (measure for complex numbers)	abs (x)
Square root: $\sqrt{x}$	sqrt (x)
Hyperbolic sine: $\sinh(x)$	sinh (x)
Hyperbolic cosine: $\cosh(x)$	cosh (x)

Suppose we want to calculate the expressions $a = e^3 + 1$ and $b = \sqrt{\frac{\sin\left(\frac{\pi}{2}\right)+\cos\left(\frac{3\pi}{2}\right)}{2}}$, and we want to store the results in two variables called a and b. Using the functions of Table B.2, the relevant commands are:

```
>> a=exp(3) + 1
a =
    21.0855
>> b=sqrt((sin(pi/2) + cos(3*pi/2))/2)
b =
    0.7071
```

B.2 Vectors and matrices in MATLAB

We have seen how variables can be used to store values. Often, it is convenient to store multiple values in a single variable. We can create lists in MATLAB that are called vectors. MATLAB is highly efficient in performing calculations involving vectors. Vectors come in two types: row vectors and column vectors.

B.2.1 Vectors

We begin by describing row vectors. A row vector is a list of values that is separated by commas or spaces. Each value within the vector is termed an element, and the total number of elements determines its length. The syntax of a row vector is $r = [a_1, a_2, ..., a_n]$, where r is the name of the vector and $a_1, a_2, ..., a_n$ are its elements. According to the previous definition, the length of r is n. The elements of a vector must always be enclosed by square brackets.

As an example, we create a vector named 'vect' containing the values 1, -5, and 12 by typing the following in MATLAB and pressing the 'Enter' key from keyboard:

```
>> vect=[1,-5,12]
vect =
    1    -5    12
```

Notice that the same result is obtained by typing the following:

```
>>vect=[1 -5 12]
vect =
     1      -5      12
```

A variable named 'vect' is now created in MATLAB's Workspace. If we double-click on the variable, we will see all the elements stored inside. Elements of a row vector are stored in the same line (row). Vectors can be used as input variables for the functions discussed earlier. For example, by typing the following in the Command Window:

```
>>abs (vect)
ans =
     1       5      12
```

it returns a new vector that contains the absolute value of each element of vect. It is also possible to access specific elements of a vector. For instance, to determine the value of the second element in vect, we can type the following:

```
>> vect (2)
ans =
    -5
```

To calculate the square root of the third value in vect and store the answer in a variable named 'a', the following command is used:

```
>>a=sqrt(vect(3))
a =
    3.4641
```

A column vector is defined like a row vector with the difference that the semicolon symbol (;) must be used to separate each element. Its syntax is: $r = [a_1; a_2; ...; a_n]$. To see the difference, we create a new vector named 'vect_2', with the same three values as vect, but this time we use semicolons between the values instead of commas or spaces. It produces the following result:

```
>> vect_2=[1;-5;12]
vect_2=
    1
   -5
   12
```

From the answer printed on the screen or by double-clicking the variable vect_2 in the Workspace, we can see that the elements of the vector are now stored in the same column. Column vectors can also be used as inputs for functions, and individual elements can be accessed in the same manner as row vectors.

The function 'length' in MATLAB is a highly useful tool for swiftly determining the number of elements in a vector. For example, to determine the number of elements in the previously used vect_2, you can enter the following command in the Command Window:

```
>> length(vect_2)
ans =
    3
```

B.2.2 Inner product

The inner product of two vectors is a scalar number and describes the projection of one vector onto the other (Figure B.4). Because the result is a scalar number, this operation is also called the scalar product. It is defined as follows:

$$\vec{a} \cdot \vec{b} = |\vec{a}||\vec{b}| \cos(\theta) \tag{B.1}$$

where $|\vec{a}|$ and $|\vec{b}|$ denote the magnitudes of vectors $\vec{a}$ and $\vec{b}$ respectively, and θ is the angle between the vectors (counterclockwise is assumed to be positive).

Assume that we have the following two vectors:

$$\vec{a} = 4\vec{i} + 3\vec{j} \quad \text{and} \quad \vec{b} = \vec{i} - 2\vec{j}.$$

If the vectors are represented in matrix form as follows:

$$\vec{a} = \begin{bmatrix} 4 & 3 \end{bmatrix} \quad \text{and} \quad \vec{b} = \begin{bmatrix} 1 & -2 \end{bmatrix}$$

then the inner product can be calculated directly through matrix multiplication:

$$\vec{a} \cdot \vec{b} = \begin{bmatrix} 4 & 3 \end{bmatrix} \cdot \begin{bmatrix} 1 & -2 \end{bmatrix}^T = \begin{bmatrix} 4 & 3 \end{bmatrix} \cdot \begin{bmatrix} 1 \\ -2 \end{bmatrix} = 4 \times 1 - 3 \times 2 = -2$$

To ensure that matrix multiplication is defined, we must transpose vector $\vec{b}$. If the vectors were written in column vectors, then we would need to take the transpose of $\vec{a}$. In MATLAB, the transpose of a vector is calculated by adding a single quote to the name of the vector:

```
>> a=[4 3];
>> b=[1 -2];
>> a*b'
ans =
    -2
```

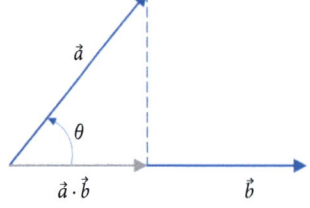

Figure B.4 The inner product of two vectors $\vec{a}$ and $\vec{b}$ is $\vec{a} \cdot \vec{b}$.

A second way to calculate the inner product of two vectors is by using the command dot (a,b), where a and b are written as column vectors or row vectors:

```
>> a=[4 3];
>> b=[1 -2];
>> dot(a,b)
ans =
     -2
```

It is important to note that in this case we did not need to utilize the transpose of either vector.

B.2.3 Cross product

The cross product (also known as vector product) of two vectors $\vec{a}$ and $\vec{b}$ is represented by $\vec{a} \times \vec{b}$ and defines a third vector $\vec{c}$ that is perpendicular to both $\vec{a}$ and $\vec{b}$. Furthermore, the measure of $\vec{c}$ is equal to the area defined by the other two vectors (Figure B.5).

The cross product of two vectors can be defined in MATLAB using the command cross (a,b), where a and b are the two input vectors. Both input vectors must be given in the form of 1×3 row vectors or 3×1 column vectors. In the previous example we had

$$\vec{a} = 4\vec{i} + 3\vec{j} \text{ and } \vec{b} = \vec{i} - 2\vec{j}$$

These vectors can be re-written as

$$\vec{a} = 4\vec{i} + 3\vec{j} + 0\vec{k} \text{ and } \vec{b} = \vec{i} - 2\vec{j} + 0\vec{k}$$

where $\vec{k}$ is the unit vector perpendicular to $\vec{i}$ and $\vec{j}$. Now we can write

$$\vec{a} = \begin{bmatrix} 4 & 3 & 0 \end{bmatrix} \text{ and } \vec{b} = \begin{bmatrix} 1 & -2 & 0 \end{bmatrix}$$

and calculate the cross product by executing:

```
>> a=[4 3 0];
>> b=[1 -2 0];
>> cross(a,b)
ans =
     0     0    -11
```

Hence, $\vec{c} = \vec{a} \times \vec{b} = 0\vec{i} + 0\vec{j} - 11\vec{k}$.

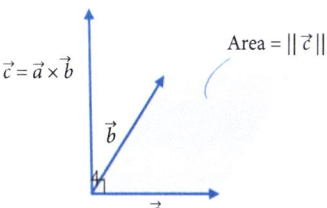

Figure B.5 The cross product of two vectors $\vec{a}$ and $\vec{b}$ is $\vec{a} \times \vec{b}$.

B.2.4 Matrices

It is possible to create a variable that consists of multiple rows and columns. This is called a matrix or a two-dimensional array. The dimensions of a matrix refer to the number of rows and columns that it contains. An $m \times n$ matrix has m rows and n columns. The number of rows always appears before the number of columns.

Because a matrix is comprised of multiple rows and columns, two indices are required to define its elements. Typically, the indices i and j are used to define the rows and the columns of a matrix, respectively (but any other letters could be used too). Using indices, it is possible to access specific elements in a matrix. In reality, a column vector is a single-column matrix and a row vector is a single-row matrix. Furthermore, a variable that contains only a single element can be considered a 1×1 matrix. A matrix that consists of the same number of rows and columns is called a square matrix.

The definition of a matrix employs the same principles discussed earlier for vectors. Square brackets are used to enclose all elements, elements in the same row are separated using commas or spaces, and elements in different rows are separated using the semicolon symbol. Elements are always typed row by row. For example, consider the matrix

$$A = \begin{bmatrix} 1 & 4 \\ -2 & 2 \\ 10 & -5 \end{bmatrix}$$

It is a 3×2 matrix (3 rows and 2 columns), while the matrix

$$B = \begin{bmatrix} 3 & 7 & 1 \\ 5 & -1 & 9 \end{bmatrix}$$

is a 2×3 matrix (2 rows and 3 columns). To define the above matrices, type the following commands in the Command Window in MATLAB:

```
>> A=[1,4; -2,2; 10,-5]
A =
    1     4
   -2     2
   10    -5
>> B=[3,7,1; 5,-1,9]
B =
    3     7     1
    5    -1     9
```

Notice how all the elements were written row by row and how the semicolon symbol was used to change rows. Using the command size (A), MATLAB returns the dimensions of a matrix (A in this case). For the previous two matrices, we have:

```
>> size(A)
ans =
    3     2
>> size(B)
ans =
    2     3
```

The first value corresponds to the number of rows while the second value refers to the number of columns in a matrix. Individual elements can be accessed using the row and column indices.

For example, the value 10 in matrix A is located at the third row and the first column. The value -1 in B is located at the first row and the third column. These values are accessed using the following commands:

```
>> A(3,1)
ans =
      10
>> B(1,3)
ans =
      1
```

Transposing a vector can change it from a column vector to a row vector (or a row vector to column vector). For example, suppose we have the following two vectors:

$$A = \begin{bmatrix} 3 & 9 & 2 \end{bmatrix}$$

$$B = \begin{bmatrix} 3 \\ 9 \\ 2 \end{bmatrix}$$

Using algebraic notation, we can write that $B = A^T$ and $A = B^T$. In MATLAB, the transpose of a matrix is calculated using single quote symbol. Hence, we can calculate vector B by defining vector A and calculating its transpose as follows:

```
>> A=[3 9 2]
A =
      3      9      2
>> B=A'
B=
      3
      9
      2
```

Extending this operation to matrices, the transpose operation exchanges the rows with the columns. This means, the first row becomes the first column, the second row becomes the second column, and so on.

For example, the transposes of the matrices

$$A = \begin{bmatrix} 2 & 5 & 7 \\ -1 & 3 & -1 \\ 2 & 6 & 9 \end{bmatrix} \text{ and } B = \begin{bmatrix} 3 & -1 & 2 \\ 5 & -1 & 6 \end{bmatrix}$$

are defined as follows:

$$A^T = \begin{bmatrix} 2 & -1 & 2 \\ 5 & 3 & 6 \\ 7 & -1 & 9 \end{bmatrix} \text{ and } B^T = \begin{bmatrix} 3 & 5 \\ -1 & -1 \\ 2 & 6 \end{bmatrix}$$

Using MATLAB, the transpose of a matrix is calculated the same way as for a vector:

```
>> A=[2 5 7; -1 3 -1; 2 6 9]          >> B=[3 -1 2; 5 -1 6]
A=                                    B=
     2    5    7                           3   -1    2
    -1    3   -1                           5   -1    6
     2    6    9                      >> B'
>> A'                                 ans=
ans=                                       3    5
     2   -1    2                           -1   -1
     5    3    6                           2    6
     7   -1    9
```

Suppose that we have two column vectors a and b. The process of summing two vectors consists of adding each element of vector a to the respective element of vector b. Obviously, the two vectors are required to have the same dimensions. The process is the same for row vectors. Column vectors and row vectors cannot be added together! Using the transpose operation, however, it is possible to turn a column vector into a row vector.

The process of subtracting two vectors is identical to that of addition. For instance, we define the following three vectors:

$$a = \begin{bmatrix} 1 & 3 & -5 \end{bmatrix}$$
$$b = \begin{bmatrix} -2 & 2 & 0 \end{bmatrix}$$
$$c = \begin{bmatrix} 4 \\ 9 \\ 2 \end{bmatrix}$$

To calculate the vectors $d = a + b$ and $f = b^T - c$:

```
>> a=[1 3 -5];
>> b=[-2 2 0];
>> c=[4; 9; 2];
>> d=a+b
d=
    -1    5   -5
>> f=b'-c
f=
    -6
    -7
    -2
```

Matrix addition

Addition and subtraction are also extended to matrices. Once more, it is essential for the two matrices to possess identical dimensions. Assuming we have the following two matrices and aim to add them together,

$$A = \begin{bmatrix} 3 & 4 & 5 \\ 6 & 7 & 8 \\ 9 & 2 & 1 \end{bmatrix}$$

$$B = \begin{bmatrix} 1 & -1 & 1 \\ -1 & 1 & -1 \\ 1 & -1 & 1 \end{bmatrix}$$

in MATLAB, we type the following:

```
>> A=[3 4 5; 6 7 8; 9 1 2];
>> B=[1 -1 1; -1 1 -1; 1 -1 1];
>> A+B
ans =
        4    3    6
        5    8    7
       10    0    3
>> A-B
ans =
        2    5    4
        7    6    9
        8    2    1
```

Matrix multiplication by a number

Sometimes we need to multiply a matrix by a scalar. The result is a new matrix, with the same dimensions as the old one. Each element of the new matrix is the product of each element of the old matrix with the scalar number. Suppose, for example, we have the following matrix A and a constant number c:

$$A = \begin{bmatrix} 1 & 2 \\ 3 & 4 \end{bmatrix} \quad \text{and} \quad c = 4$$

Then, the product $(c \cdot A)$ will be: $c \cdot A = 4 \begin{bmatrix} 1 & 2 \\ 3 & 4 \end{bmatrix} = \begin{bmatrix} 4 & 8 \\ 12 & 16 \end{bmatrix}$.

In MATLAB it would be as follows:

```
>> A=[1  2;  3  4];
>> C=4;
>> C*A
ans =
        4    8
       12   16
```

Matrix multiplication

Matrix multiplication is considered a rather complex operation. It is defined only when the number of columns of the first matrix is the same as the number of rows of the second one. The resulting matrix will have the same number of rows as the first matrix, and the same number of columns as the second matrix. For example, suppose A and B are matrices with dimensions $m \times n$ and $k \times l$ respectively. The matrix $C = A \cdot B$ can only be defined if $n = l$ and will have dimensions $m \times l$. The elements of C can be calculated using the following expression:

$$c_{ij} = a_{i1} \cdot b_{1j} + a_{i2} \cdot b_{2j} + a_{i3} \cdot b_{3j} + \ldots + a_{in} \cdot b_{nj} \tag{B.2}$$

Using this expression, each individual element of C can be calculated by accounting for all the values of i and j in the equation. Take, for example, the matrices

$$A = \begin{bmatrix} 1 & 2 & 3 \\ 1 & 3 & 2 \end{bmatrix} \quad \text{and } B = \begin{bmatrix} 2 & 2 \\ 3 & 1 \\ 1 & 2 \end{bmatrix}$$

The dimensions of A and B are 2×3 and 3×2, respectively, and hence matrix multiplication can be performed. The resulting matrix will be of dimension 2×2. Matrix C is constructed by multiplying each row of matrix A by each column of matrix B (this is an inner product).

In MATLAB, matrix multiplication is denoted using the multiplication symbol '*'. Hence, C can be calculated by typing the following in MATLAB:

```
>> A=[1  2  3; 1  3  2];
>> B=[2  2; 3  1; 1  2];
>> C=A*B
C=
      11    10
      13     9
```

The same applies for vectors as well. Suppose we have the following two vectors:

$$A = \begin{bmatrix} 1 & 5 & 2 \end{bmatrix} \text{ and } B = \begin{bmatrix} -2 \\ 4 \\ 3 \end{bmatrix}$$

The dimensions of the vectors are 1×3 and 3×1 for A and B, respectively. Looking at the dimensions of the two matrices given, the product $C = BA$ will generate a matrix with dimensions 3×3 while the product $D = AB$ will return a single value as a result (a 1×1 matrix). This can be verified in MATLAB as follows:

```
>> A=[1  5  2];
>> B=[-2; 4; 3];
>> C=B*A
C=
      -2   -10   -4
       4    20    8
       3    15    6
>> D=A*B
D =
      24
```

Elementwise matrix multiplication (Hadamard product)

Suppose now that we have two matrices A and B with the same dimensions $m \times n$. Multiplying each element of matrix A by the respective (same indices) element of matrix B is called elementwise multiplication. Hence, elementwise multiplication is the procedure of taking the product of corresponding elements. The same procedure is also extended to element-by-element division (elementwise division). This operation must not be mistaken for the standard matrix multiplication. In MATLAB, the symbols '.*' and './' are used to denote elementwise multiplication and elementwise division respectively. Take, for example, the following matrices:

$$A = \begin{bmatrix} 3 & 4 & 5 \\ 6 & 7 & 8 \\ 9 & 2 & 1 \end{bmatrix} \text{ and } B = \begin{bmatrix} 1 & -1 & 1 \\ -1 & 1 & -1 \\ 1 & -1 & 1 \end{bmatrix}$$

The elementwise product and division between A and B are calculated in MATLAB as follows:

```
>> A=[3  4  5; 6  7  8; 9  1  2];
>> B=[1  -1  1; -1  1  -1; 1  -1  1];
>> A.*B
ans =
     3    -4     5
    -6     7    -8
     9    -1     2
>> A./B
ans =
     3    -4     5
    -6     7    -8
     9    -1     2
```

Obviously, the same also applies to vectors.

Assume that A is a matrix; the elementwise multiplication A.*A can be simplified in MATLAB as A.^2. Similarly, the elementwise multiplication A.*A.*A can be simplified to A. ^3. The symbol '.^' is called elementwise power and returns the power of each element in the matrix. If B is also a matrix, A.^B will return each element of A to the power of the corresponding element of B.

For example, if we have

$$A = \begin{bmatrix} 1 & 4 \\ 3 & 2 \end{bmatrix} \text{ and } B = \begin{bmatrix} 1 & 0 \\ 2 & 2 \end{bmatrix}$$

we can use MATLAB to quickly calculate A.^2 and A.^B by typing the following in the Command Window:

```
>> A=[1  4; 3  2];
>> B=[1  0; 2  2];
>> A.^2
ans =
     1    16
     9     4
>> A.^B
ans =
     1     1
     9     4
```

Square matrices, trace, and determinant

We have already stated that if both dimensions of a matrix are equal, then the matrix is called a square matrix. The elements of a square matrix that have the same indices form the principal diagonal of the square matrix. If all the elements of the matrix are zero except for those in the principal diagonal, then the matrix is called a diagonal matrix. If, additionally, the elements in the diagonal are all equal to 1, then the matrix is called the identity matrix or unit matrix and is denoted by I. An example of a 3×3 unit matrix is

$$I = \begin{bmatrix} 1 & 0 & 0 \\ 0 & 1 & 0 \\ 0 & 0 & 1 \end{bmatrix}$$

In MATLAB, we can swiftly create a unit matrix using the command eye (x). This command returns a square unit matrix with dimensions x × x. For example, a 3 × 3 unit matrix is created by executing:

```
>> I=eye(3)
I=
     1     0     0
     0     1     0
     0     0     1
```

In matrix multiplication, the unit matrix is analogous to the number 1 in the familiar multiplication from elementary algebra. For example, consider the matrices

$$A = \begin{bmatrix} 3 & 4 & 5 \\ 6 & 7 & 8 \\ 9 & 2 & 1 \end{bmatrix} \text{ and } I = \begin{bmatrix} 1 & 0 & 0 \\ 0 & 1 & 0 \\ 0 & 0 & 1 \end{bmatrix}$$

Then we can say that $A \cdot I = A$. We can verify this in MATLAB by typing the following lines in the Command Window:

```
>> A=[3  4  5; 6  7  8; 9  1  2];
>> I=eye(3);
>> A*I
ans =
     3     4     5
     6     7     8
     9     1     2
```

The trace of a square matrix is the sum of all the elements in the main diagonal. If A is an $n \times n$ matrix, its trace is defined as

$$\text{tr}(A) = a_{11} + a_{22} + ... + a_{nn} \tag{B.3}$$

In MATLAB, the trace of a matrix is calculated using the command trace (X), where X is the name of the matrix. For example, consider the matrix B,

$$B = \begin{bmatrix} 1 & -1 & 1 \\ -1 & 4 & -1 \\ 1 & -1 & 2 \end{bmatrix}$$

Its trace is calculated as: tr $(B) = 1 + 4 + 2 = 7$. In MATLAB, this calculation is performed by typing the following lines in the Command Window:

```
>> B=[1  -1  1; -1  4  -1; 1  -1  2];
>> trace(B)
ans =
      7
```

325 VECTORS AND MATRICES IN MATLAB

Using MATLAB, it is possible to quickly calculate the determinants of large matrices. The determinant can be evaluated using the command det (X), where X is the input square matrix. Assume that the following matrix is given:

$$A = \begin{bmatrix} 1 & -2 & 3 & 7 \\ -1 & 2 & 7 & 5 \\ 2 & 3 & 5 & 1 \\ 1 & 8 & -1 & 3 \end{bmatrix}$$

Because A is a 4×4 matrix, evaluating its determinant by hand requires long calculations that can easily lead to errors. In MATLAB, on the other hand, the determinant of A can be found by typing the following commands:

```
>> A=[1  -2  3  7; -1  2  7  5; 2  3  5  1; 1  8  -1  3];
>>det(A)
ans =
    -1.2360e+03
```

Inverse matrix

If the determinant of a matrix is not zero, then the matrix is called invertible or nonsingular. If the determinant of a matrix is zero, then the matrix is called noninvertible or singular. If a matrix A is square and nonsingular, then we can calculate its inverse (i.e., A^{-1}). Note that the superscript -1 does not denote power here but is a notation formalism to indicate the inverse of a matrix. The following property holds for any invertible matrix:

$$A^{-1} \cdot A = A \cdot A^{-1} = I \tag{B.4}$$

Note that I is the unit matrix discussed earlier. In MATLAB, the inverse of a matrix is calculated using the inv (X) command, where X is the input square invertible matrix. In this example, the inverse of A is calculated by typing the following:

```
>> A=[1  -2  3  7; -1  2  7  5; 2  3  5  1; 1  8  -1  3];
>> inv(A)
ans =
    0.1537    -0.2718    0.2880    -0.0016
   -0.0696     0.0388    0.0065     0.0955
   -0.0437     0.0825    0.0971    -0.0680
    0.1197     0.0146   -0.0809     0.0566
```

We can verify the property $A \cdot A^{-1} = I$ by typing the following command line:

```
>> A*inv(A)
ans =
    1.0000    -0.0000    0.0000    -0.0000
    0.0000     1.0000   -0.0000     0.0000
   -0.0000    -0.0000    1.0000    -0.0000
   -0.0000     0.0000    0.0000     1.0000
```

REMARK A more efficient way to calculate the inverse of A is by using the \ operator in the form A\eye(size(A)). However, to solve for large scale linear systems using the inverse is not advisable. Different methods are used in such cases.

Eigenvalues and eigenvectors

The eigenvalue problem for a matrix A is written as

$$(A - \lambda I) \cdot x = 0 \tag{B.5}$$

where A is a matrix, I is the unit matrix with the same dimensions as A, λ are the eigenvalues of matrix A, and x are the eigenvectors of A. First, we will work towards finding the characteristic polynomial of a matrix. Analytically, this can be found by evaluating the determinant $\det(A - \lambda I)$ from the eigenvalue problem. To find the characteristic polynomial and the eigenvalues of a matrix (like matrix A) in MATLAB, we can use the command poly (A). Assume that we have the matrix

$$A = \left[\begin{array}{cc} 1 & -1 \\ 1 & 5 \end{array} \right]$$

and we want to find its characteristic polynomial. Analytically, this is done as follows:

$$\det(A - \lambda I) = 0 \rightarrow \lambda^2 - 6\lambda + 6 = 0.$$

Using the command poly (A), MATLAB returns the vector of coefficients for the characteristic polynomial (see Section B.3.2). In this example, we type the following:

```
>> A=[1 -1; 1 5];
>> p=poly(A)
p =
    1.0000  -6.0000  6.0000
```

where we stored the result in a vector called p. Obviously, the coefficients correspond to the characteristic polynomial we calculated analytically. The eigenvalues of the matrix can be found by finding the roots of the characteristic polynomial. Solving analytically the quadratic equation $\lambda^2 - 6\lambda + 6 = 0$, we find that the roots are

$$\lambda_1 = 1.2679 \text{ and } \lambda_2 = 4.7321$$

Using the command poly (A) in MATLAB, the vector of coefficients was obtained above (i.e., 1, −6, and 6). This vector can be then inserted directly into the command roots (p) to find the roots of the characteristic polynomial. The roots correspond to the eigenvalues of matrix A. In the Command Window of MATLAB, we type the following:

```
>> r=roots(p)
r=
    4.7321
    1.2679
```

where we stored the result in a vector called r. The elements of r are the eigenvalues.

In order to find the eigenvectors of A, we will use the command eig (A). This command is different to what we have seen so far because it returns two results. The full syntax of the command is:

$$[\text{eigenvectors, eigenvalues}] = \text{eig}(A)$$

Hence, we must provide two variable names on the left-hand side of the above equation. The eigenvectors and the eigenvalues of A will be stored in the first and the second variable, respectively. Assume that we want to store the eigenvalues of A in a variable called l and the eigenvectors in a variable called x. Obviously, variable names are only indicative. We type the following commands in the Command Window:

```
>> A=[1  -1; 1  5];
>> [x,l] = eig(A)
x =
    -0.9659    0.2588
     0.2588   -0.9659
l=
     1.2679         0
          0    4.7321
```

Each column of x corresponds to an eigenvector and each element in the diagonal of l corresponds to the respective eigenvalue.

Simultaneous equations

Assume we have the following two equations and we need to solve them simultaneously:

$$y = 3x + 5, \quad y = -x + 2$$

These equations describe two straight lines in the xy plane. To find whether there is a point of intersection for the above two equations, we must solve the simultaneous system and find the x and y that satisfy both equations. The above system can be rewritten as follows:

$$\begin{Bmatrix} -3x + y = 5 \\ x + y = 2 \end{Bmatrix}$$

By doing so we merely gathered on the left-hand side of the system all the unknowns with their coefficients and on the right-hand side all the constants. We can now rewrite the system in the following matrix form:

$$\begin{bmatrix} -3 & 1 \\ 1 & 1 \end{bmatrix} \begin{bmatrix} x \\ y \end{bmatrix} = \begin{bmatrix} 5 \\ 2 \end{bmatrix} \quad \text{or} \quad A \cdot x = B$$

where A is the matrix that contains the coefficients of the unknown variables, B is the column vector that contains the constant values, and x is the column vector that contains the unknowns (bold notation is used for the column vector x to distinguish it from the variable x).

The solvability of the above system can be investigated by calculating the determinant of A:

- If the determinant is nonzero, then A is invertible and the system has a unique solution.
- If the determinant of A is zero, then the system is either impossible or has infinite solutions and A cannot be inverted.

In the above example, we have $\det(A) = -4$ and hence A is invertible. To find the unknown values x and y inside the column vector x, we premultiply the above equation by A^{-1} to get $A^{-1} \cdot A \cdot x = A^{-1} \cdot B$. Using the property $A^{-1} \cdot A = I$, we have $x = A^{-1} \cdot B$. The last step can be easily performed in MATLAB using the command inv (A). To find the unknowns we execute the following:

```
>> A=[-3  1; 1  1];
>> B=[5; 2];
>> det(A)
ans=
    -4
>> inv(A)*B
ans =
       -0.7500
        2.7500
```

where we can also see that the value of $\det(A)$ is calculated as -4. It is important to notice that B was defined as a column vector. Furthermore, $A^{-1} \cdot B \neq B \cdot A^{-1}$ so the order of calculation is important. From the result, we see that

$$x = \begin{bmatrix} -0.75 \\ 2.75 \end{bmatrix} = \begin{bmatrix} x \\ y \end{bmatrix}$$

and hence $x = -0.75$ and $y = 2.75$. This means that the two lines intersect at the point with coordinates $(x, y) = (-0.75, 2.75)$.

Finding the inverse to solve a system of simultaneous equations is not very efficient, particularly when the number of equations and unknowns increases. MATLAB has the 'back-slash' command for solving simultaneous equations. This command is simply x = A\B and will produce the same solution as we found above. It is recommended that the backslash command is always used for solving simultaneous equations.

B.3 Plotting functions and finding roots in MATLAB

B.3.1 Plotting functions

The MATLAB command 'plot' is valuable for graphically representing results. Points are defined using pairs of coordinates (x, y) and are joined together with straight lines. The basic syntax of the command is plot(x_values, y_values), where x_values and y_values are vectors that define the pairs of coordinates of each point. It is important that the two vectors have an equal number of elements, otherwise the command will not work.

Suppose we want to make a plot of the straight line represented by $y = x+1$ in the domain $-1 \leq x \leq 1$. Since it is a straight line, we can define it using only two points. For $x = -1$ and $x = 1$, the coordinates of the two points are $(-1, 0)$ and $(1, 2)$, respectively. We can define the vector x_vect = [-1; 1] that contains the x coordinates of the two points and the vector y_vect = [0; 2] that contains the y coordinates. Here, we used column vectors for the coordinate pairs. Each line of the two vectors defines one point. The same can be achieved using row vectors but then the column of each pair will define each point. The plot can be created by executing plot(x_vect, y_vect). The Command Window should look like the following:

```
>> x_vect=[-1; 1];
>> y_vect=[0;2];
>> plot(x_vect, y_vect)
```

Following the last command, a new window will appear containing the plot specified with the two vectors. Figure B.6 illustrates the plot created. Depending on the default setting on a particular computer, the plot might appear with slight variations.

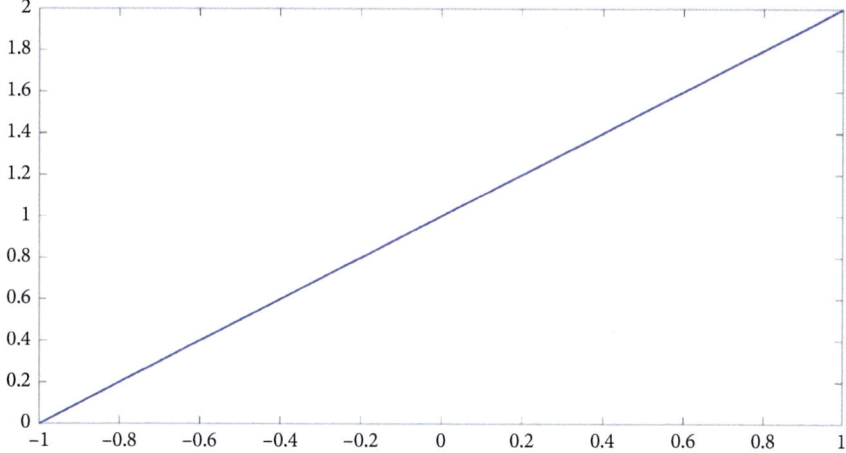

Figure B.6 Plot of the line segment represented by $y = x + 1$ for the interval $-1 \le x \le 1$.

The previous example was a simple case as only two points were adequate to represent the graph. In most cases, additional points are needed to improve the accuracy and quality of the graphs. Now we consider an example with $y = \sin(2\pi x)$ and $0 \le x \le 2$. To illustrate the need for additional points, we create the plot using six points. Following the same steps as before, we first create the x and y coordinate vectors and then plot the points using plot. This should generate the plot shown in Figure B.7.

```
>> x=[0  0.4  0.8  1.2  1.6  2.0];
>> y=sin(2*pi*x);
>> plot(x,y)
```

Using such a small number of points, the graph in Figure B.7 hardly resembles a sine curve. Increasing the number of points will also increase the resolution. Manually typing each one of the coordinate points can become quite tedious; thus we need an automated method. This can be achieved using the 'linspace' command. The syntax of the command is linspace (a, b, N), which generates N equally spaced points

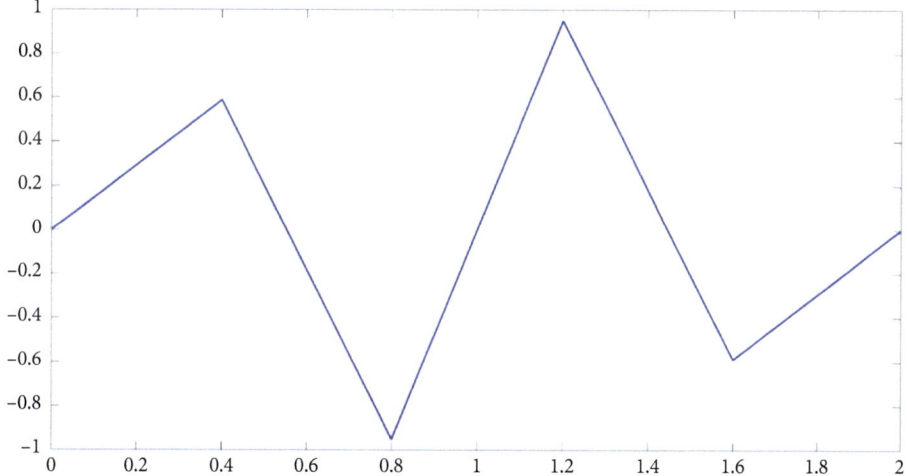

Figure B.7 Plot for the equation $y = \sin(2\pi x)$ in the domain $0 \le x \le 2$ using six points.

in the closed interval $[a, b]$. Here, we use this command to generate 200 points on the x axis, using the following line in MATLAB's Command Window:

```
>> x=linspace(0,2,200);
```

Notice in the Workspace that a vector x has been created containing 200 equidistant elements in the closed interval $[0, 2]$. Following the same steps as before the refined plot can now be created whose outcome is shown in Figure B.8:

```
>> x=linspace(0,2,200);
>> y=sin(2*pi*x);
>> plot(x,y)
```

There are various formatting options within the plot command. An extensive list of all options can be found by searching 'plot' in the help window at the top right of the MATLAB desktop (Figure B.2). Here, a few useful options are given. We define a title to the figure as well as name the two axes of the plot. A figure can be titled using the command 'title' after a plot command. The syntax is title ('...'), where the string is the text to be used as a title. The axes are titled using the xlabel and ylabel commands, which have a similar syntax. As an example, by specifying the following lines in MATLAB:

```
>> x=linspace(0,2,200);
y=sin(2*pi*x);
plot(x,y)
>>title('plot of y=sin(2πx)');
>>xlabel('x axis');
>>ylabel('y axis');
```

the titled plot in Figure B.9 is produced. Additional formatting tools can be accessed by clicking the Plot Tools icon in MATLAB's figure window.

Assume now that we do not want to specify the exact number of points within the interval, but we want to specify their distance. Using the command: $a = (start : step : end)$, MATLAB automatically creates a vector of elements in the interval $[start, end]$. The last element will be as close as possible

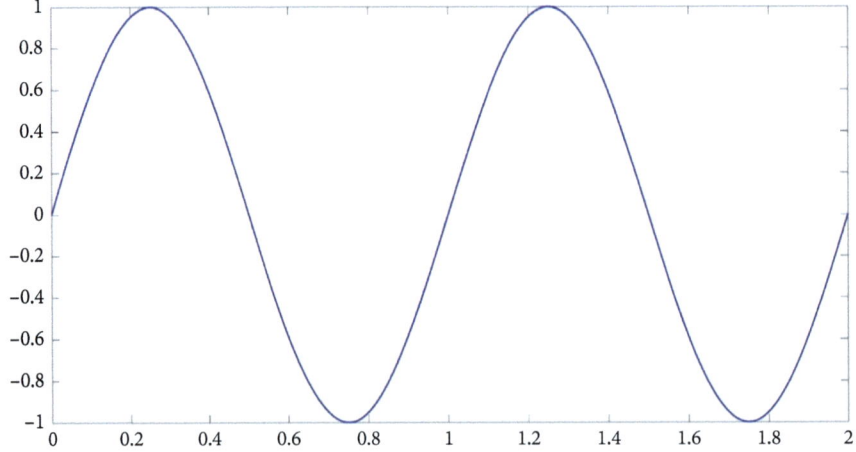

Figure B.8 Plot for the equation $y = \sin(2\pi x)$ in the domain $0 \le x \le 2$ using 200 points.

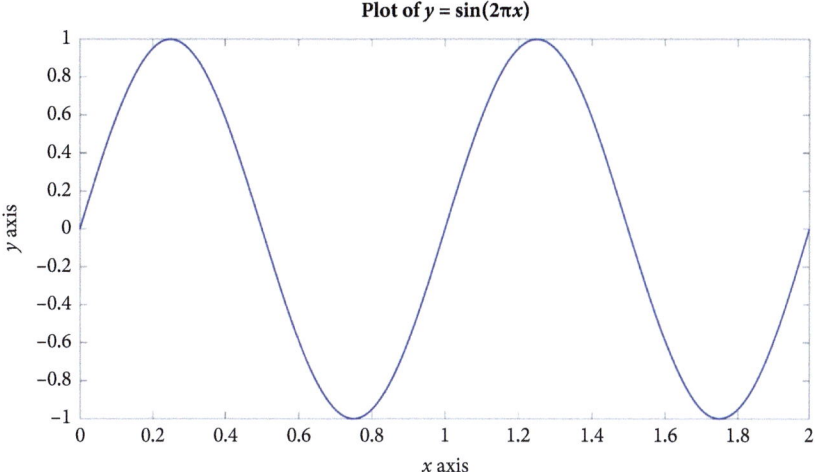

Figure B.9 Plot with title and axis labels.

to the value *end*, depending on the specified step. If no step is specified, it is assumed to be one by default (i.e., *step* = 1). For example, if we execute a = (1 : 0.5 : 5), MATLAB creates a vector with nine elements that start at one and end at five. If, however, we execute b = (1 : 0.9 : 5), the vector contains five elements, and the last one is 4.6:

```
a =                 b=
   1.0000              1.0000
   1.5000              1.9000
   2.0000              2.8000
   2.5000              3.7000
   3.0000              4.6000
   3.5000
   4.0000
   4.5000
   5.0000
```

Elementwise operations are very convenient when we want to plot functions in MATLAB. Assume that we want to create the plot of the following function in the given domain:

$$y = x \cdot \sin(2\pi x) \text{ for } 0 \le x \le 2$$

Here we use 500 points in the interval $x \in [0, 2]$ (remember, this can be done using the command linspace). The plot is created by executing the following commands and is shown in Figure B.10:

```
x=linspace (0, 2, 500);
y=x.*sin (2*pi*x);
plot (x, y)
xlabel ('x')
ylabel ('y')
```

Notice that to multiply x (which is a vector) by sin (2*pi*x) (also a vector) we used the symbol '.*'. On the other hand, the multiplication of x by 2*pi is performed with the symbol '*' since it is a scalar. The resulting plot can be found in Figure B.10.

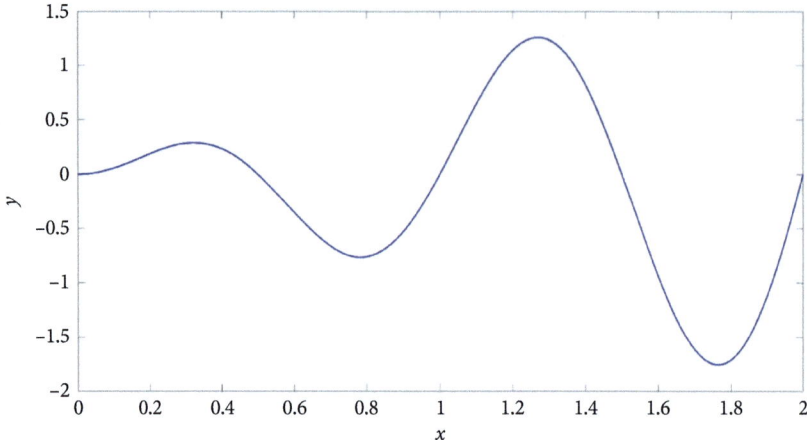

Figure B.10 Plot of $y = x \sin(2\pi x)$ in MATLAB.

MATLAB also includes plot commands for functions of several variables, enabling the plotting of surfaces and volumes.

B.3.2 Roots of polynomials

The general form of a polynomial in one variable (also known as a univariate polynomial) can be given as

$$p = c_n x^n + c_{n-1} x^{n-1} + \ldots + c_2 x^2 + c_1 x + c_0 \tag{B.6}$$

Where the coefficients c_i are constants and x is a variable. The highest power that appears in the polynomial (n) defines its degree. In many problems, one needs to find the roots of a polynomial (meaning solve for $p = 0$). For example, an expression like $c_2 x^2 + c_1 x + c_0 = 0$ is a quadratic equation and its roots can be conveniently determined using the formula $x = \frac{-c_1 \pm \sqrt{c_1^2 - 4c_2 c_0}}{2c_2}$. Similarly, solving $x^3 + x^2 - 6x = 0$ is also possible as it can be factorized to $x(x-2)(x+3) = 0$. Unless the polynomial is of order $n \leq 3$, or the polynomial can be simplified to a special case, finding its roots is not straightforward.

MATLAB has a built-in function for finding the roots of a polynomial. The syntax of the function is roots (p), where $p = [c_n, c_{n-1}, \ldots, c_2, c_1, c_0]$ is a vector containing the coefficients of a single-variable polynomial containing nonnegative exponents.

As an example, we find the roots of $x^3 + x^2 - 6x = 0$ (discussed earlier) using MATLAB. From its factorized form, we can easily see that the roots of the polynomial are 0, 2, and -3. To use the roots command in MATLAB, we must first write the polynomial in the form $a_n x^n + \cdots + a_3 x^3 + a_2 x^2 + a_1 x + a_0 = 0$. In this case, we have $1 \cdot x^3 + 1 \cdot x^2 - 6x + 0 = 0$ with $c_3 = 1$, $c_2 = 1$, $c = -6$, and $c_0 = 0$. Hence, the vector containing the coefficients of the polynomial is $p = [1, 1, -6, 0]$. We can now find the roots by typing the following commands in MATLAB:

```
>> p=[1  1  -6  0];
>> roots(p)
ans =
        0
       -3
        2
```

B.4 The MATLAB Symbolic Toolbox

Various toolboxes are available in MATLAB that provide additional capabilities. The Symbolic Math Toolbox allows the user to solve problems using symbolic variables, much like writing down mathematical equations. This way, one can solve systems of equations, integrals, differential equations, and many other mathematical problems.

B.4.1 Introduction to symbolic calculations

To better understand the use of the Symbolic Toolbox, we start by defining a simple function $f(x) = x^2$ and evaluating it for different values of x. First, we must define in MATLAB that x is now a symbolic variable. This can be done by using the command syms. The syntax is simple: syms Var_1 Var_2 ... Var_N, where the Var_i are the variable names that we want to define as symbolic variables. Remember, the syntax and functionality of various commands can always be verified by conducting searches in the Help window. To access detailed information about a specific function, enter the following command:

```
>> syms  x
```

Now the symbolic variable x appears in the Workspace (notice how the icon and value field are different). It is possible now to define the symbolic function $f(x) = x^2$. This is done by typing the following:

```
>> f(x)=x^2;
```

We can now evaluate the value of $f(x)$ by inserting values directly into the symbolic function. Assume that we want to find $f(x)$ for $x = 2$; then we type the following in the Command Window:

```
>> f(2)
ans =
4
```

MATLAB understands that the symbolic variable x takes the value of 2 in the expression of the symbolic function $f(x)$ and evaluates it. We can also use vectors to quickly evaluate a symbolic function for multiple values. Assume that we have a function $g(x) = 3x^4 + 7x - 3$ and we want to evaluate it for $x = 1, 2, 3, 4, 5, 6, 7, 8, 9$, and 10. Assume that we also want to store the result for each value of x. We can do this task one value at a time as before, but this means we should execute the command many times, as well as creating many variables to store the results. It is easy to see how this can become problematic, especially for large-scale calculations.

Using vectors, however, we need execute the command only once and all the results will be stored in single variable that is also a vector. The first step is to always define our symbolic variable and function (in the previous example the symbolic variable x had already been created, so this step can be avoided here, but for completeness we assume that we start from the beginning). Next, we define the vector containing all the values for which we want to evaluate $g(x)$. For this problem the name of the vector can be selected arbitrarily as a = [1, 2, 3, 4, 5, 6, 7, 8, 9, 10]. Finally, we evaluate g (a) and store the results in a variable named b. The Command Window should look like the following:

```
>> syms x
>> g(x)=3*x^4+7*x-3;
>> a=[1  2  3  4  5  6  7  8  9  10];
>> b=g(a)
b =
[7, 59, 261, 793, 1907, 3927, 7249, 12341, 19743, 30067]
```

Symbolic functions can also be used for the roots of polynomials. Let's assume we have a function $f(x) = x^3 + 4x^2 - 15x - 18$ and we want to solve it for $f(x) = 0$. We will use the command 'solve' to directly write the polynomial in MATLAB. The syntax of the command is solve (*equation, variable*). Here, for *equation* we write the expression that we want to solve in the manner we would normally write it, the only exception being that the symbol (==) needs to be used to indicate equality. For *variable*, we need to specify the symbolic variable used. To solve the above example, we type the following commands:

```
>> syms x
>> f(x) = x^3 + 4*x^2 - 15*x - 18;
>> solve(f(x)==0,x)
ans =
      -6
      -1
       3
```

B.4.2 Derivatives and differentiation

One of the biggest advantages of this toolbox is that it allows the user to write equations directly, just like we would write them in mathematics. We will now study how we can write expressions and evaluate their derivatives. Assume, for instance, the function $f(x) = x^2$. Using the differentiation rule $(x^n)' = nx^{n-1}$ we can calculate the derivative of f analytically as

$$\frac{df}{dx} = f' = 2x$$

In MATLAB, we can evaluate the derivative of a symbolic function using the command diff (*fun, var*), where in place of *fun* we write our function and in place of *var* we define the differentiation variable (it has to be a symbolic variable). To find the derivative of $f(x) = x^2$, we have to first define a symbolic variable (for compatibility we will use x), then we have to define the symbolic function f. Consequently, we can evaluate the derivative using the command diff (*fun, var*). We execute the following lines in the Command Window:

```
>> syms x
>> f(x)=x^2
f(x) =
x^2
>> diff(f,x)
ans(x) =
2*x
```

Alternatively, we can define a variable to store the derivative of f. For example:

```
>> f_prime=diff(f,x)
f_prime(x) =
2*x
```

Similarly, given $g(x) = 10x^5$, the derivative g' is calculated as follows:

```
>> g(x)=10*x^5;
>> g_prime=diff(g,x)
g_prime(x) =
50*x^4
```

We can see that using MATLAB's Symbolic Toolbox we can find the analytical derivative of a function. Of course, the rules of derivatives are followed during the solution. For example, the derivatives of $h(x) = 3x^2 + 2x^4$ and $k(x) = x \cos x$ can be calculated as

$$\frac{dh}{dx} = h' = \left(3x^2\right)' + \left(2x^4\right)' = 6x + 8x^3$$

$$\frac{dk}{dx} = k' = (x)' \cos x + x(\cos x)' = \cos x - x \sin x$$

In MATLAB, we can evaluate the above derivatives using the following commands:

```
>> syms x
>> h(x)=3*x^2+2*x^4;
>> g(x)=x*cos(x);
>> h_prime=diff(h,x)
h_prime(x) =
8*x^3 + 6*x
>> g_prime=diff(g,x)
g_prime(x) =
cos(x) - x*sin(x)
```

The same commands can be used for higher-order derivatives. Recall the function $f(x) = x^2$ we used earlier. In mathematics, the second derivative of f is evaluated as

$$\frac{d^2f}{dx^2} = \frac{d}{dx}\left(\frac{df}{dx}\right) = f'' = \left(\left(x^2\right)'\right)' = (2x)' = 2$$

In MATLAB, we can calculate higher-order derivatives by using the command diff(*fun, var*) multiple times. The second derivative of f is evaluated by executing the following:

```
>> syms x
>> f(x)=x^2;
>> diff(diff(f,x),x)
ans(x) =
2
```

Alternatively, we can evaluate each derivative separately and store the result in a variable. We first calculate the derivative of f using the command diff(fun, var) and store the result in a variable named f_prime. Consequently, we calculate the derivative of the derivative (i.e., the second derivative) by using the command diff(fun, var) once more, but this time we give as an input the derivative of f. Execute the following in the Command Window:

```
>> syms x
>> f(x)=x^2;
>> f_prime=diff(f,x);
>> f_prime_f_prime=diff(f_prime,x)
f_prime_prime(x) =
2
```

We now discuss some common methodologies for the numerical approximation of the derivative of a function. To enhance visualization of the concepts discussed here, we will use the example of finding the slope of the tangent line at a point on a curve. Consider a function $f(x)$ as illustrated in Figure B.11. At x_0, there is a line that touches the curve only at point $(x_0, f(x_0))$, i.e., a line that is tangent to the curve. From calculus we know that the slope of the tangent at x_0 is given by the derivative of f evaluated at x_0.

Assume a second point on the curve located at $(x_0 + \Delta x, f(x_0 + \Delta x))$ as illustrated in Figure B.12. If we connect the two points on the curve with a straight line, then the slope of the new line is calculated as

$$\tan \theta' = \frac{f(x_0 + \Delta x) - f(x_0)}{x_0 + \Delta x - x_0} = \frac{f(x_0 + \Delta x) - f(x_0)}{\Delta x} \tag{B.6}$$

The value of Δx was arbitrarily selected. Comparing Figures B.11 and B.12, as the value Δx becomes small, the second line tends to approximate the tangent at x_0. If an even smaller value of Δx was selected, the accuracy of the approximation would be better. We see that as Δx becomes a small value an approximation of the derivative can be found from

$$f'(x_0) \approx \frac{f(x_0 + \Delta x) - f(x_0)}{\Delta x} \tag{B.7}$$

In the previous method, a point was selected in front of x_0 at a distance Δx. For this reason, the method is called the 'forward Euler' method. A point could be selected behind x_0 at a distance Δx from it. This case is illustrated in Figure B.13. In this case the approximation would be written as

$$f'(x_0) \approx \frac{f(x_0) - f(x_0 - \Delta x)}{x_0 - (x_0 - \Delta x)} = \frac{f(x_0) - f(x_0 - \Delta x)}{\Delta x} \tag{B.8}$$

Because we now selected a point behind x_0, this method is called the 'backward Euler' method.

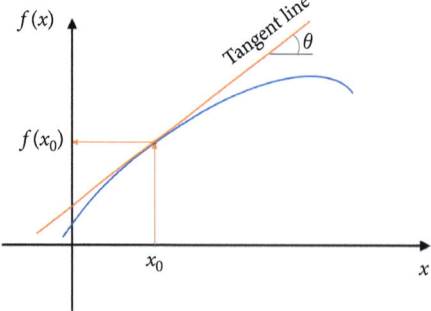

Figure B.11 Tangent of a curve at the point x_0.

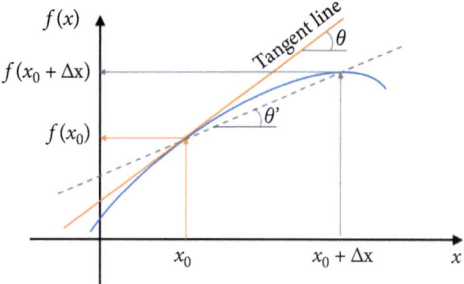

Figure B.12 Line defined from the points $(x_0, f(x_0))$ and $(x_0 + \Delta x, f(x_0 + \Delta x))$ of the curve.

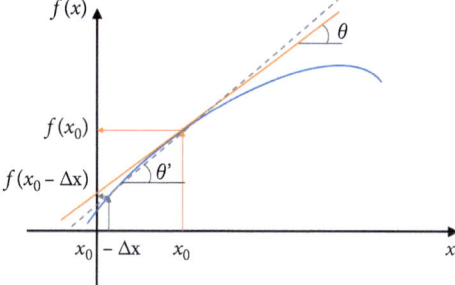

Figure B.13 The backward Euler method.

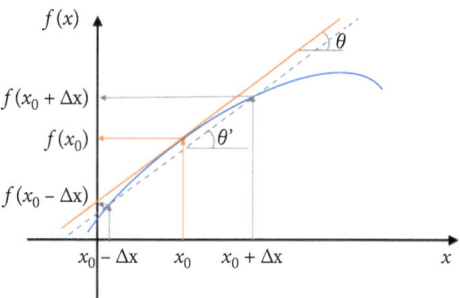

Figure B.14 The central difference approximation.

It is also possible to use a point behind x_0 and a point in front of it at the same time to approximate the derivative at x_0. This is called the 'central difference' method, and it is illustrated in Figure B.14. In this case the approximation is written as

$$f'(x_0) \approx \frac{f(x_0 + \Delta x) - f(x_0 - \Delta x)}{x_0 + \Delta x - (x_0 - \Delta x)} = \frac{f(x_0 + \Delta x) - f(x_0 - \Delta x)}{2\Delta x} \tag{B.9}$$

Assume, for example, that we have the function $f(x) = \sin(x)$ and we want to approximate its derivative at $x_0 = \pi/6$ for $\Delta x = [0.1, 0.01, 0.001, 0.0001]$ using the approximation methods presented earlier.

First, we calculate the value of $f'(x_0)$ analytically so we can compare with the numerical results:

$$f'(\pi/6) = \cos(\pi/6) = 0.866025403784439$$

For $\Delta x = 0.1$ the forward approximation can be calculated by executing the following command:

```
>> x0=pi/6;
dx=0.1;
fw_euler=(sin(x0+dx)-sin(x0))/dx
fw_euler =
    0.839603576017623
```

For $\Delta x = 0.1$ the backward approximation can be calculated by executing the command:

```
>> bw_euler=(sin(x0)-sin(x0-dx))/dx
bw_euler =
    0.889561923237365
```

For $\Delta x = 0.1$ the central difference approximation can be calculated by executing the following command:

```
>> cntr_diff=(sin(x0+dx)-sin(x0-dx))/(2*dx)
cntr_diff =
    0.864582749627494
```

For $\Delta x = 0.1$, the central difference method provides the best approximation for this example. We can repeat the commands for all values of Δx to compare the results. For convenience, we introduce an error measure defined as

$$\text{relative \% error} = \left| \frac{\text{actual value} - \text{approximation}}{\text{actual value}} \right| \cdot 100\% \tag{B.10}$$

The above example can be calculated using the following commands:

```
x0=pi/6;
exact=cos(x0);
dx=[0.1  0.01  0.001  0.0001];
fw_euler=(sin(x0+dx)-sin(x0))./dx;
bw_euler=(sin(x0)-sin(x0-dx))./dx;
cntr_diff=(sin(x0+dx)-sin(x0-dx))./(2*dx);
error_fw=abs((exact-fw_euler)/exact)*100;
error_bw=abs((exact-bw_euler)/exact)*100;
error_cntr=abs((exact-cntr_diff)/exact)*100;
```

The results are presented in Table B.3.

B.4.3 Integrals and integration

In general, integration of a function can be considered as the opposite procedure to differentiation. An essential distinction between definite and indefinite integrals needs to be emphasized at this point.

Table B.3 The results of error calculations for estimating a derivative using the forward, backward, and central Euler schemes.

	Error %			
	$\Delta x = 0.1$	$\Delta x = 0.01$	$\Delta x = 0.001$	$\Delta x = 0.0001$
Forward Euler	3.05059	0.2903	0.0289	0.0029
Backward Euler	2.7178	0.2870	0.0289	0.0029
Central difference	0.16666	0.0017	1.667E−05	1.667E−07

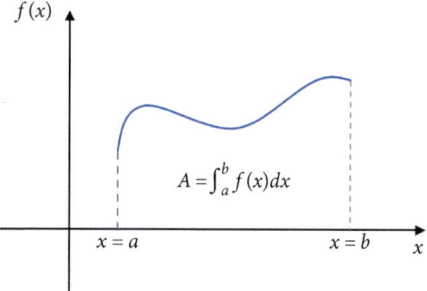

Figure B.15 Illustration of the area defined by a definite integral.

An indefinite integral of $f(x)$ is a new function that answers the question, what function do we differentiate to get $f(x)$? The indefinite integral is denoted as

$$\int f(x)\,dx \tag{B.11}$$

where x is the differentiation variable. Take, for instance, the function $f(x) = 3x^2$. The integral of f is

$$\int f(x)\,dx = \int 3x^2\,dx = x^3 + C$$

where C is a constant called the 'integration constant'. We can easily verify that

$$(x^3 + C)' = \left(x^3\right)' + (C)' = 3x^2 = f(x)$$

The definite integral of $f(x)$ from $x = a$ to $x = b$ is a number and corresponds to the area under the curve bounded by the values a and b. The definite integral is denoted as

$$\int_a^b f(x)\,dx \tag{B.12}$$

Figure B.15 illustrates the area defined by a definite integral.

Similar to the case of derivatives, integrals can be evaluated either analytically or numerically. In certain cases, the analytical solution of complex integrals is not always possible, and they can be evaluated only numerically.

Using MATLAB's Symbolic Toolbox, integration can be performed with the command int (*fun, var*), which takes the same inputs as the command diff (*fun, var*) used earlier. Using this command, we can swiftly find the analytical solution of an integral. Consider our example function, $f(x) = 3x^2$. To use the Symbolic Toolbox, we must first define a symbolic variable and construct our symbolic function using this variable. We can then evaluate its integral by executing the following commands:

```
>> syms x
f(x)=3*x^2;
int(f,x)
ans(x) =
x^3
```

Notice how MATLAB cannot add the integration constant at the end of the result. If we want to add the constant, we can do this separately and we would have to execute:

```
>> syms x C
f(x)=3*x^2;
g(x)=int(f,x)+C
g(x) =
x^3 + C
```

Notice also that we have now added the constant C in the list of symbolic variables during the execution of the first command. MATLAB follows the integration rules from mathematics to arrive at the solution. Consider the function $g(x) = x \cos x$. The integral of g can be evaluated using the integration by parts rule:

$$\int g(x)\, dx = \int x \cos x\, dx = \int x(\sin x)'\, dx = \int (x \sin x)'\, dx - \int (x)'\, \sin x\, dx =$$

$$= x \sin x - \int (-\cos x)'\, dx + C = x \sin x + \cos x + C$$

In MATLAB, the above integral can be evaluated using the command int (*fun, var*) by typing the following lines:

```
>> syms x C
>> g(x)=x*cos(x);
>> int(g,x)+C
ans(x) =
C + cos(x) + x*sin(x)
```

The examples we have seen so far refer to indefinite integrals. Assume that we want to calculate the area below the function $f(x) = x^2 + x - 1$ between the values $a = 1$ and $b = 0$. We can do this analytically:

$$\int_a^b f(x)\, dx = \int_1^2 x^2 + x - 1\, dx = \left(\frac{2^3}{3} + \frac{2^2}{2} - 2\right) - \left(\frac{1^3}{3} + \frac{1^2}{2} - 1\right) = 2.8333$$

In MATLAB, we can evaluate definite integrals by using a modified syntax of the command int (*fun, var*). This time we must provide four inputs in total. The command is written as int (*fun, x, a, b*) where *fun* is the function we want to integrate, *x* is the integration variable, and *a* and *b* are the integration limits. Using this syntax, we can evaluate the above example by using the following commands:

```
>> syms x
>> f(x)=x^2+x-1;
>> A=int(f,x,1,2)
A =
17/6
```

Notice now that the result is a number, while earlier the result was a function.

Earlier we saw how we can numerically calculate the derivative of a function. Remember that numerical methods approximate the result and do not give exact solutions. Engineers are often required to solve problems that lead to complex integrals. These integrals are not always possible to calculate analytically. Consider the function $f(x) = e^x\sqrt{1 + \cos^2 x}$. Using what we did in the previous paragraph, the two integrals $\int f(x)\,dx$ and $\int_1^2 f(x)\,dx$ can be evaluated as follows:

```
>> syms x
>> f(x)=exp(x)*sqrt(1+cos(x)^2);
>> int(f,x)
ans(x) =
int(exp(x)*(cos(x)^2 + 1)^(1/2),x
>> int(f,x,1,2)
ans =
int(exp(x)*(cos(x)^2 + 1)^(1/2), x, 1, 2)
```

Notice how MATLAB did not actually perform any calculations and only returned the question without answering it. In such cases we must use numerical methods and simply approximate the result. Here we study two common procedures for numerical integration. To be able to compare the approximations with the actual result, we will use a simpler example. Consider the function $f(x) = x^2 + 3$ where we want to calculate the area below the curve between the values $a = -3$ and $b = 3$. We can do this first analytically to find the exact solution:

$$\int_{-3}^{3} x^2 + 3\,dx = \frac{x^3}{3} + 3x\Big|_{-3}^{3} = 36$$

The function f, along with the integration limits, is plotted in Figure B.16. As a first approximation, we divide the area below the curve into six rectangular shapes. The subdivision is illustrated in Figure B.17. The base of each rectangle is $\Delta x = \frac{b-a}{6} = 1$. At the middle of the base, we take a point on the x axis and evaluate the function at that point. The area of each element can then be calculated from $A_i = \Delta x f(x_i)$. It is clear from Figure B.17 that each element does not fit exactly. However, the total area can then be approximated by taking the sum of all the elements: $A \approx \sum A_i$.

To calculate the above approximation, we need to first define the coordinates of the points $(x_i, f(x_i))$ and to calculate the area for each element. Notice that the first and last points are located at a distance $\frac{\Delta x}{2}$ from the integration limits. We can use the command linspace(*start, finish, points*) to create the x_i. Then, we calculate $f(x_i)$ from the expression of the function and we calculate the area of each element. Lastly, we use the command sum to add together all the areas. Execute the following lines in the Command Window:

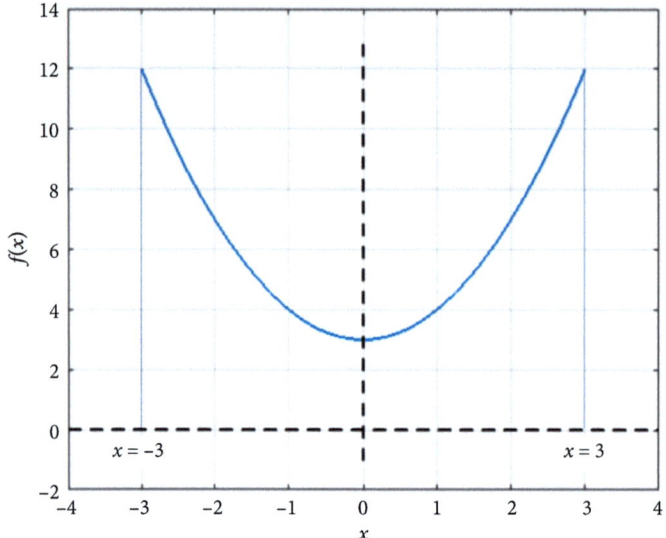

Figure B.16 Plot of $f(x) = x^2 + 3$ in the domain $-3 \le x \le 3$.

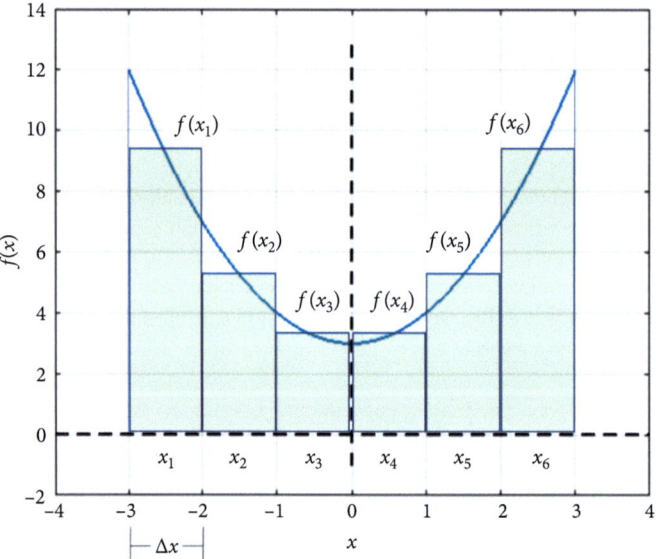

Figure B.17 Approximation of the area below the function $f(x) = x^2 + 3$ in the domain $-3 \le x \le 3$ using six rectangular areas.

```
>> a=-3;
b=3;
no_of_points=6;
dx=(b-a)/no_of_points;
x=linspace(a+dx/2,b-dx/2,no_of_points);
fx=x.^2+3;
Ai=fx*dx;
A=sum(Ai)
A =
   35.500000000000000
```

Similar to the numerical integration, by increasing the number of elements (hence decreasing the distance Δx), the accuracy of the approximation will improve. We try the above process once again using 10 elements:

```
>> a=-3;
b=3;
no_of_points=10;
dx=(b-a)/no_of_points;
x=linspace(a+dx/2,b-dx/2,no_of_points);
fx=x.^2+3;
Ai=fx*dx;
A=sum(Ai)
A =
   35.820000000000000
```

We try again, now using 100 elements:

```
>> a=-3;
b=3;
no_of_points=100;
dx=(b-a)/no_of_points;
x=linspace(a+dx/2,b-dx/2,no_of_points);
fx=x.^2+3;
Ai=fx*dx;
A=sum(Ai)
A =
   35.998200000000011
```

A second approach to approximating the area underneath a curve is by employing trapezoidal elements instead of rectangular. For this reason, this method is called the trapezoidal rule. Like the previous method, the area is approximated by the sum of the areas of all the trapezoids. MATLAB has a built-in command to apply the trapezoidal rule. The command is written as: trapz $(x_i, f(x_i))$, where x_i and $f(x_i)$ are vectors containing the coordinate points as indicated in Figure B.18.

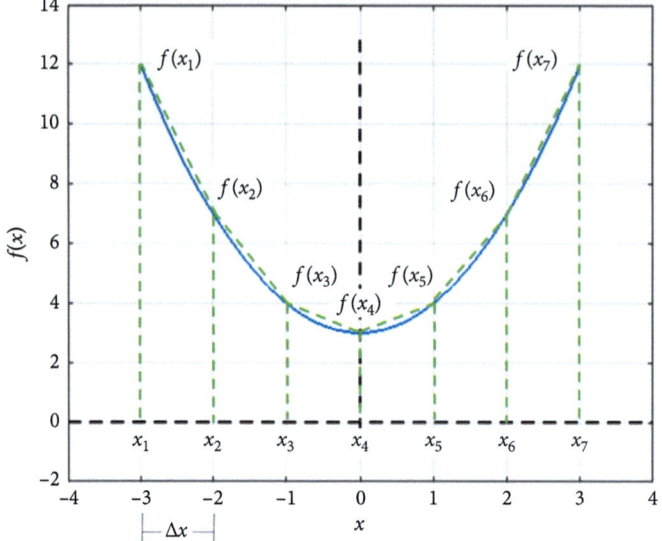

Figure B.18 Approximating the area beneath a curve using the trapezoidal rule in MATLAB.

Here we apply the trapezoidal rule and approximate the integral using six trapezoids. In total, we create seven points to define the six trapezoids. We evaluate the integral by executing the following commands:

```
>> clear
>> x=linspace(-3,3,7);
>> fx=x.^2+3;
>> trapz(x,fx)
ans =
      37
```

Using 10 elements we would get the following result:

```
>> x=linspace(-3,3,11);
>> fx=x.^2+3;
>> trapz(x,fx)
ans =
  36.359999999999999
```

And with 100 elements, the result is:

```
>> x=linspace(-3,3,101);
>> fx=x.^2+3;
>> trapz(x,fx)
ans =
  36.003600000000006
```

B.5 Programming in MATLAB

MATLAB can be used for high-level programming and solving large-scale and complex engineering problems. To generate our own script in MATLAB, the New Script option needs to be selected. This is denoted with a yellow frame in Figure B.19. In the following, a simple script is presented to introduce some of the basic programming commands in MATLAB. It is recommended to create scripts when using MATLAB, as this allows the saving of commands and calculations, facilitating easy recall when needed. Consequently, it is advisable to generate a separate script for each MATLAB example from the preceding sections of this chapter.

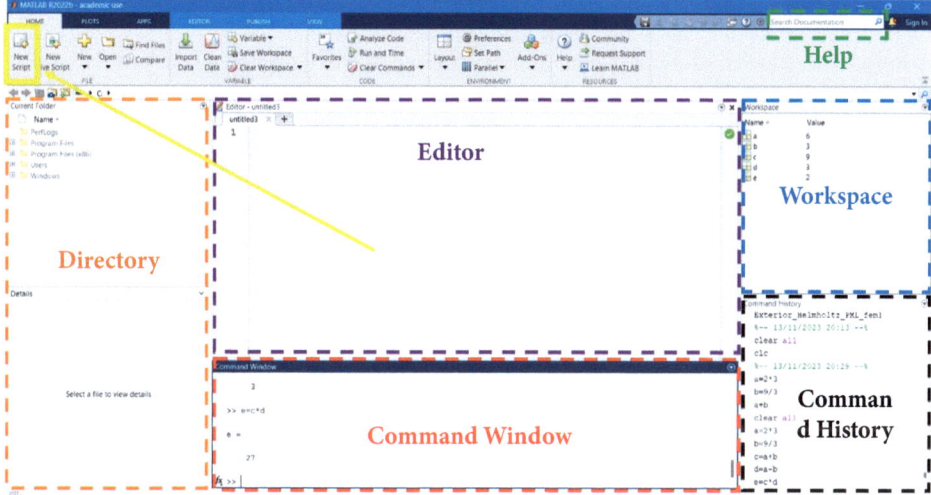

Figure B.19 Selecting the New Script feature in the MATLAB desktop to create a MATLAB script.

We aim to find the sum of the squares of all natural numbers from 1 to 10, excluding 3, i.e.,

$$S = 1^2 + 2^2 + 4^2 + \cdots + 10^2$$

The MATLAB script in Code B.1 can perform this calculation. This code and other MATLAB scripts presented in this book can be downloaded at: http://www.oup.com/AdvancedMathematicalModelling.

Code B.1 *MATLAB code to calculate the sum of the squares of all natural numbers from 1 to 10, excluding 3. This code can be downloaded at: http://www.oup.com/AdvancedMathematicalModelling.*

```matlab
% Calculation of S
clear all
clc


S=0; % initial value for S


for i=1:10 % perform a loop over the natural numbers from 1 to 10
    if i~=3 % exclude 3 but not calculating when i is not equal to 3


        S=S+i^2; % update sum S by adding in each loop the next square


    end % close the if statement
```

```
end % close the for loop over i

S % show on command window the value of S
```

The % symbol is used to introduce comments in the script for our convenience. Everything written after this symbol appears in a green font and is not recognized as a command to be executed by MATLAB. These comments are added to explain the details of the code to a user and help them understand the code. It is advisable to add lots of comments while developing code in MATLAB.

We initialize the sum variable S to zero and perform a loop, using 'for' over the index i from 1 to 10. Each time we add the value of the index squared to account for all the contributions. We use an 'if' statement to disregard the case i = 3. For each statement and loop to terminate, we use 'end'. Note that for, if, and end all appear in a blue font. Finally, we type S without a semicolon to see the value of the sum when calculated in the Command Window. The result is S = 376.

It is possible to generate code for the solution of the preceding problem without the need to use for, if, and end. It is advisable in MATLAB to employ object-oriented programming techniques for increased efficiency, aiming to minimize or avoid the use of for and if loops. The MATLAB script in Code B.2 reproduces this case.

Code B.2 *MATLAB code to calculate the sum of the squares of all natural numbers from 1 to 10 excluding 3 by avoiding for, and if loops. This code can be downloaded at: http://www.oup.com/AdvancedMathematicalModelling.*

```
% Calculation of S
clear all
clc

i=1:10; % generate a vector with natural numbers from 1 to 10

S=sum(i.^2); % calculate S as the sum of the vector values squared

S=S-3^2 % update S subtracting the square of 3 and show on command window
```

The result obtained employing Code B.2 is S = 376, which is the same as in the previous calculations with Code B.1. MATLAB offers several commands that can perform operations with vectors and matrices. The use of such commands is always recommended as they speed up all calculations and save computational time.

B.6 Further reading

The interested reader can find more information in several relevant books, such as Kalechman (2009) and Nagar (2017).

References

Heidarzadeh, M., Harada, T., Satake, K., Ishibe, T., and Takagawa, T. (2017). Tsunamis from strike-slip earthquakes in the Wharton Basin, northeast Indian Ocean: March 2016 Mw 7.8 event and its relationship with the April 2012 Mw 8.6 event. *Geophysical Journal International*, 47(3), 1601–1612. doi: 10.1093/gji/ggx395

Papathanassiou, T.K., Filopoulos, S.P., and Tsamasphyros G.J. (2011) Optimization of composite patch repair processes with the use of genetic algorithms. *Optimization and Engineering*, 12, 73–82.

MathWorks Inc. (2022). *MATLAB version 9.13.0 (R2022b)*. https://www.mathworks.com

Kalechman, M. (2009). *Practical MATLAB Basics for Engineers*. CRC Press.

Nagar, S. (2017). *Introduction to MATLAB for Engineers and Scientists*. Apress.

Index